CISM International Centre for Mechanical Sciences

Courses and Lectures

Volume 609

For more than 40 years the book series edited by CISM, "International Centre for Mechanical Sciences: Courses and Lectures", has presented groundbreaking developments in mechanics and computational engineering methods. It covers such fields as solid and fluid mechanics, mechanics of materials, micro- and nanomechanics, biomechanics, and mechatronics. The papers are written by international authorities in the field. The books are at graduate level but may include some introductory material.

Katalin Bagi · Maurizio Angelillo
Editors

Discrete Computational Mechanics of Masonry Structures

Springer

Editors
Katalin Bagi
Department of Structural Mechanics
Budapest University of Technology
and Economics
Budapest, Hungary

Maurizio Angelillo
Dipartimento di Ingegneria Civile
University of Salerno
Fisciano, Italy

ISSN 0254-1971 ISSN 2309-3706 (electronic)
CISM International Centre for Mechanical Sciences
ISBN 978-3-031-32475-8 ISBN 978-3-031-32476-5 (eBook)
https://doi.org/10.1007/978-3-031-32476-5

This Springer imprint is published by the registered company Springer Nature Switzerland AG
The registered company address is: Gewerbestrasse 11, 6330 Cham, Switzerland

Preface

The material composing traditional masonry structures is a "structured" material, that is, a collection of individual bodies in dry or mortared mutual contact among each other, rationally assembled in such a way to form a load-bearing structure. These techniques were used by mankind since the most ancient times and have evolved through history into sophisticated forms. The majority of our more precious architectural heritage and a great portion of contemporary civilian residences and transportation infrastructures are made of masonry.

The assessment of the structural integrity of such structures poses serious challenges, since, due to their discrete built-up, standard continuum-based calculation methods are often incapable to correctly reflect their mechanical behavior. The most common manifestations of masonry deformation—to be understood as a change of geometry with respect to an ideal "perfect" design form—are sustained by cracks opening up between voussoirs, sliding among the blocks at the block interfaces and cracks developing inside the individual blocks. Transient failures due to actions produced by the environment—such as large settlements, earthquakes, and anthropic activities—entail the kind of deformation described above and are usually not a problem of strength of materials but rather a matter of a geometrical transition from one to another state of equilibrium. The mechanics of these problems may better be captured by methods that consider the structure as a collection of discrete bodies. An effective alternative is represented by suitably discretized continua, with some kind of no-tension material or no-tension interfaces.

Such methods are available in a wide variety today. The aim of the present book is to give an overview to the theoretical fundaments, advantages, and preferable fields of application of the most important methods, but also calling the attention to their limitations so that the readers would build up a critical view of the choices they have when attacking a problem concerning real masonry structures. The book gives a basis for the readers to become able to develop their own methods, inspired either by classical Graphic Statics or by any modern technique they find promising.

Among the methods we adopt here, we can distinguish between two main approaches. The first one is the equilibrium approach, that is, the assessment of the possibility to construct a class of bearable states of equilibrium inside the structure under the given forces, and the kinematic approach, that is, the evaluation of the displacements that a structure can exhibit as a result of the external actions. These two different areas are dual problems in the discretized-continuum approach, and are coupled in the Discrete Element Method (DEM). Both the equilibrium and the kinematic approach are tools commonly used in Limit Analysis which remains the basic method to assess the safety of masonry structures.

Chapter 1 explains why is the mechanical behavior of masonry structures so different from modern building materials like steel or concrete, and what are the main approaches of mechanical analysis that reflect those peculiarities. This chapter calls the attention to common mistakes in strengthening and renovating that originate from the poor understanding of the discrete nature of masonry: distrust in masonry structures standing for many centuries often lead to too heavy interventions, sometimes even causing damages themselves. Fundaments of methods of analysis like Graphic Statics, Limit Analysis, and the Discrete Element Method are summarized.

Chapter 2 focuses on Graphic Statics, a commonly applied method of analysis in the nineteenth and early twentieth centuries. Graphic Statics is the traditional way to assess structures with the equilibrium method and attracts particular interest today due to its high potential in computerized applications. Old techniques for arches, vaults, and domes are introduced and applications of their modern computerized versions are presented. TNA, which is essentially a graphical equilibrium method which has been developed as an efficient computer program for the assessment of vaults and domes of any geometry, is treated in Chap. 4.

Chapter 3 is concerned with Continuum models incorporating the unilateral assumptions of Heyman. In the first part of Chap. 3, the model is formulated mathematically and the main theorems of Minimum Potential and Complementary energy are introduced. Reference to the Theorems of Limit Analysis is also made. Singular strains, that is, cracks, and singular stresses are discussed. By admitting singular stresses, admissible stress states in masonry structures can be seen as compressive networks within the structure geometry. This theory forms the basis for the applications which are shown in the second part of Chap. 3, dedicated to the implementation of the kinematical and equilibrium approach into computer programs. Chapter 4 describes TNA a "new" methodology which represents a modern 3d development of traditional Graphic Statics and can also be seen within Limit Analysis, as an equilibrium method for 3d structures composed of unilateral material and exhibiting singular stresses.

Chapter 5 introduces the most important available versions of the Discrete Element Method (DEM). Special attention is paid on 3DEC, the most widespread commercial code for masonry analysis. After introducing the theoretical fundaments, several applications are presented about quasi-static and dynamic problems of masonry structures.

The authors hope that this book will inspire young researchers to devote their energy, fresh ideas, and enthusiasm to the field of the mechanics of masonry structures.

Budapest, Hungary
Fisciano, Italy

Katalin Bagi
Maurizio Angelillo

Contents

Chapter 1
Discrete Computational Mechanics of Masonry Structures

An Introduction

Maurizio Angelillo and Katalin Bagi

1.1 Masonry as a Structural Material

The Architect/Engineer education. "Engineers, called on to advise on the repair of an old building—a cathedral, say—will usually have learned their skills in the design of modern buildings using modern materials—steel and reinforced concrete. Care must be taken in transferring those skills to ancient structures." This sentence, taken from a recent paper by Heyman (2018), brings our attention on two critical matters concerning the interventions on old and ancient masonry structures, that is the difference in behaviour between new and old structures and the lack of training and syllabi on old construction techniques in the theory and practice of modern schools of Engineering and Architecture.

We will discuss the first issue at length in what follows; for what concerns the second issue, there are historical reasons for such a condition.

By the end of the nineteenth century, the advent of new materials accompanied by a sense of superiority of the new technologies over the old construction rules, and lately, the so-called Modern Architecture of Bauhaus dramatically broke with the millenary tradition of vaulted construction: masonry walls, arches and vaults were considered old-fashioned determining a progressive fading in the knowledge of the old construction techniques. Currently, there is a deficiency of such teaching in Schools which is difficult to compensate for with specialised education. Where

M. Angelillo
Department of Civil Engineering, University of Salerno, Fisciano, Italy

K. Bagi (✉)
Department of Structural Mechanics, Budapest University of Technology and Economics, Budapest, Hungary
e-mail: bagi.katalin@emk.bme.hu

K. Bagi and M. Angelillo (eds.), *Discrete Computational Mechanics of Masonry Structures*, CISM International Centre for Mechanical Sciences 609,
https://doi.org/10.1007/978-3-031-32476-5_1

can this vast heritage of knowledge be re-learned? In recent years we have assisted in many initiatives aimed at this objective, such as open-air laboratories (Carola, 2019), fieldwork experiences (Fuentes and Huerta, 2014), practical construction stages (Al Asali and Ramage, 2020; Workshop, 2020; Summer Schools, 2022; Angelillo et al., 2022), but it is obvious that a true inversion of this trend can be driven only by a *political* decision.

The threat that this lack of competence may represent for conserving and preserving the historic built heritage is now briefly described.

Renewal. Old masonry structures do not usually come to us untouched, and, even if they do, they will inevitably be changed due to the new needs and requirements of modern life.

In this activity, besides the guide of skilful Engineers, experienced stone masons are needed, and, nowadays, most existing construction firms lack such competent figures. This insensitivity of the ordinary technicians, at all levels, toward the comprehension of the inherent structural behaviour of masonry, has produced and will continue to produce damage to the ancient and old built environment, some of which are shown in Fig. 1.1, such as:

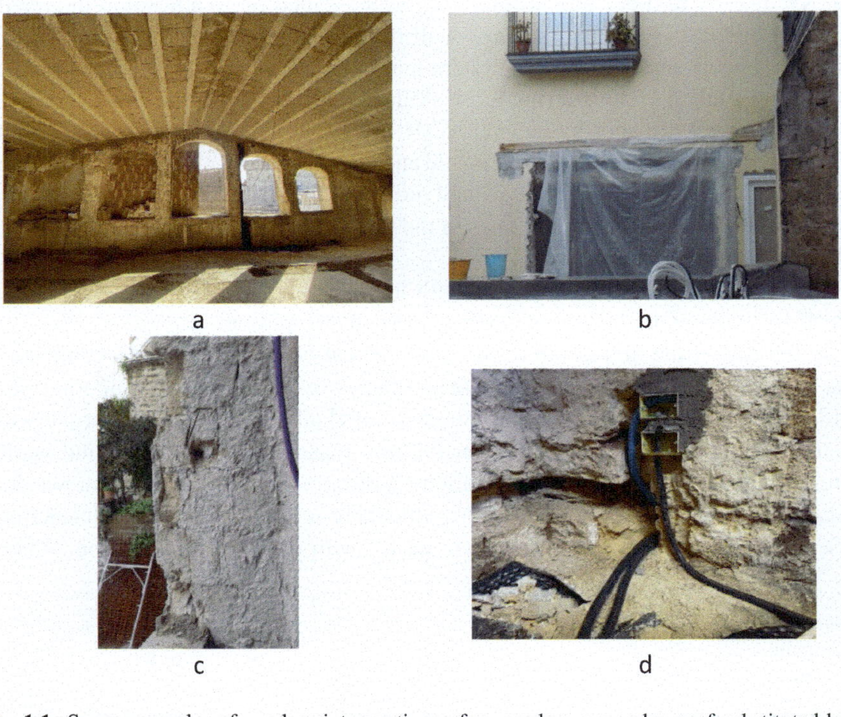

Fig. 1.1 Some examples of careless interventions of renewal. **a**: a wooden roof substituted by a concrete slab; **b** The realisation of a misaligned large opening in a bearing wall; **c** well-constructed window corners messed up to put in place new window frames; **d** tunnelling and wrecking of walls to install new electrical systems

- the removal of partition walls
- the demolition of vaults,
- the substitution of wooden roofs and ceilings with concrete slabs,
- the execution of large or disordered new openings in load-bearing walls,
- the modification of existing openings to put in place new windows;
- the drilling, tunnelling and wrecking of walls to put new technological installations such as plumbing, heating, electrical and air conditioning systems.

Retrofitting. Old masonry structures, as all human artefacts, deteriorate and need maintenance. The problem arises of intervening to preserve the structural integrity of these constructions. In modern times, a general feeling of suspect and distrust on the safety of masonry structures has pervaded the technicians who, due to their poor knowledge on the actual way of functioning of masonry structures and to their scarce familiarity with traditional maintenance techniques, believe that the use of new techniques and materials would necessarily ameliorate the structural response of masonry constructions.

Despite the extreme stability and robustness that ancient buildings have proved by surviving severe challenges, such as earthquakes, strong winds, floods, settlements, lack of maintenance, and man attack (bombing, despoliation, modern transformations), masonry structures are generally perceived as unsafe. This feeling of distrust has produced, mainly after the first world war, heavy interventions of retrofitting that, far from improving the performance of the buildings, deprived these constructions of their original firmness, demonstrated, if nothing else, by centennial experience. Some examples of this strategy based on the idea of transferring the bearing function of walls and vaults to a hidden reinforced concrete structure, and for which Antonino Giuffrè coined the expressive definition of "the Monsters" (see Giuffrè, 1988), are shown in Fig. 1.2.

A new strategy, originating from the same frame of mind, is based on the application on the masonry surface of walls, arches and vaults, of plastic or mortar matrix

a b

Fig. 1.2 The *Monsters*. In **a**, **b**, removal of the filling and realisation of concrete vaults reinforced with a double mat of steel bars, on top of perfectly constructed and conserved masonry vaults (S.Basilio convent interventions, after the L'Aquila earthquake of 2009)

a b

Fig. 1.3 The new monsters. In **a**, a wall sandwiched between two nets of plastic. In **b**, strips of composite material applied at the extrados of cross vaults

composites with fibres (Fig. 1.3). The effect of these retrofitting on masonry components is justified by showing (either through experiments or often through naïve numerical checks) the increase of the strength of such elements with respect to the effect of large local actions, that no one intends to apply. Indeed, this is not the goal of a safe masonry design. The robustness of masonry structures is mainly due to the availability of various load paths as a response to changes of the load distribution and boundary conditions. Speaking about robustness to deterioration of masonry, a kind of structure for which the self-load dominates the effect of the accidental loads, the most natural definition states that robustness is taken to imply tolerance to damage from extreme geometrical changes, human error and deterioration.

As the example shows, offering a quality support to restoration people has become a challenge for structural Engineers. Correct interventions should be more sensible and in keeping with the construction language, and based on the knowledge of old construction techniques and design rules.

The "genetical" evolution of Masonry. Masonry is constructed with the rules of art. Where did these rules of art come from?

Masonry design evolved before Mechanics was invented, but the evolution was slow and took millennia so that a sort of "genetical" selection of construction techniques occurred. This genetic evolution can be explained as the selection of best solutions for the given materials under the laws of Statics and Dynamics that rule the Physics close to the earth's surface.

The first traces of the use of fire are dated 1 million years ago. The birth of an organised form of agriculture and then of the first permanent human artificial settlements is dated approximately 12,000 years ago. The first stone remains (at Stonehenge) are dated around 3000 years ago. 476 AC is the end of the West Roman Empire, that is, of the ancient era, and the conventional start of the post-classical and modern era, which brought the explosion of Romanic and Gothic masonry technology. 1850 is the conventional start date of the second industrial revolution, the beginning of the late modern era and the swan song of masonry architecture.

a b

Fig. 1.4 The adobe brick technology in ancient Mesopotamia (2500 BC): **a** Similar technique still used (1960’s) in North Africa: **b** from Besenval (1984)

The images shown in Figs. 1.4, 1.5, 1.6 and 1.7, which come from a presentation by Santiago Huerta (whose courtesy in sharing them is acknowledged here) show some examples of the evolution of masonry technology through history.

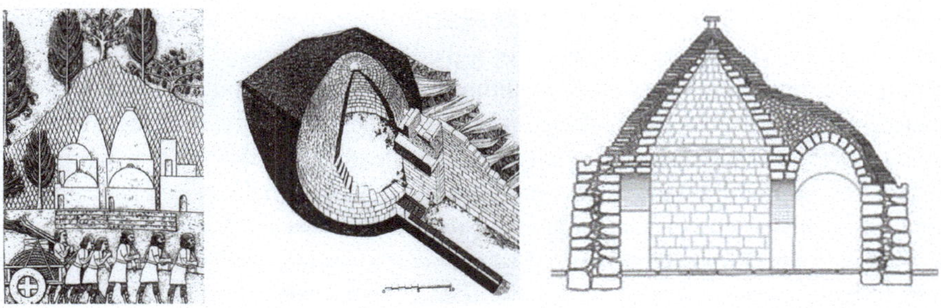

a b c

Fig. 1.5 Tholos vaulting, developed between 3000–1000 BC. In **a**, a Neo-Assyrian bas-relief (8th Century BC) depicting old vault buildings in a Syrian village (from Besenval, 1984). In **b** the treasury of Atreus in Mycenae (1250 BC, from Sanpaolesi, 1989). **c** “modern” tholos, a typical section of a Trullo in Alberobello (Brindisi, Italy)

a b c

Fig. 1.6 Vaults and domes from the Roman times to the nineteenth century. **a** the temple of Minerva Medica (Roma fourth century); **b** the ribbed cross vault of Gothic times (thirteenth century); **c** the dome of the church of San Francesco di Paola in Napoli (nineteenth century, from Cennamo et al., 2018)

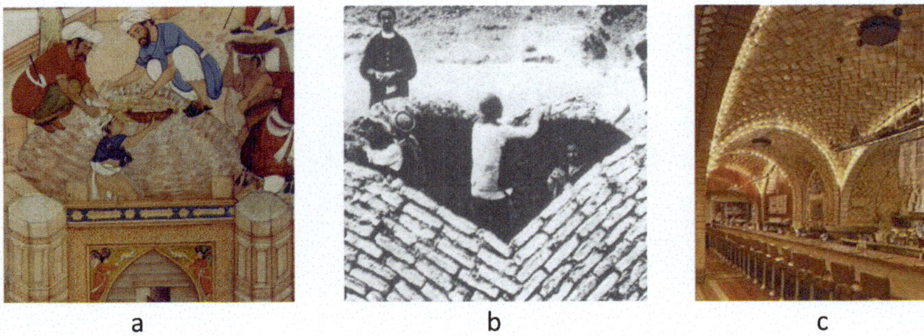

Fig. 1.7 Self-supporting vaults from antiquity to modernity. In **a** vault construction in India (sixteenth century); in **b**, similar technology used in North Africa in the early 60ties of the twentieth century. Tile vaulting of the Guastavino in Grand Central Oyster Bar and Restaurant, New York (1913)

The peculiarities of masonry material. Why masonry is a peculiar *material*? masonry is an assembly of "small" stiff blocks, usually weakly connected. How walls, arches and vaults are done? Masonry is designed to be compressed and the rules of art exploit compression in order to avoid sliding among blocks through different mechanisms, such as friction, interlocking and confinement. Typically, these construction tricks are realised through stereotomy -that is, the size and the form in which the blocks are cut- and how the blocks are assembled. These tricks can be appreciated more easily in the numerous examples of dry stone walls used as fences in sheep farming and agriculture worldwide. In the example shown in Fig. 1.8, we can see some of these aspects, such as stacking sequence, through stones, and corner details.

Masonry is not a material in a conventional sense, but a structure; the strength of the individual blocks is less important than their geometry and assembly. Besides,

Fig. 1.8 Michael Fearnhead giving advice on dry stone wall building https://www.youtube.com/watch?v=mtQChQXKrkA

most of the deformations that masonry constructions exhibit are associated with cracks and produced by the movement of the surrounding posts, therefore they are far from being a direct effect of elasticity. All modern Construction Rules are based on strength and elasticity and freeing from the stamp of this frame of mind is difficult.

Approaches to model masonry. Then, masonry needs special models to be analysed and since most of the existing tools to analyse structures neglect contact, fracture and friction, special computational tools need to be developed. The two most important ways to approach the problem are Heyman's Limit State Analysis for masonry structures, and the Distinct Element Method (DEM). DEM, aims at modelling each block of the structure as an individual body in mutual unilateral contact with the other blocks and interacting with them through special interface laws. Heyman's theory is usually less detailed and consists in a set of crude material restrictions that try to catch at the macroscale the basic ingredients of masonry behaviour, namely unilaterality, compressive strength and friction. The two methods have been adopted with success in recent years and can be used in parallel to assess real masonry buildings.

A common feature of the two methods is that they are in keeping with the traditional Graphic Static approaches, and may be used to complement them and extend their reach by automating the search for equilibrium states. In this book we have all these lines of thinking represented. Traditional Graphic Statics is considered by Danila Aita, its 3d generalisation and implementation and integration with computer graphical tools is demonstrated by Philippe Block. DEM is the research area of José Lemos (who actually was a pioneer of the method) and Katalin Bagi, who will give an overview of both DEM and Heyman's theory—among other approaches—in the subsequent Sections. Maurizio Angelillo has been working for decades on Heyman's model for masonry and, in recent years, has given impulse to the development of new energy-based methods to analyse equilibrium and collapse of masonry structures, here presented by Andrea Montanino and by Antonino Iannuzzo. Iannuzzo recently contributed to implement one of these methods on the COMPAS platform of the ETH.

Heyman's model. The essential ingredients of masonry behaviour are condensed into the material restrictions first foreseen by some old pioneers such as Coulomb, Couplet and Mascheroni and later on formulated in modern terms by Jacques Heyman in the seminal paper Heyman (1966) and in the more recent book Heyman (1995).

The key aspect of masonry behaviour is unilaterality, that is the extreme weakness under tensile loads. Some small tensile stresses are actually carried by most masonry structures at distances of the size of a few blocks, and some large structures, besides those special elements that are explicitly conceived for carrying large tensile forces (such as cantilevering elements), may well live for years under relevant internal tensile forces. The problem is that for such masonry structures we cannot depend on the tensile strength, since any small movement of the supports or any other changes of the environment may produce fractures and the consequent sudden disruption of their state of metastable equilibrium. Then, the existence of a nearby state of equilibrium of pure compression is needed to ensure stability.

Heyman's book of 1995 (Heyman, 1995) contains a synthetic, but deep and sophisticated treatment of what is the essence of masonry behaviour. The flood of multi-parameter constitutive assumptions that have been lately proposed to model masonry materials, including elastic deformation, brittleness, friction, damage and degradation, with the excuse of being closer to reality, and with few exceptions, fail to catch this essence, missing the main point: masonry behaviour is ruled primarily by mass, geometry and unilateral behaviour (see Huerta, 2006)). The model of Heyman (put forward in Heyman 1966), simple as it may seem, penetrates the very idea behind old masonry design, and incorporates those assumptions that were the goals of any master builder. Indeed, to them, giving the construction a geometry and to the stones an arrangement such that the structure could remain, with limited cracks and scant sliding, comfortably compressed, whatever be the external actions that may affect it, were their main concerns (see Huerta, 2008). The material restrictions discovered/invented by Heyman are the essential ingredients for the theorems of Limit Analysis to be valid, and in the present course, we will exploit both the safe and kinematic theorems for assessing the equilibrium of masonry structures.

It is curious to observe how there is virtually no trace of such clever ideas and good sense understanding in any of our modern rules of practice. These rules, though based on Limit States, are often so confused in defining them, that a proper priority of criteria, specific to each single material, hardly emerges. And, in the end, the fact that the ultimate goal of any modern structural analysis should be the assessment of Limit States for which the structure loses equilibrium by developing a mechanism, is kind of lost in the mind of the reader.

Indeed, this last criterion, strength being not the primary concern, is actually, in a very specific way, the most important (if not the sole) for masonry, and the intuition of Heyman that Limit Analysis can still be used in this new context, is one of the turning points of the twentieth century Building Construction Mechanics.

Limit Analysis. Heyman's material restrictions put the theory of masonry within the realm of Perfect Plasticity. It is difficult for a student of Strength of Materials to accept that masonry, a material that is not at all ductile under tensile loads, and shows no yielding due to atomic plane shearing, that is, lacks of all the physical ingredients that motivate the very adoption of perfect plasticity, can be modelled as such. The secret of this lies in the low compressive stress level (i.e., low in comparison to crushing strength) characterising masonry construction and on the unilateral assumption that cuts away all the energy required to activate cracks. Indeed, the material restrictions defining the so-called Normal Rigid No-Tension (NRNT) material (see Angelillo, 2014; Angelillo et al., 2014 for a thorough description of these) described for completeness in Chap. 3, and among which the normality assumption (that is the no sliding assumption of Heyman) plays a crucial role, are sufficient to prove the two theorems of Limit Analysis. A rigorous mathematical proof of the theorems can be found in Del Piero (1998), and Lucchesi et al. (2012), but elementary proofs which disregard the complexity of the function spaces inherent to the NRNT assumptions, can be managed with simple mathematical tools.

Bounded Variation functions. The mathematics of Normal Rigid No Tension (NRNT) materials is non-standard, the function spaces involved are not Sobolev spaces, the ones typical for example of linear and non-linear elasticity and visco-elasticity. Sobolev spaces are vector spaces of functions that have weak derivatives. Intuitively, a Sobolev space is a space of functions which are sufficiently regular, and equipped with a norm that measures both the size and regularity of a function. Motivation for studying these spaces is that solutions of most partial differential equations, when they exist, belong naturally to Sobolev spaces: weak solutions of some important partial differential equations exist in appropriate Sobolev spaces, even when there are no strong solutions in spaces of continuous functions with the derivatives understood in the classical sense. A Sobolev space X is reflexive, meaning that the if X' denotes the dual space of X, (X')' coincides with X. This property is one of the key ingredients to prove the existence of the solution with the "classical" theorems.

The natural spaces of functions describing a typical Boundary Value Problem (BVP) for NRNT materials are not reflexive, and require therefore a more sophisticated treatment. They are actually rather rich and interesting from a purely mathematical point of view. The most interesting aspect from an engineering point of view is the fact that both the strain and the stress can be singular (or, more precisely, they are bounded measures), meaning that both the displacement and the stress vector can be discontinuous, namely, in the mathematical terminology, functions of Bounded Variation (Ambrosio et al., 2000; Bouchitte & Dal Maso, 1993). BV function can be rather complicated, in the sense that the support of their jump set can be geometrically weird and difficult to be segmented and approximated with simple geometrical objects such as lines or surfaces. But, if we are not worried by the completeness and by the closure of our working spaces, we may well restrict ourselves to smaller and more familiar classes of BV functions, namely the BV functions whose derivatives are line or surface Dirac deltas.

Singular strain and stress. Singular strains, that is deformations concentrated on lines or surfaces, are familiar in perfect plasticity. An easy way to construct kinematically admissible mechanisms, to be used as test functions for the Kinematic (upper bound) Theorem of Limit Analysis, is to consider rigid blocks moving the one relative to the other and producing concentrated strains at their interfaces.

Singular stresses are less common. Their possible use in the context of Masonry Mechanics was discovered by Lucchesi et al. (2005), but a possible smart application of concentrated stresses in elasticity was proposed in the seminal paper by Fraternali et al. (2002) where the Lumped Stress Method is introduced. LSM is a mathematically rational way to admit singular approximations for the stress in elasticity, obtained by relaxing the potential energy functional. This is actually one of the few existing stress-based elastic approximations that can be easily implemented on a computer. More recently force based computer codes, created for approximating singular stresses in masonry-like materials, have been introduced by Montanino et al. (2022).

Structures inside structures: strut-and-tie. The potential usefulness of singular stresses can be more appreciated if we look at the support of stresses concentrated on lines or surfaces as to 1d and 2d structures that we envisage inside our 3d continuum.

This, in the same spirit of the strut-and-tie method adopted to apply the safe theorem to reinforced concrete structures modelled as perfectly plastic bodies. We may think that the surface tractions and body forces applied to a 3d body are transferred to such structures through regular stresses having jumps across the lines and surfaces that define them. Or, simply, we may lump such forces to these ideal structures disregarding the detailed way in which the theoretical surface tractions and body forces are transmitted to them, in much the same way it is done with strut-and-tie methods.

Strut-and-tie methods are based on the so-called truss analogy; the truss model helps to understand how forces are transferred throughout a structural member. The well-known truss analogy proposed by Ritter (1899) and Morsch (1909) for the calculation of shear reinforcements in reinforced concrete beams is the oldest example of reasoning by means of "strut-and-tie models".

More recently the strut-and-tie method was revisited and formulated by Schlaich et al. (1987). Designing members by means of the strut-and-tie method belongs to the domain of the plastic design philosophy, and more in particular to the domain of (lower bound) equilibrium methods. Indeed, the design is based on a truss model for which admissible stresses are not exceeded in the members of the truss, and is thus based on a statically admissible system, which may transfer the loads with sufficient safety.

According to the lower bound theorem, the structural member does not fail if for the imposed load combination, a stress distribution can be identified for which: equilibrium is guaranteed (the system of forces is in equilibrium with a given set of loads), and the threshold for plastic behaviour of materials is not exceeded. In this respect, Thrust Network Analysis (TNA), which is described in this book in Chap. 4, is a striking example of a lower-bound limit-analysis-based method to assess vaulted masonry structures. With TNA, admissible equilibrium states correspond to compressive force networks within the structural geometry.

Brittle fracture. Modelling fracture propagation and nucleation in solids is one of the frontiers of modern Continuum Mechanics. In Fig. 1.9a the effect of a transversal impact point load applied to a glass plate is shown. The beautiful crack pattern that we see in Fig. 1.9a, representing the collapse mechanism of a glass plate composed of elastic brittle material and loaded by a transverse point load, though presents some resemblance with the elastic stress trajectories before collapse takes place, have features that are hardly reproducible with any existing numerical simulation.

Low toughness brittle fracture. The situation is somehow different for materials which are weak in tension, and actually one of the main merits of strut-and-tie methods is that with them one can follow the crack patterns that are seen in the mechanisms forming when approaching collapse, or conversely, predict the correct crack pattern.

In Fig. 1.9b, the crack pattern which forms in a concrete beam submitted to a four-point bending test, is reported. The beam is reinforced with inferior longitudinal rebars only, without stirrups. By looking at Fig. 1.9b, the first consideration that one can make is that the crack pattern that we see has only a very feeble resemblance with the principal stress trajectories predicted by an elastic analysis. This is rather typical of materials with low toughness: the state of pure compression that is reached when

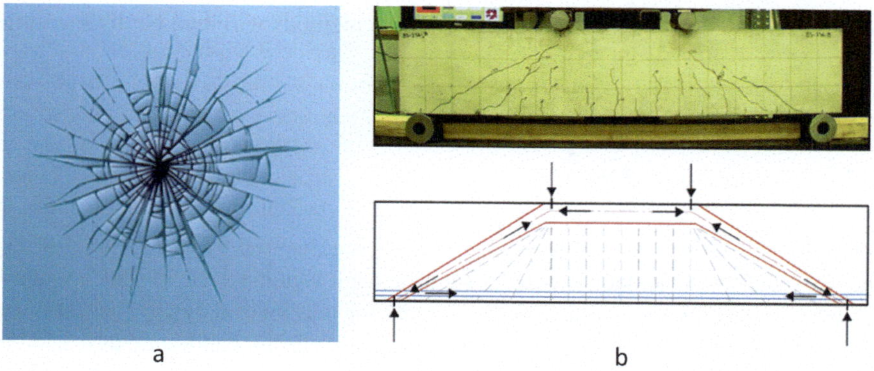

Fig. 1.9 In **a**, crack pattern in a glass plate loaded by a transverse point load. In (**b** up), crack pattern in a concrete beam subject to a four-point bending test. The beam is reinforced only with inferior rebars. In (**b** down), truss analogy: concentrated stress field balancing the load at the verge of collapse for the beam depicted in (b up)

the tensile threshold is overcome and a mechanism is formed, is a state of minimum for the potential energy that usually has nothing to do with the elastic solution.

The crack pattern shown in Fig. 1.9b is predicted with good accuracy, by adopting the strut-and-tie method and considering the concentrated stress field depicted in Fig. 1.9b.

Freeing from the stamp of elasticity. One of the difficulties in understanding and using in all its generality the concept of devising structures inside structures, which is inherent to the possibility of considering concentrated stresses on filaments (tendons/linear arches) and membranes inside a 3d body, is due to the stamp of the theory of elasticity. One has to free his mind from lines of axis going through centroids in beams, middle surfaces for wall beams, plates and shells, and from the idea of Saint Venant distance of extinction. Lines of axis are useful when elastic force–displacement decoupling is sought, and Saint Venant extinction is a result of the inherent ellipticity of the elastic problem. But for NRNT materials, the elastic deformation effect is typically irrelevant and in most of the non-trivial cases the equilibrium equations are non-elliptic. Moreover, one should be ready to generalise the theorem of Cauchy to calculate emerging tractions at the boundary, generalise the Gauss theorem transforming volume into surface integrals, and be able to impose displacement boundary conditions for displacement fields that can be discontinuous.

1.2 Discrete Computation Methods

As explained above, computational methods reflecting the discrete nature of masonry and/or the unilaterality of the material are in the focus of the interest of this book, and hence this section gives an overview on the theoretical fundaments of those methods available today for the public. The details of the methods will then be discussed in Chaps. 2, 3, 4 and 5 of this book.

1.2.1 Limit State Analysis

After the pioneering work of Heyman (1966) being inspired by Kooharian (1952), the static and the kinematic theorem of plastic limit state analysis has been widely applied for developing computational methods for masonry analysis. The masonry material can either be considered as a unilateral, non-sliding continuum as it will be done in Chap. 3, or as a collection of separate blocks where each individual block is considered separately.

According to the classic approximations given by Heyman (1966), the masonry material is a continuum that is assumed to obey the following criteria:

(1) **The stone material has no tensile strength.** (This practically means a no-tension continuum so that when the individual voussoirs of a masonry shell are very small in comparison to the overall dimensions of the structure, this is a proper approximation. Note that the crosswise tension resistance due to the interlocking of voussoirs of bricks is neglected in this case, which is a safe approximation; the issue of crosswise tensile resistance will be discussed in Chap. 5 where the phenomenon is sanalysed with the help of DEM-simulated experiments.)

(2) **The compressive strength of stone is effectively infinite.** (While the heights of usual masonry structures are not that large that the engineer should worry about too large compressions due to self-weight, stress concentrations may occur at singularities, and it may lead to local crushing for too large compressions. So this assumption is on the unsafe side.)

(3) **Sliding cannot occur.** (This assumption allows the engineer to use "associated flow rules" for the no-tension continuum, which makes the mathematical background far simpler as if non-associated flow rules should be applied. However, this assumption is the most questionable one among Heyman's three approximations. Though master masons protected their structures with different methods against sliding of the stones or bricks among each other (Beatini et al., 2019), sliding is dangerous as it often occurs unexpectedly and may lead to sudden catastrophic failure. Heyman (1966) and Mark (1990) both hypothesise that sliding played an important role in the famous collapse of Beauvais cathedral in 1284.)

It is also assumed by Heyman, though not stated explicitly, that the material is undeformable for compression: if the geometry of a masonry shell or structure is known, the same geometry can be assumed under any loads, unless cracks open due to not resisting tension.

Considering the masonry as a collection of individual solid bodies, the basic assumptions (Boothby, 2001; D'Ayala and Tomasoni, 2008) are as follows:

(a) the masonry **units** are **infinitely rigid**;
(b) the masonry **units** are **infinitely strong**;
(c) the masonry units **do not slide at their joints**;
(d) the **joints transmit no tension**.

The two theorems of classic Limit State Analysis can be formulated in both cases as follows:

Safe ("Static") Theorem:

If a force system can be found for the given set of external loads which satisfies the above material criteria and equilibrates the given external loads, then the structure with the given geometry is safe under those loads.

The Safe Theorem gives the theoretical fundament for those classic or modern computerised graphic statics methods where the aim is not to find "the" actual force system of the structure, but to find any force system to equilibrate the loads not violating the above assumptions. Huerta (2001) refers to it as "the equilibrium approach" where "The task of the analyst is not to find the actual equilibrium state, but to find reasonable states of equilibrium for the structure under study. In fact, this has been the approach of all the great architects and engineers. It was implicit in the "geometrical design" of the master builders. It was explicit in the work of Maillart, Torroja, Nervi, Candela or Gaudi, to cite only a few great engineers and architects". Also, this theorem explains how can those 19-twentieth century analyses based on very different alternative assumptions how the forces "flow down" in ribbed cross vaults equally lead to correct results (Huerta, 2009).

However, the Safe Theorem may fail, even in lack of frictional sliding. The issue was discussed in Bagi (2014) where different examples were given for structures being unsafe in spite of having force systems satisfying the material assumptions. Figure 1.2.1 shows two elementary cases; more practical examples are given in Bagi (2014). To resolve the problem, the possibility of any tangential relative displacements had to be excluded, and alternative formulations of the Safe Theorem were suggested (Fig. 1.10).

Unsafe ("Kinematic") Theorem:

If a virtual displacement system can be found for the given set of external loads which satisfies the material criteria and produces non-negative work with the given external loads, then the structure will collapse under these loads.

The Unsafe Theorem can also fail if the possibility of sliding or the tangential relative translations of contacting blocks are not a priori excluded from the analysis. The elementary example given by Drucker (1954) is shown in Fig. 1.10, right.

Duality gap. This theoretical phenomenon, well known in plastic limit state analysis, occurs in case of non-associated flow rules, i.e. when the normality condition fails. In this case the static and kinematic theorems may lead to contradicting results.

Fig. 1.10 Elementary examples for the failure of LSA theorems. Left and middle: unsafe structures having equilibrated force systems; Right: safe structure having a displacement system that produces positive work

The structural behaviour becomes history dependent and without such a history analysis, the behaviour cannot reliably be assessed. The problem of duality gap does not occur if the tangential relative translations between the blocks are completely excluded from the analysis, or if the friction angle and the dilation angle between the blocks are equal. However, in masonry structures the joints approximately obey the Coulomb friction law: their frictional sliding angle is usually around 35–50°, while the dilation angle in the joints is only a few degrees or practically zero. Hence a realistic contact failure law is non-associated and the phenomenon of duality gap occurs.

About associated versus non-associated LSA methods see Casapulla and Argiento (2018), and the references therein. Repeated attempts occur in the literature to overcome the issue of duality gap. Gilbert et al. (2006) introduced a method containing an iterative scheme that works as if incorporating a history dependence into the calculations. Casapulla and Argiento (2018) developed a technique for finding the "exact" solution for non-associated flow rule (Coulomb friction with zero dilatancy) for walls with running bond pattern; they found realistic failure modes for in-plane shear, which were successfully compared to the outcomes of experiments.

The most important LSA codes for masonry arch bridges are Archie-M (based on the static theorem) that excludes tangential relative translations, and Limit State: RING based on combining the two theorems) that includes the possibility of frictional sliding by assuming an associated contact flow rule.

1.2.2 Graphic Statics and Computerised Graphic Statics

Graphical Statics was a fundamentally important analysis tool for structural engineers in the XIXth and early XXth century, aiming at finding a statically admissible force system for the structure considered. In the second half of the twentieth century though, due to the rapid development of continuum-mechanics-based numerical techniques and Limit State Analysis methods, graphical statics methods gradually lose their practical importance in structural mechanics. The latest decades brought a renaissance to their application in masonry analysis: powerful computerised methods were developed, part of which are widely used today for different problems (e.g. Thrust Network Analysis (TNA) of Block and Ochsendorf (2007), inspired by O'Dwyer (1999), and part of which are used as research tools (e.g. Aita et al., 2019). Some of the modern computerised graphic statics methods are based directly on those well-known classical techniques like the old Eddy-Lévy, Wolfe, Durand-Claye etc. algorithms, while some of them are novel, based on computer graphics and modern optimisation, like the Thrust Network Analysis method.

These methods search for possible equilibrium force systems to the given loads, hence they theoretically belong to LSA techniques as they are applications of Heyman's static theorem, some of them including restrictions due to friction limits. However, due to their particular importance in the history of structural analysis as well as among modern computational methods, they deserve to be considered a separate group.

Graphic statics methods will be introduced and discussed in detail in Chap. 2 of this book. TNA is illustrated by Avelino & Block in Chap. 4

1.2.3 Methods Based on Energy Principles

Section 1.1 already shortly called the attention to the possibilities of those methods that are based on energy principles. Chapter 3 will introduce the most important versions of these methods in detail, together with practical applications, but the theoretical fundaments are briefly summarised here. Further explanations can be found in Richards (1977) or Reddy (2002).

Theorem of potential energy. In order to be able to assign a potential energy to the structural problem under analysis, two important conditions have to be satisfied: (1) the material of the system cannot be dissipative; and (2) the loads that act have to be conservative. The first condition is satisfied if the material is either elastic, or consists of perfectly rigid blocks (the internal potential is zero in this case). The second requirement is automatically satisfied in case of gravity loads like self-weight for example.

The theorem states that among all kinematically admissible displacement-and-deformation systems of the structure with a given initial geometry, that one leads the structure to the equilibrium state corresponding to the given loads, for which the potential energy is stationary (i.e. its every first variation is zero). This theorem is practically the sufficient and necessary condition of equilibrium. The displacement system leading to the equilibrium may be small or large; and the potential energy function at the stationarity point can have a local minimum, maximum, saddle point, constant value in a finite region etc.

The complete potential energy is the sum of the external and the internal potential. The external potential is the opposite of the work done by the external static loads from the given initial geometry to the equilibrium state. The internal potential energy is the same as the strain energy in the equilibrium state; for systems consisting of perfectly rigid blocks only, the internal potential is zero. Note that for structures being statically determinate or indeterminate (i.e. over-constrained), the solution is unique: the structure has only one equilibrium state belonging to the given loads.

The equilibrium state may be stable, unstable or neutral; this issue belongs to the field of stability analysis where higher variations of the potential energy are analysed. Stable equilibrium state is recognised from the potential energy having a local minimum.

Theorem of the complementary potential energy. In order to be able to have a complementary potential energy, three conditions have to be satisfied: (1) the material has to be non-dissipative; (2) the displacements have to be small; and (3) the structure has to be statically indeterminate (over-constrained). The theorem gives the compatibility condition of the displacements and deformations leading to the equilibrium with the given kinematic loads, e.g. prescribed support displacements (zero or nonzero). It states that among all those statically admissible force systems (including the static loads as well as the internal forces), that one will belong to the compatible displaced state for which the complementary potential energy is stationary. It can be proved that the stationarity point can only be a local minimum point for elastic materials, there is no possibility for saddle point, local maximum etc.

The external complementary potential is the complementary work of the prescribed support displacements on the corresponding reactions. The internal complementary potential is the dual quantity of the internal potential energy, and it is also zero for structures consisting of perfectly rigid blocks only. Again, the uniqueness of the stationary point can be proved.

Masonry analysis using energy principles. A new simple method to detect fractures and predict crack patterns due to horizontal actions or to settlements, named PRD (Piecewise Rigid Displacement method) and based on the minimisation of the potential energy, has recently been proposed and developed (Iannuzzo et al., 2018a; Angelillo et al., 2018). The most recent results obtained with this kinematic approach are rather promising and encouraging (Iannuzzo, 2019; Iannuzzo et al., 2019, 2020, 2021a), and represent the dual counterpart of the force approach which is also described below.

Another method, the CDF method, which is also displacement based but for which the fractures appear as diffused rather than concentrated, has been put forward in Iannuzzo et al., (2018a, 2018b), Iannuzzo (2019), and fully developed in Iannuzzo et al. (2021b).

A very promising and efficient force-based approach consisting in the minimisation of the complementary energy has been also recently proposed by Montanino et al. in Montanino et al. (2022). This is a unique method which allows to obtain automatically, balanced and admissible stress fields, even in cases in which stress singularities are expected. This method is from the safe side in a Limit Analysis perspective and produces a clear picture of the possible flow of forces which can arise inside masonry structures due to the effect of different kinematical data.

1.2.4 The Discrete Element Method

Discrete Element Method (DEM) is a phrase denoting those numerical models that consider the simulated material (e.g. a sand pile) or structure (e.g. a masonry arch) as an assembly of separate solid bodies that are able to partly or completely detach from their contacting neighbours, slide along each other, and to form new contacts when the large displacements of the solid bodies lead to the rearrangement of contact topology. The first DEM code was born at the end of the 1960ies (Cundall, 1971), for fractured rock soils. Unlike in the Finite Element method (FEM) where the mechanical model is continuous and the discretisation is done for numerical purposes only, DEM is able to fully reflect the discrete built-up of masonry systems.

There are many terminological confusions about DEM and "discrete models" in general. A numerical technique belongs to the Discrete Element Method if the following three criteria are satisfied:

(i) the model consists of finite-sized separate elements (these are either rigid or deformable bodies) each having their own shape, size and position;
(ii) the elements are able to do finite (i.e. large) displacements, independently of each other (they can translate, rotate, and—in most codes—even deform);
(iii) the modifications of the contact topology are automatically kept track by the code: in particular, the new contacts are automatically detected and taken into consideration after they are formed (which is the computationally most demanding part of DEM calculations).

All existing DEM codes simulate the mechanical behaviour of the modelled system in a step-by-step manner, in small consecutive displacement increments. Different DEM codes apply different mathematical tools for finding these increments.

From the late 1990s on—when computer hardware development reached the level that practical problems could already be simulated within reasonable computation times—DEM has widely been applied to masonry constructions (Lemos, 2007; Tóth et al., 2009; Rafiee and Vinches, 2013; Rizzi et al., 2014; McInerney & DeJong, 2015; Forgacs et al., 2017 etc.). In this CISM course, the DEM software 3DEC by Itasca (2016) was explained and used.

In DEM, the mechanical model is composed of separated blocks interacting with each other at the interfaces: the blocks and the joints are equally important. The blocks can be considered rigid or deformable, plastic or even breakable in some models. Many different joint models are available in the codes; most commonly, no-tension joints with Mohr–Coulomb type failure criteria, often amended with dilation, cohesion, softening etc. are used. Since the blocks can be considered rigid and the contact interfaces with no tensile strength, this method is suitable for the investigation of the behaviour of unreinforced masonry structures.

Further details (classification of DEM software according to how they determine the displacement increments, what degrees of freedom they assign to the solid bodies etc.) and an overview of the most important DEM software packages that are available today will be introduced in Chap. 5. by Lemos and Bagi.

1.2.5 Block-By-Block Models Based on Stiffness Matrix

The major drawback of DEM is that the repeated analysis of whether new contacts are born somewhere in the system is very time consuming. A possible way to avoid this disadvantage is simply neglect the possibility of new contacts: assume that the contact state and topology changes only because of existing contacts opening up or reaching the friction limit, and compile the stiffness matrix of the system, taking into consideration the elasticity of the blocks and the joints (Baraldi & Cecchi, 2017). From the resultant forces reduced to the properly chosen nodes of the system, the usual equilibrium equation system of displacement method can be solved by inverting the stiffness matrix, perhaps directly, perhaps with a more sophisticated technique without even compiling the whole matrix, like in the frontal solver techniques (Bagi, 1993) or with successive approximation (Kishino, 1988).

This approach can be useful for problems when rearrangements of the topology are not expected, but the discrete built-up of the system is relevant (e.g. formation of partial cracks). These methods can be considered being in-between DEM and nonlinear FEM. Such software packages are still not available commercially, but they have a potential for future developments and practical applications.

References

Al Asali, M. W., & Ramage M. H. (2020). Learning through building: Approaches to craft training in thin-tile vaulting, *Journal of Traditional Building, Architecture and Urbanism*, 1.

Ambrosio, L., Fusco, N., & Pallara, D. (2000). *Functions of bounded variation and free discontinuity problems*. Oxford, UK: Clarendon Press.

Angelillo, M. (2014). Practical applications of unilateral models to masonry equilibrium. In M. Angelillo (Ed.), *Mechanics of masonry structures* (pp. 109–210). Vienna: Springer.

Angelillo, M., Fortunato, A., Montanino, A., & Lippiello, M. (2014). Singular stress fields in masonry structures: Derand was right. *Meccanica, 49*(5), 1243–1262.

Angelillo, M., Fortunato, A., Gesualdo, A., Iannuzzo, A., & Zuccaro, G. (2018). Rigid block models for masonry structures. *International Journal Masonry Research Innovation, 3*(4), 349–368.

Angelillo, M., De Felice, G., & Huerta, S. (2022). The international summer school on historic masonry structures. *Structural, 241*, 1–10.

Bagi, K. (1993). A quasi-static numerical model for micro-level analysis of granular assemblies. *Mechanics of Materials, 16*(1–2), 101–110.

Bagi, K. (2014). When Heyman's safe theorem for rigid block systems fails: Non-Heymanian collapse modes of masonry structures. *International Journal Solids and Structures, 51*, 2696–2705.

Baraldi, D., & Cecchi, A. (2017). A full 3D rigid block model for the collapse behaviour of masonry walls. *European Journal of Mechanics - A/Solids, 64*, 11–28. https://doi.org/10.1016/j.euromechsol.2017.01.012

Besenval, R. (1984). Technologie de la voute dans l'Orient Ancien, Paris

Boothby, T. E. (2001). Analysis of masonry arches and vaults. *Progress in Structural Engineering and Materials, 3*, 246–256.

Bouchitté, G., & Dal Maso, G. (1993). Integral representation and relaxation of convex local functionals on BV(Ω). *Annali della Scuola Normale Superiore di Pisa - Classe di Scienze, 4, 20*(4), 483–533.

Carola, F. (2019). Neagorà sette piazze. Retrieved from https://ilgiornaledellarchitettura.com/2019/01/30/fabrizio-carola-1931-2019/

Casapulla, C., & Argiento, L. U. (2018). In-plane frictional resistances in dry block masonry walls and rocking-sliding failure modes revisited and experimentally validated. *Composites, Part B: Engineering, 132*, 197–213.

Cennamo, C., Cusano, C., Fortunato, A., & Angelillo, M. (2018). A study on form and seismic vulnerability of the dome of San Francesco di Paola in Naples. *Ingegneria Sismica, 35*(1), 88–108.

Cundall, P. A. (1971). A computer model for simulating progressive large scale movements in blocky rock systems. In: Proceedings of the Symposium of the International Society of Rock Mechanics, Nancy, France.

D'Ayala, D. F., & Tomasoni, E. (2008). The structural behaviour of masonry vaults: limit state analysis with finite friction. In D D'Ayala, E Fodde (Eds.), *Proceedings of the 6th International Conference on Structural Analysis of Historic Construction, Bath, UK. Taylor and Francis Group, London* (pp. 47–61)

Del Piero, G. (1988). Limit analysis and no-tension materials. *International Journal of Plasticity, 14*(1–3), 259–271.

Drucker, D. C. (1954). Coulomb friction, plasticity and limit loads. *Journal of Applied Mechanics, 21*, 71–74.

Fearnhead, M. Giving advice on dry stone wall building. Retrieved from https://www.youtube.com/watch?v=mtQChQXKrkA

Forgács, T., Sarhosis, V., & Bagi, K. (2017). Minimum thickness of semi-circular skewed masonry arches. *Engineering Structures, 140*, 317–336.

Fraternali, F., Angelillo, M., & Fortunato, A. (2002). The lumped stress method and the discrete-continuum approximation. *International Journal of Solids and Structures, 39*, 6211–6240.

Fuentes, P., & Huerta, S. (2014). Bóvedas sin cimbra: La construcción tabicada. Memoria académica del taller experimental de Historia de la Construcción del curso 2013/2014. Instituto Juan de Herrera

Giuffrè, A. (1988). Restauro e sicurezza in zona sismica, la Cattedrale di Sant'Angelo dei Lombardi. *Palladio, 1*, 1–26

Heyman, J. (1966). The stone skeleton. *International Journal of Solids and Structures, 2*(2), 249–279.

Heyman, J. (1995). *The stone skeleton: Structural engineering of masonry architecture*. Cambridge University Press.

Heyman, J. (2018). The structural engineer's view of ancient buildings. *Journal Mechanics Materials and Structures, 13*(5), 609–615.

Huerta S. (2001). Mechanics of masonry vaults: The equilibrium approach. In P.B. Lourenco, & P. Roca (Eds.), *Historical constructions* (pp. 47–69). University of Minho

Huerta, S. (2006). Galileo was wrong: The geometrical design of arches. *Nexus Network Journal, 8*(2), 25–52.

Huerta, S. (2008). The analysis of masonry architecture: a historical approach. *Architectural Science Review, 51*(4), 297–328

Huerta, S. (2009). The debate about the structural behaviour of gothic vaults: From Viollet-le-Duc to Heyman. In: *Proceedings Third International Congress on Construction History, Cottbus, May 2009* (pp. 837–844)

Iannuzzo, A., Angelillo, M., De Chiara, E., De Guglielmo, F., De Serio, F., Ribera, F., & Gesualdo, A. (2018). Modelling the cracks produced by settlements in masonry structures. *Meccanica, 53*, 1857–1873.

Iannuzzo, A., De Serio, F., Gesualdo, A., Zuccaro, G., Fortunato, A., & Angelillo, M. (2018). Crack patterns identification in masonry structures with a C° displacement energy method. *International Journal Masonry Research Innovation, 3*(3), 295–323.

Iannuzzo, A. (2019). Energy based fracture identification in masonry structures: The case study of the church of "Pieta dei Turchini." *Journal of Mechanics of Materials and Structures, 14*, 683–702.

Iannuzzo, A., Olivieri, C., & Fortunato, A. (2019). Displacement capacity of masonry structures under horizontal actions via PRD method. *Journal of Mechanics of Materials and Structures, 14*, 703–718.

Iannuzzo, A., Van Mele, T., & Block, P. (2020). Piecewise Rigid Displacement (PRD) method: A limit analysis-based approach to detect mechanisms and internal forces through two dual energy criteria. *Mechanics Research Communications*, 107.

Iannuzzo, A., Dell'Endice, A., Van Mele, T., & Block, P. (2021a). Numerical Limit Analysis-based modelling of masonry structures subjected to large displacements. *Computers & Structures*, 242(2)

Iannuzzo, A., Block, P., Angelillo, M., & Gesualdo, A. (2021). A continuous energy based numerical approach to predict fracture mechanisms in masonry structures: CDF method. *Computers and Structures, 257*, 1–25.

Itasca Consulting Group, Inc. (2016). 3DEC, Three-Dimensional Distinct Element Cod, Ver. 5.2. Minneapolis: Itasca.

Kishino, Y. (1988). Disc model analysis of granular media. In M. Satake, J.T. Jenkins (Eds.), *Micromechanics of granular materials, Elsevier* (pp. 143–152)

Kooharian, A. (1952). Limit analysis of voussoir (segmental) and concrete arches. *Journal of American Concrete Institute, 4*(4), 317–328.

Lemos, J. V. (2007). Discrete element modeling of masonry structures. *International Journal Architectural Heritage, 1*(2), 190–213.

Lucchesi, M., Silhavy, M., Zani, N. (2005). In M. Frémond, & F. Maceri (Eds.), *Singular equilibrated stress fields for no-tension panels. Lecture notes in applied and computational mechanics* (Vol. 23, pp. 255–265) Springer

Lucchesi, M., Silhavy, M., & Zani, N. (2012). On the choice of functions spaces in the limit analysis for masonry bodies. *Journal of Mechanics of Materials and Structures, 7*(8–9), 795–836

Mark, R. (1990). *Light, wind, and structure: The mystery of the master builders.* MIT Press. ISBN: 026263158X (ISBN13: 9780262631587)

McInerney, J., & DeJong, M. J. (2015). Discrete element modeling of groin vault displacement capacity. *International Journal of Architectural Heritage, 9*(8), 1037–1049.

Montanino, A., De Gregorio, D., & Iannuzzo, A. (2022). The continuous Airy-based for stress-singularities (CASS) method: an energy-based numerical formulation for unilateral materials, Preprint

Morsch, E. (1909). Concrete-steel construction. McGraw-Hill Book Company (Translation from third edition of Der Eisenbetonbau, first edition 1902)

O'Dwyer, D. (1999). Funicular analysis of masonry vaults. *Computers and Structures, 73*, 187–197

Reddy, J. N. (2002). *Energy principles and variational methods in applied mechanics.* Wiley

Richards, T. H. (1977). *Energy methods in stress analysis.* Wiley

Ritter, W. (1899). Die Bauweise Hennebique. *Schweizerische Bauzeitung, 33*, 59–61.

Rizzi, E., Rusconi, F., & Cocchetti, G. (2014). Analytical and numerical DDA analysis on the collapse mode of circular masonry arches. *Engineering Structures, 60*, 241–257.

Sanpaolesi, P. (1989). Strutture a cupola autoportanti. *Palladio, 21*, 3–64.

Schlaich, J., Schäfer, K., & Jennewein, M. (1987). Toward a consistent design of structural concrete. *PCI Journal, 32*(3), 74–150.

Summer School on Historic Masonry Structures (2018–2022). Retrieved from https://www.himass.org

Tóth, A. R., Orbán, Z., & Bagi, K. (2009). Discrete element analysis of a stone masonry arch. *Mechanics Research Communications, 36*(4), 469–480.

Workshop: VIII Taller de Bóvedas Tabicadas – MUPAAC. (2020). Retrieved from http://www3.uah.es/mupaac/viii-taller-de-bovedas-tabicadas/

Chapter 2
Graphical Statics

Danila Aita

Abstract In this chapter fundamentals of graphical statics are recalled in order to introduce those techniques (Méry, Durand-Claye, Eddy, Lévy, Wolfe) that inspired modern computerized versions for the equilibrium analysis of masonry arches and domes. The historical contributions are critically examined, by highlighting their application potential. Particular emphasis is given to the Durand-Claye method, a graphical procedure able to check the existence of admissible solutions by taking into account the compressive strength of masonry. Applications to masonry mechanics issues and on historic structures are finally shown.

2.1 Fundaments of Graphical Statics

Some brief historical notes. The basis of graphical statics lies in the identification of a force as an oriented segment (Benvenuto, 1981).

From the 3rd Century b.c. to the 17th a.d. the concept of force was not precisely defined. The various schools of thought tried to define the force by means of different interpretations of this concept: *dynamics*, *vis*, potential, *momentum* (inspired by Aristotle), action, energy, stress (Benvenuto, 1991). The start of a paradigm shift took place only between the end of the fifteenth century and the beginning of the 16th with the funicular devices introduced by Leonardo: the rope becomes the materialization of the tension that acts along it; moreover, it allows the deviation of the line of action of the force. The funicular machines are also studied later by Stevin (1608). But it is Varignon in 1687 (Varignon, 1725) that, analyzing the equilibrium configuration reached by a rope that supports weights, introduces for the first time, although empirically, some key tools of graphical statics, such as the funicular polygon, corresponding to the equilibrium configuration of the rope, and the force polygon, i.e., the polygon of the supported weights or the forces applied to the rope (Sinopoli, 2012).

D. Aita (✉)
Dipartimento di Ingegneria Civile e Ambientale (DICA) - Department of Civil and Environmental Engineering, Politecnico di Milano, Milan, Italy
e-mail: danila.aita@polimi.it

K. Bagi and M. Angelillo (eds.), *Discrete Computational Mechanics of Masonry Structures*, CISM International Centre for Mechanical Sciences 609,
https://doi.org/10.1007/978-3-031-32476-5_2

In its contemporary meaning, the roots of graphical statics can be placed around 1835 (Culmann, 1866), when Poncelet (1835) published his method on the *Solution graphique des principales questions sur la stabilité des voûtes*, systematically applying graphic methods inspired by the *Géometrie descriptive* proposed by Monge (1794).

In that period, the works by Möbius (1833) and Maxwell (1864, 1870) formalized the properties of reciprocal figures in space and their orthographic projections, which emerge in the analysis of a system of forces in equilibrium.

In the following years, through the fundamental contributions by Bresse (1880), Culmann (1866, 1875, 1880), Rankine (1864), Ritter (1888) and Cremona (1890), graphical statics assumed the dignity of an autonomous discipline able to describe the properties of one-, two- and three-dimensional geometric entities, used in the applications concerning structural calculation. Graphical statics allows for defining both the rules of the algebraic structure of such entities and their properties of reciprocity.

The algebraic structure of a vector and the funicular polygon. In this section a brief review of vector algebra is proposed, in order to theoretically frame the graphical procedures within vector analysis (the reference texts are Olivito, 2009; Roman, 2005; Sinopoli, 2012, 2021).

A vector can be defined as an entity of algebraic nature. More precisely, a vector is an element of the linear vector space $\boldsymbol{V}^n$, i.e., the set of the *n*-tuple $\boldsymbol{v} = v_1, v_2, v_3, \ldots, v_n$, defined in the field of real numbers $\boldsymbol{R}$, for which it is possible to establish two operations:

(1) Vector addition $+ : \boldsymbol{V}^n \times \boldsymbol{V}^n \rightarrow \boldsymbol{V}^n$. This operation takes any two vectors $\boldsymbol{u}$ and $\boldsymbol{v}$ and assigns to them a third vector which is commonly written as $\boldsymbol{u} + \boldsymbol{v}$, and is the sum of these two vectors:

$$\forall \boldsymbol{u}, \boldsymbol{v} \in \boldsymbol{V}^n \rightarrow \exists \boldsymbol{w} \in \boldsymbol{V}^n / \boldsymbol{u} + \boldsymbol{v} = \boldsymbol{w},$$

with $\boldsymbol{w} = (w_1, w_2, w_3, \ldots, w_n) = (u_1 + v_1, u_2 + v_2, u_3 + v_3, \ldots, u_n + v_n)$.

(2) Scalar multiplication $\cdot : \boldsymbol{R} \times \boldsymbol{V}^n \rightarrow \boldsymbol{V}^n$. This operation takes any scalar a and any vector $\boldsymbol{v}$ and gives another vector $a\boldsymbol{v}$:

$$\forall \boldsymbol{u} \in \boldsymbol{V}^n \wedge \forall a \in \boldsymbol{R} \rightarrow \exists \boldsymbol{v} \in \boldsymbol{V}^n / a \cdot \boldsymbol{u} = \boldsymbol{v},$$

with $\boldsymbol{v} = (v_1, v_2, \ldots, v_n) = a \cdot (u_1, u_2, \ldots, u_n) = (a \cdot u_1, a \cdot u_2, \ldots, a \cdot u_n)$

A set V^n can be defined as a vector space if the operations of addition and multiplication adhere to the following axioms:

(1) commutativity of addition: $\forall \boldsymbol{u}, \boldsymbol{v} \in \boldsymbol{V}^n \rightarrow \boldsymbol{u} + \boldsymbol{v} = \boldsymbol{v} + \boldsymbol{u}$
(2) associativity of addition:

$$\forall \boldsymbol{u}, \boldsymbol{v}, \boldsymbol{w} \in \boldsymbol{V}^n \rightarrow \boldsymbol{u} + \boldsymbol{v} + \boldsymbol{w} = (\boldsymbol{u} + \boldsymbol{v}) + \boldsymbol{w} = \boldsymbol{u} + (\boldsymbol{v} + \boldsymbol{w})$$

(3) identity element of addition (zero vector):

$$\forall \boldsymbol{v} \in \boldsymbol{V}^n \rightarrow \exists 0 \in \boldsymbol{V}^n / 0 + \boldsymbol{v} = \boldsymbol{v}$$

(4) inverse elements of addition: $\forall \boldsymbol{v} \in \boldsymbol{V}^n \rightarrow \exists(-\boldsymbol{v}) \in \boldsymbol{V}^n / (-\boldsymbol{v}) + \boldsymbol{v} = 0$

(5) compatibility of scalar multiplication with field multiplication:

$$\forall \boldsymbol{v} \in \boldsymbol{V}^n \wedge \forall a, b \in \boldsymbol{R} \rightarrow a \cdot b \cdot \boldsymbol{v} = a \cdot (b \cdot \boldsymbol{v}) = (a \cdot b) \cdot \boldsymbol{v}$$

(6) identity element of scalar multiplication: $\forall \boldsymbol{v} \in \boldsymbol{V}^n \rightarrow 1 \cdot \boldsymbol{v} = \boldsymbol{v}$

(7) distributivity of scalar multiplication with respect to vector addition:

$$\forall \boldsymbol{u}, \boldsymbol{v} \in \boldsymbol{V}^n \wedge \forall a \in \boldsymbol{R} \rightarrow a \cdot (\boldsymbol{u} + \boldsymbol{v}) = a \cdot \boldsymbol{u} + a \cdot \boldsymbol{v}$$

(8) distributivity of scalar multiplication with respect to field addition:

$$\forall \boldsymbol{v} \in \boldsymbol{V}^n \wedge \forall a, b \in \boldsymbol{R} \rightarrow (a + b) \cdot \boldsymbol{v} = a \cdot \boldsymbol{v} + b \cdot \boldsymbol{v}$$

Let us introduce the Cartesian coordinate system $\{O, x, y, z\}$ and the orthonormal basis $(\boldsymbol{i}, \boldsymbol{j}, \boldsymbol{k})$, with $\boldsymbol{i} = (1, 0, 0), \boldsymbol{j} = (0, 1, 0)$, and $\boldsymbol{k} = (0, 0, 1)$, which consists in a system of unit vectors, each of them associated with the corresponding dimension of the space.

The components of $\boldsymbol{u}$ can be interpreted as the coordinates of the point P, the extreme of the oriented segment $(P - O)$, which represents the diagonal of the parallelepiped defined by the three orthogonal sides corresponding to the components of the vector (Fig. 2.1).

The generic vector $\boldsymbol{u}$ in the three-dimensional Euclidean space can be interpreted as an entity characterized by three components, u_x, u_y, u_z, corresponding to three oriented segments along the directions defined by one of the coordinate axes: $\boldsymbol{u} = (u_x,\ u_y,\ u_z) = (u_x, 0, 0) + (0, \mathrm{u}_y, 0) + (0, 0, \mathrm{u}_z) = u_x \cdot \boldsymbol{i} + u_y \cdot \boldsymbol{j} + u_z \cdot \boldsymbol{k} = \boldsymbol{u}_x + \boldsymbol{u}_y + \boldsymbol{u}_z$, where $\boldsymbol{u}_x$, $\boldsymbol{u}_y$ and $\boldsymbol{u}_z$ are vectors applied at the origin O, each of them parallel to one of the coordinate axes.

Fig. 2.1 The generic vector $\boldsymbol{u}$ in the three-dimensional Euclidean space

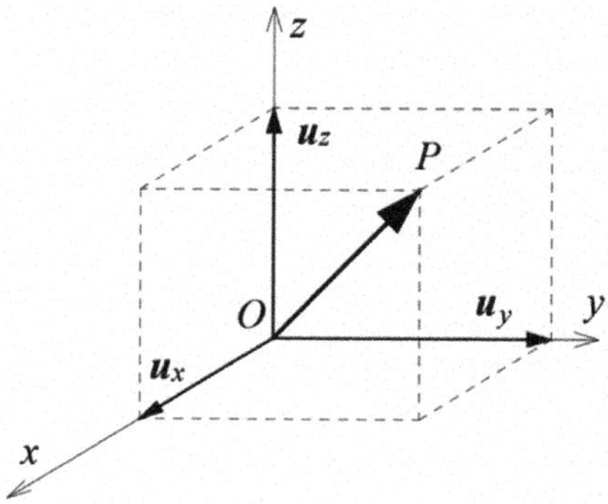

Scalar product. The algebraic structure of a linear vector space to which an orthonormal basis system is associated can be expanded introducing the scalar product:

$$\forall \boldsymbol{u}, \boldsymbol{v} \in \boldsymbol{V}^n \rightarrow \exists c \in \boldsymbol{R}/\boldsymbol{u} \times \boldsymbol{v} = c, \text{ with } c = \sum_{i=1}^{n} u_i \cdot v_i.$$

Vector product. A further extension is represented by the definition of vector product which, for the vectors of the three-dimensional Euclidean space, is written as follows:

$\forall \boldsymbol{u}, \boldsymbol{v} \in \boldsymbol{V}^3 \rightarrow \exists \boldsymbol{w} \in \boldsymbol{V}^3/\boldsymbol{u} \bigwedge \boldsymbol{v} = \boldsymbol{w}$, with

$$\boldsymbol{w} = \begin{vmatrix} \boldsymbol{i} & \boldsymbol{j} & \boldsymbol{k} \\ u_x & u_y & u_z \\ v_x & v_y & v_z \end{vmatrix}$$

The addition of a set of m vectors. Consider now a set of m vectors $(P_i - A_i)$, with $i = 1, 2, 3, \ldots, m$, laying in a plane π. As a consequence of the properties related to addition, it is possible to identify different ways to perform addition. The fact that the vectors are applied at points A_i is irrelevant—for the very definition of vector—with respect to the addition of a set of m vectors:

$$\boldsymbol{R} = \sum_{i=1}^{m} (P_i - A_i),$$

which represents the resultant vector of the system.

The so-called parallelogram law gives the rule for vector addition of two vectors (Fig. 2.2a).

The polygon of vectors (Fig. 2.2b) is a graphic construction which allows for finding the resultant vector, $\boldsymbol{R}$, of a system of m vectors. The m sides of this polygon, whose vertexes are $Q_0, Q_1, Q_2, \ldots, Q_m$, are parallel to the assigned vectors $(P_i - A_i)$ and their length represents the magnitude of the corresponding vectors, measured to a suitable scale. The resultant vector, $\boldsymbol{R}$, is given by the oriented segment $(Q_m - Q_0)$.

The opposite vector of $\boldsymbol{R}$, characterized by the same magnitude as $\boldsymbol{R}$ but with reverse direction, identifies the anti-resultant vector.

Moment of force. With reference to Fig. 2.3, the moment $\boldsymbol{M}(O)$ of a force $(P - A)$ about a point O is a vector obtained by calculating the vector product $(A - O) \bigwedge (P - A)$, where $(A - O)$ is the distance vector between the point A, at which the force $(P - A)$ is applied, and the pole O:

$$\boldsymbol{M}(O) = (A - O) \wedge (P - A) = |A - O| \cdot |P - A| \sin\alpha \cdot \boldsymbol{n}.$$

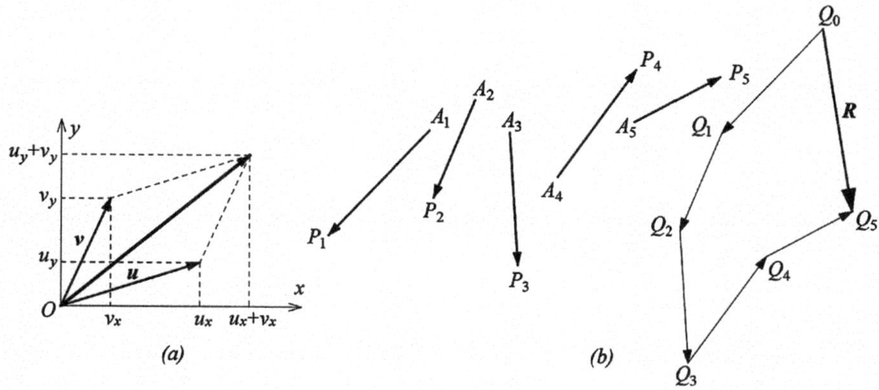

Fig. 2.2 The parallelogram rule (**a**); system of $m = 5$ vectors and corresponding vector polygon (**b**)

Fig. 2.3 The moment $\boldsymbol{M}(O)$ of a force $(P - A)$ with respect to the point O

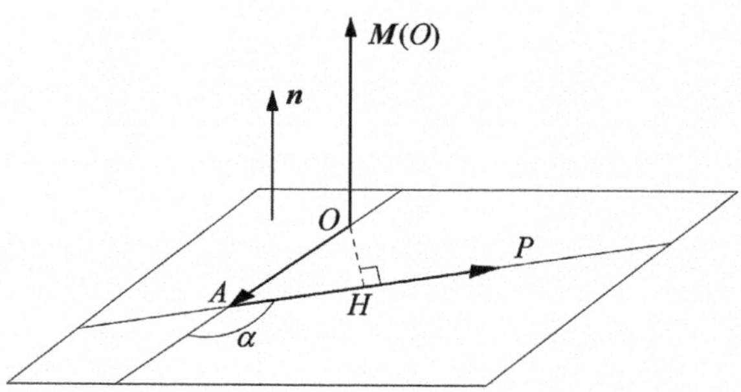

The magnitude of the moment is given by $|H - O| \cdot |P - A|$. The segment OH is the moment arm or lever arm, i.e., the perpendicular distance between the line of action of the force and the base point, O.

Equivalent systems of forces. In many problems concerning the analysis of the equilibrium conditions of rigid bodies subject to the action of a system of forces, it is often convenient to transform the system of forces in an equivalent system, consisting of a minimum number of vectors.

Equilibrium condition for a rigid body. For a rigid (or slightly deformable) solid body the necessary and sufficient condition for equilibrium is given by:

$$\boldsymbol{R} = 0$$

$$\boldsymbol{M}(O) = 0$$

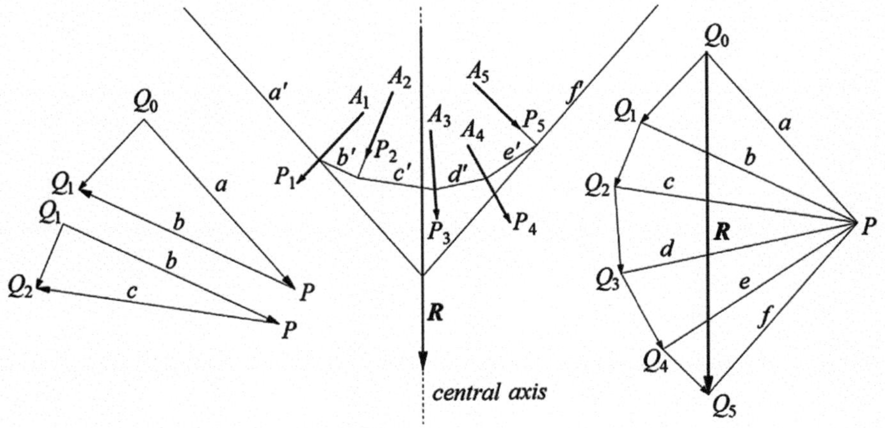

Fig. 2.4 The funicular polygon and the determination of the central axis

The funicular polygon. Consider now a system of m planar forces $(P_i - A_i)$, applied at points A_i, belonging to a plane π, with $i = 1, 2, 3, \ldots, m$. The resultant vector, $\boldsymbol{R}$, can be graphically identified by moving each vector parallel to its line of action, so that the vectors $(P_i - A_i)$, after their translation, are represented by $(Q_i - Q_{i-1})$ (Fig. 2.4).

In order to transform the original system of forces into one equivalent to it, consisting of the resultant vector $\boldsymbol{R}$ applied along the central axis, a point P (called pole) is arbitrarily chosen in the plane π.

Through an operation that is invariant with respect to addition, the vector $(Q_1 - Q_0)$ can be decomposed into the sum of two vectors:

$$(Q_1 - Q_0) = (P - Q_0) + (Q_1 - P).$$

Analogously, the vector $(Q_2 - Q_1)$ can be decomposed as follows:

$$(Q_2 - Q_1) = (P - Q_1) + (Q_2 - P).$$

Analyzing the two relations, it is possible to verify that the second addend, $(Q_1 - P)$, of the first vector, $(Q_1 - Q_0)$, is opposite to the first addend, $(P - Q_1)$, of the second vector, $(Q_2 - Q_1)$, so that they constitute a couple with null lever arm and their sum vanishes; this property always holds in the decomposition of two contiguous vectors.

More generally, the generic vector $(Q_i - Q_{i-1})$ can be expressed as the addition of two vectors:

$$(Q_i - Q_{i-1}) = (P - Q_{i-1}) + (Q_i - P), \text{ with } i = 1, 2, \ldots, m. \tag{2.1}$$

The segments corresponding to $(P - Q_{i-1})$ and $(Q_i - P)$ are called projecting rays of the vector polygon associated with the vector $(Q_i - Q_{i-1})$.

Equation (2.1) represents the ∞^2 possible decompositions of the generic vector $(Q_i - Q_{i-1})$ that can be obtained by varying the position of the pole P in the plane π, position that determines both the direction and the magnitude of the vectors addends along the corresponding projecting rays.

From the above, also the resultant vector, $\boldsymbol{R}$, of the system of forces can be decomposed into the sum of two vectors:

$$\boldsymbol{R} = \sum_{i=1}^{m}(Q_i - Q_{i-1}) = (Q_m - Q_0) = (P - Q_0) + (Q_m - P) \qquad (2.2)$$

since, with the exception of the first vector, $(P - Q_0)$, and of the last vector, $(Q_m - P)$, along the projecting rays only couples with nil lever arm are obtained.

With reference to Fig. 2.4, where a system of five forces is represented, the funicular polygon a'-b'-c'-d'-e'-f' is identified by tracing a' parallel to a, b' parallel to b and so on, by starting from a point arbitrarily chosen along the line of action of the first vector. The funicular polygon corresponds to the transformation of the assigned system of forces in a system consisting of only two forces, acting along the straight lines a' and f', and equivalent to the given system, since it is obtained through invariant operations with respect to the sum of vectors. The procedure allows for determining the central axis of the vector system by considering the straight lines a' and f': the line of action of $\boldsymbol{R}$ will pass through their intersection point.

Graphical determination of the resultant moment. Consider now a system of m planar forces $(P_i - A_i)$, applied at points A_i, belonging to the plane π, with $i = 1, 2, 3, \ldots, m$.

By constructing the force polygon, which gives the resultant $\boldsymbol{R}$, and the funicular polygon associated with the projecting pole P, which identifies the central axis, the given system of forces is transformed into one equivalent to it consisting of the resultant vector, $\boldsymbol{R}$, applied to any point of the central axis.

With reference to Fig. 2.5, related to a system of five forces, the resultant moment with respect to the generic point O can be expressed as:

$$\boldsymbol{M}(O) = (A - O) \wedge \boldsymbol{R},$$

where A is a generic point belonging to the central axis.

Let now consider the force polygon and denote as h the distance of the projecting pole, P, from the line Q_0Q_5, representing the resultant vector $\boldsymbol{R}$. Then, by considering the funicular polygon, draw the parallel r to the line of action of $\boldsymbol{R}$, so that it passes through point O; the first and last side of the funicular polygon, intersecting each other at point X, intercept two points on the line r, $0''$ and $5''$, which together with X define the triangle $0''5''X$ similar to the triangle $Q_0Q_5\,P$ of the vector polygon. By denoting as d^* the distance of the central axis from the point O (centre of moments), it follows that:

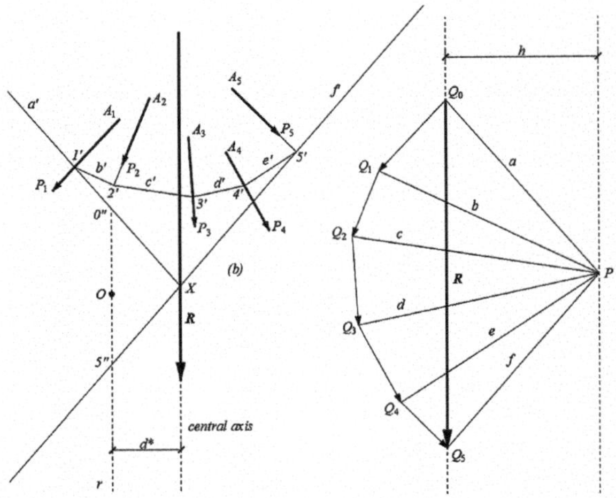

Fig. 2.5 Graphical determination of the resultant moment

$$|\boldsymbol{R}| : \overline{0''5''} = h{:}\ d^*.$$

Thus, the magnitude of the moment is measured by:

$$|M(O)| = |R| \cdot d^* = \overline{0''5''} \cdot h.$$

Properties of the funicular polygon. Observe that the tracing of the funicular polygon of a system of plane vectors depends on the choice in the plane of the arbitrary position of the projecting pole, *P*, and on the arbitrary choice of point *1'* along the line of action of the first vector; it follows that the funicular polygon of a system of plane vectors is three times indeterminate, since that there are ∞^3 funicular polygons that lead to the identification of the only central axis of the vector system.

It may be interesting, then, to analyze how to draw funicular polygons so that they are associated with a given system of forces and subject to two or three conditions, compatible and independent. To this purpose, Culmann's Theorem is briefly recalled.

Culmann's Theorem. Given a vector system $\boldsymbol{v}_1$, $\boldsymbol{v}_2$, ..., $\boldsymbol{v}_m$, consider the polygon p' which is obtained from the generic pole P'. Now consider a new pole, P'', and the new funicular polygon p'' which is thus obtained. The homologues sides of the two polygons meet at points $C_1,\ldots, C_{m+1}$, which are aligned on a straight line, called Culmann's axis, parallel to that joining the two poles P' and P'' (Fig. 2.6).

Funicular polygon passing through two points. Culmann's theorem allows for drawing a funicular polygon passing through two points, *A* and *B*. The procedure is described in the following, by means of an example. A funicular polygon p', corresponding to the pole P', is drawn so that the first side passes through point *A* (Fig. 2.7).

A second funicular polygon p'' has to satisfy the following condition: in addition to passing through point *A* (first side), also it must pass through point *B* (third side).

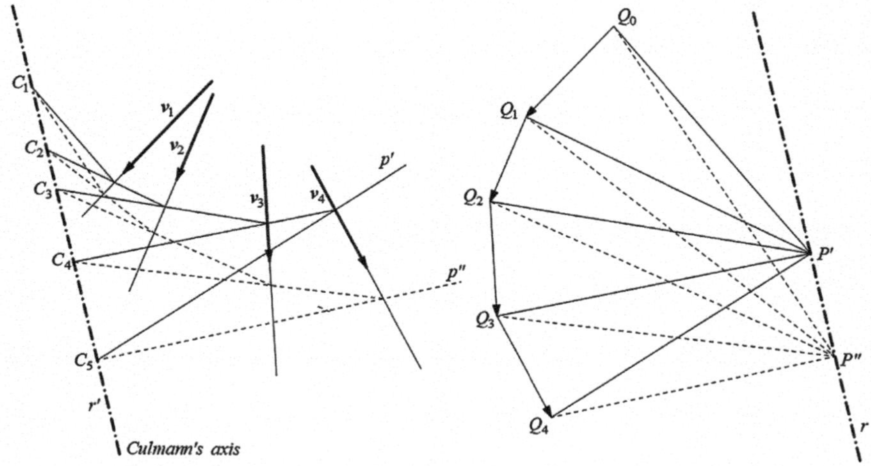

Fig. 2.6 Culmann's theorem: determination of Culmann's axis

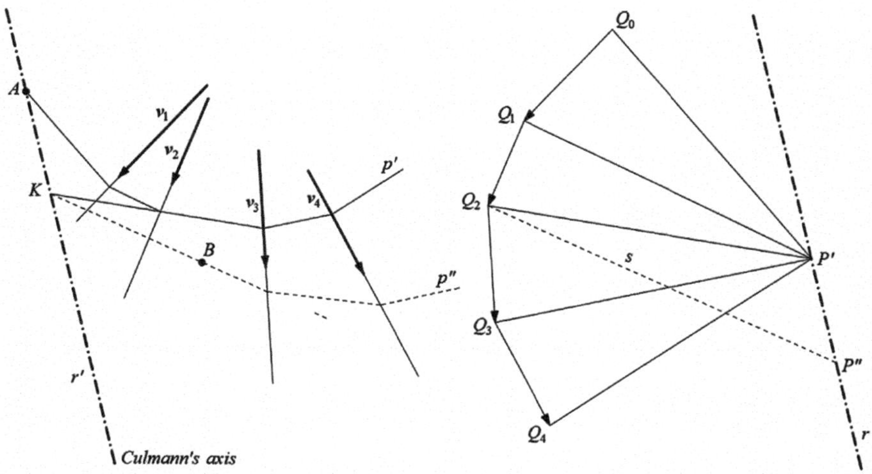

Fig. 2.7 Funicular polygon passing through two points: determining pole P''

Since this funicular polygon p'' must first pass through A, the first side meets the homologous side (i.e., the first side) of the first funicular polygon right at point $A = C_1$. More in detail, point C_1 is the point where the first sides of the two funicular polygons intersect.

Then consider the third side of the second funicular polygon p'': this side must pass through B. Draw this side arbitrarily starting from B and let $K = C_3$ the point where it meets the homologous side of the first polygon p' (i.e., the third side). The straight line joining A with K, then, is the Culmann axis r'.

Fig. 2.8 Funicular polygon passing through two points, *A* and *B*

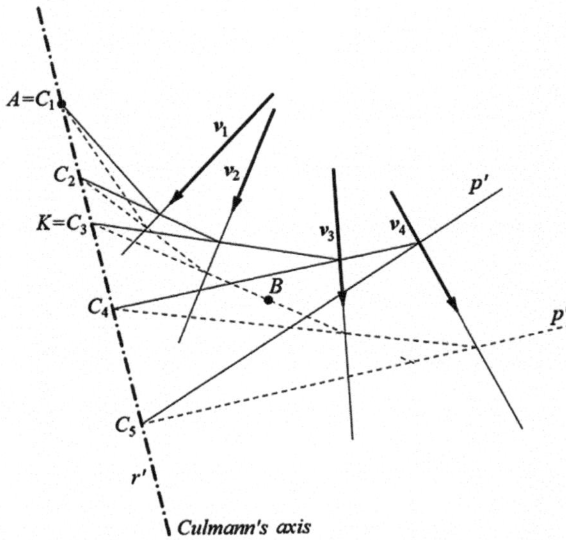

The two polygons p' and p'' have their respective poles P' and P'' on a straight line r parallel to r'. Since the straight line *BK* is the third side of the funicular polygon, this must connect two points belonging to the lines of actions of vectors $\boldsymbol{v}_2$ and $\boldsymbol{v}_3$ and therefore must be parallel to the straight line joining the new pole P'' with the extreme Q_2 of the segment corresponding to vector $\boldsymbol{v}_2$.

By drawing the parallel to *BK* from point Q_2 of the vector polygon, a straight line s is obtained on which the pole P'' must be located. Since P'' must also lie on r, the pole P'' is determined at the intersection of s and r.

Thus, the funicular polygon p'' is identified. Given the pole P'', it is easy to draw the corresponding funicular polygon, p'', starting from point *A* (see Fig. 2.8).

Funicular polygon passing through three points. Consider now a further condition, i.e., the fifth side of a new funicular polygon, p''' is imposed to pass through a third point, *C*.

Since p''' must also pass through points *A* and *B*, as imposed by the assigned constraints, the Culmann axis related to the funicular polygons p'' and p''' is the straight line t' joining *A* with *B*. Therefore the pole P''' must be on the straight line t parallel to the straight line t' and passing through P'' (Fig. 2.9).

According to Culmann's theorem, if the pole P''' lies on the straight line t, the homologues sides of p'' and p''' must meet on the straight line t'; moreover, p'' and p''' will certainly pass through *A* and *B*.

The side of p''' that is imposed to pass through *C* is the fifth. Therefore the fifth side of p''' must pass through the point *H* at which the fifth side of p'' intersects the axis t'. Since the fifth side of p''' must also pass through *C*, this side will be identified by the straight line joining *H* with *C*.

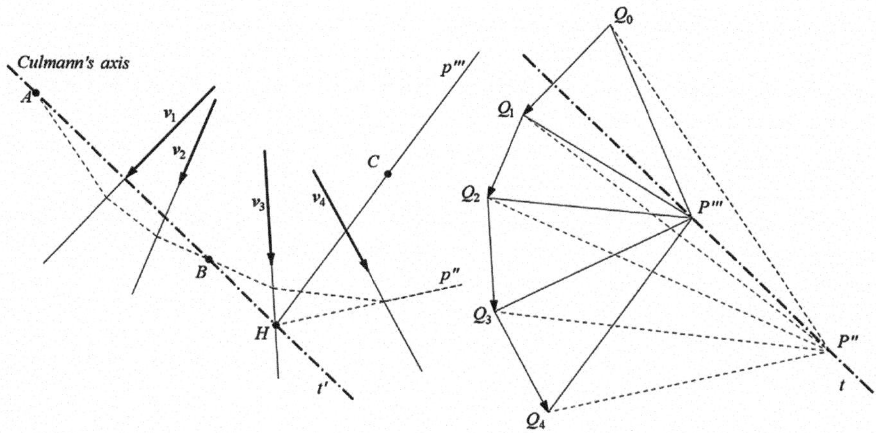

Fig. 2.9 Funicular polygon passing through three points: determining pole *P*"'

In the current example, the fifth side of p''' is also the last. So it corresponds to the projection of point Q_4 in the force polygon. By tracing the straight line parallel to HC from point Q_4, this straight line must contain the pole P''' of the funicular polygon p'''. Since P''' must also lie on t, the pole P''' is determined at the intersection of the two straight lines.

Thus, it is possible to draw the funicular polygon that passes through A, B and C, by assuming P''' as the pole (Fig. 2.10).

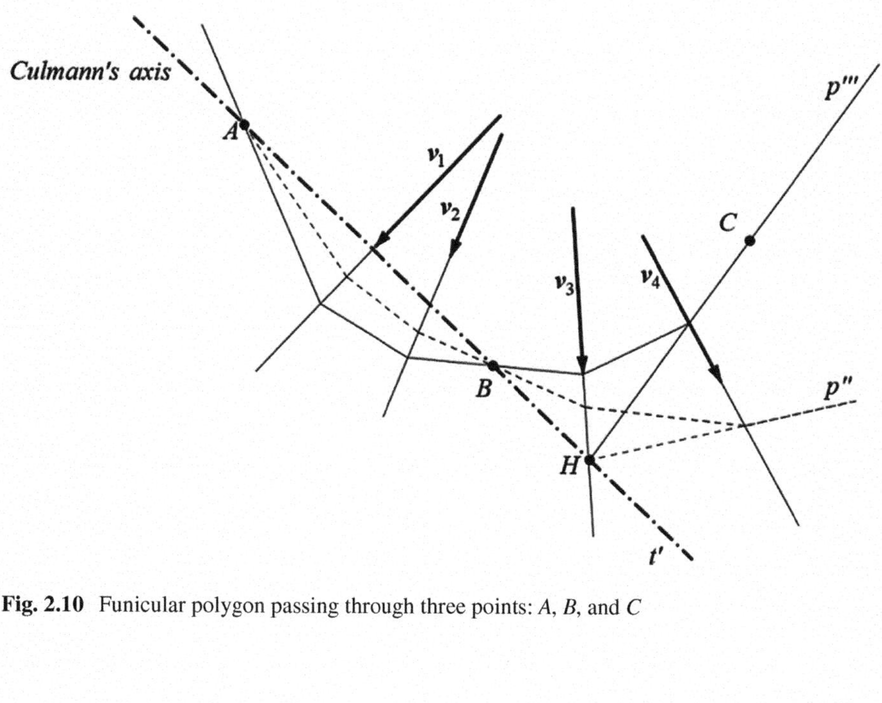

Fig. 2.10 Funicular polygon passing through three points: *A*, *B*, and *C*

Re-reading the treatise by Wolfe (1921). The treatise *Graphical Analysis. A text book on graphic statics* by Wolfe (1921) is distinguished for clarity and completeness. The interested reader is referred to this text for insights on the applications of graphical statics, which are presented by starting from the fundamentals through a graphic approach. In the following paragraphs an interesting application of graphical statics to the mechanics of masonry arches is introduced by referring to Wolfe's text.

The subject under examination is a special case of non-concurrent coplanar forces, i.e., a system of parallel forces. The line of action of their resultant can be found by exploiting the funicular polygon construction.

As an example, in Fig. 2.11, *left* (Fig. 32 from the original), the force polygon is drawn for the seven parallel forces shown in Fig. 2.11, *right* (Fig. 31 from the original) whose magnitude, direction, and action line is defined. The force polygon is drawn by starting at point 1 with the vector representing F_1. Observe that the sense of F_4 and F_6 is up. The magnitude of the resultant is obtained by considering the distance from point 1 to point 8, i.e., the end of the last vector. Through this construction, the resultant and the central axis are determined.

Starting from the construction described in Fig. 2.11, Wolfe presents a fundamental application of graphic statics, i.e., determining the line of pressure in a masonry arch. Although this concept was introduced and developed by some previous scholars (among them, Méry, 1840), it is interesting to recall the treatment proposed by Wolfe in order to position its application potential within the context of graphical statics.

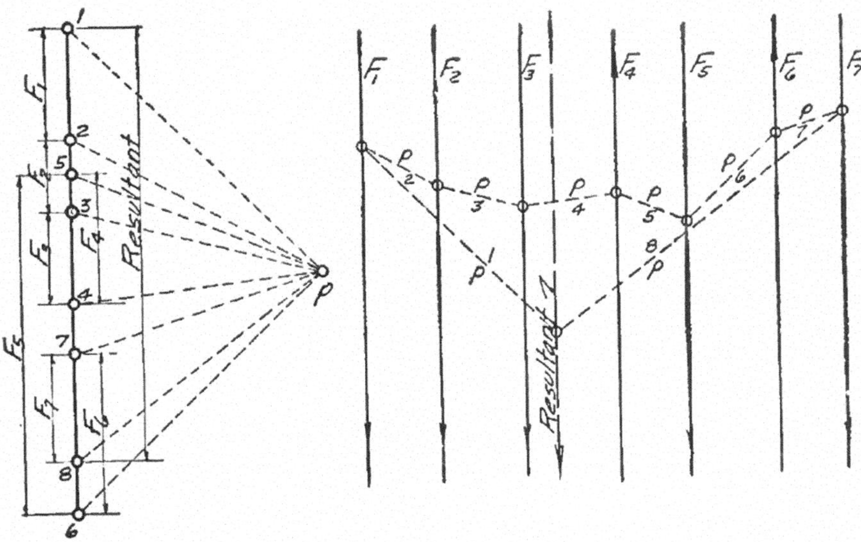

FIG. 32. FIG. 31.

Fig. 2.11 Resultant of a system of parallel forces (from Wolfe, 1921)

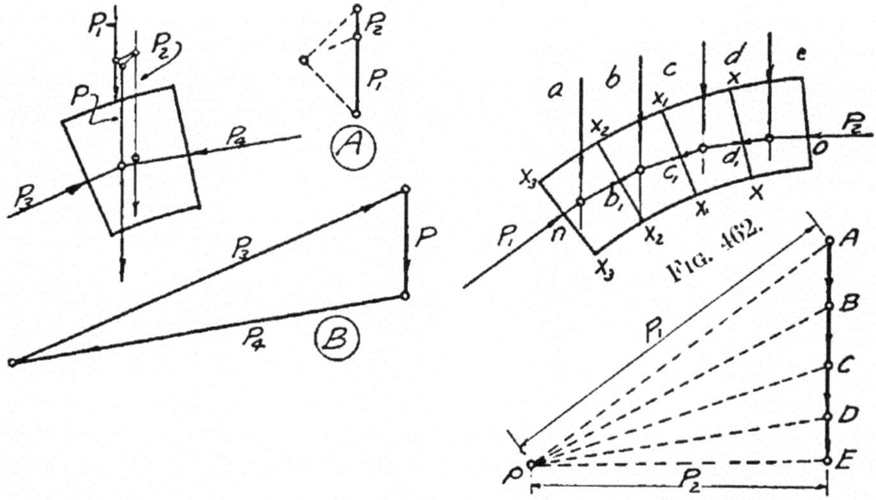

FIG. 461. FIG. 463.

Fig. 2.12 Line of pressure in a masonry arch (from Wolfe, 1921)

Figure 2.12 (Fig. 461 from the original) shows a *voussoir* belonging to a masonry arch. The force P represents the resultant of the load, P_1, and the weight, P_2. The force polygon A and the corresponding funicular polygon allow for determining the action line of P.

This force, P, is equilibrated by forces P_3 and P_4, which are the resultant forces exerted on the *voussoir* by the portion of the arch at its left and at its right, respectively. These three forces must be concurrent in order to guarantee the equilibrium.

In Fig. 2.12 (Fig. 462 from the original) a portion of the arch is shown, subdivided into a number of *voussoirs*. Below the force polygon is shown (Fig. 463 from the original): the segments A-B, B-C, C-D, etc., correspond to the resultant vertical loads acting on each subdivision, including the weight of each *voussoir* (similarly to the force P). With reference to Fig. 2.12, suppose that force P_2 is represented in magnitude and direction by the line E-P in Fig. 463 (from the original) and passes through point o in Fig. 462. The resultant of the two forces D-E and P_2 is represented by D-p in Fig. 463 and its action line passes through the intersection point of forces D-E and P_2. This resultant acts on the section x-x (see Fig. 462) from the right and intersects x-x at point d_1. In order to guarantee equilibrium, there must be an opposite reaction on x-x from the left. Then, the magnitude and direction of the resultant pressure on the section x-x is given then by line D-p in Fig. 463; this internal force acts at d_1. The procedure can be repeated for sections x_1-x_1, x_2-x_2, etc.

Wolfe gives, then, his definition of line of pressure (Fig. 462): "The broken line n-b_1-c_1-d_1-o may be called the line of pressure for the portion of the arch shown in the figure, but the student should keep in mind that the true line of pressure for a distributed load is a smooth curve tangent to the various lengths of the broken line, on the inside" (Wolfe, 1921, p. 231).

Taking into account his definition, Wolfe observes that "the broken line n-b_1-c_1-d_1-o is nothing more than a funicular polygon for the forces *A*-*B*, *B*-*C*, etc., passing through the points *o* and *n*, with *p* of Fig. 463 used as the pole. The resultant force acting on any section like x_1-x_1 or x_2-x_2 is given in magnitude and direction by the corresponding ray in the force polygon, and it acts at the place where the section cuts the line of pressure. Having located the point where the line of pressure intersects the section, the eccentricity may be measured, and since the magnitude of the force is known, the maximum unit pressure on the section can be found and also the distribution of stress" (Wolfe, 1921, p. 231).

Before proceeding further, in the Sect. 2.2 some remarks on the concept of line of pressure will be recalled, by starting from a re-reading of some historical contributions. This will allow for making some clarifications on the definition provided by Wolfe, which overlaps the definition of line of pressure with that of line of resistance (Gerstner, 1831–1834).

2.2 Traditional Methods and Their Novel Versions for Masonry Arches and Domes

2.2.1 *Coulomb's Method of* "Maximis et Minimis"

The thrust line method can be seen as a graphical reinterpretation of the method proposed by Coulomb in his *Essai sur une application de règles de maximis et minimis à quelques problèmes de Statique, relatifs à l'Architecture*, dedicated to the equilibrium of masonry vaults. It is therefore appropriate to briefly recall in the following Coulomb's fundamental contribution. The subject of Coulomb's treatment is a symmetric arch (or vault), whose intrados curve *aB* and extrados curve *Gb* are assigned (Fig. 2.13); the joints *Mm* are assumed to be perpendicular to the intrados curve. The purpose of the study is to define the limits of the internal horizontal force applied at point *f* of the key joint able to guarantee the equilibrium of the arch, by assuming that the structure is subjected to its own weight. The constitutive hypotheses on masonry are the presence of finite friction and cohesion between the joints.

An important point on Coulomb's treatment, is the fact that he acknowledges the role of internal reactions for the equilibrium of the arch.

More in detail, due to the symmetry of both geometry and loads, he analyses the equilibrium of the *voussoir* extending from the crown section *aG* to a generic joint, *Mm*, as a function of the unknown thrust, *P*, applied at any point *f* of the key joint (Fig. 2.14).

In the following Coulomb's treatment is described in a simplified way, by supposing that the friction is finite, the compressive strength of the material is infinite, the tensile strength is nil. The following paragraphs are re-elaborated from (Aita &

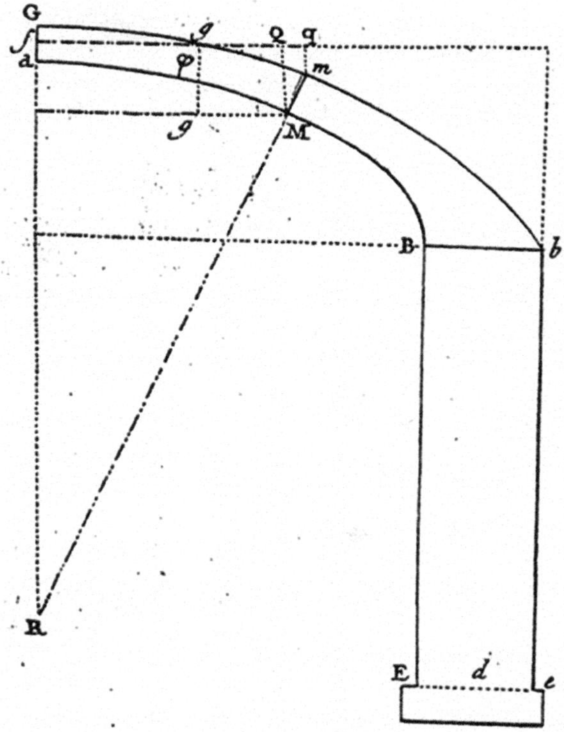

Fig. 2.13 The Coulomb method (*maximis et minimis*) (from Coulomb, 1776)

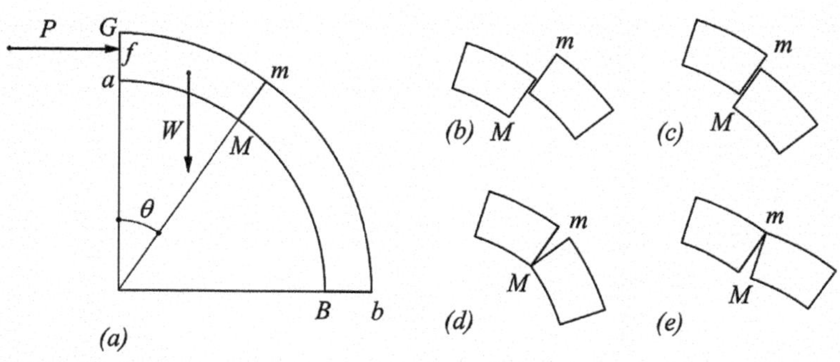

Fig. 2.14 Coulomb's scheme for the equilibrium analysis of the arch

Sinopoli, 2020), to which the interested reader is addressed (for further insights, see also Heyman, 1966; Sinopoli, Corradi & Foce, 1997).

Let R_n and R_t denote the normal and tangential component of the resultant reaction $\boldsymbol{R}$ acting at joint Mm, measured with respect to local unity vectors at joint θ, and let $\boldsymbol{M}$ be the corresponding moment with respect to the joint centre (Fig. 2.15). In accordance with the assumed mechanical model, the application point of reaction $\boldsymbol{R}$ must be internal to the joint thickness with extreme positions at the extrados, M, and intrados, m, respectively. Thus, if h represents the arch thickness at joint θ, the necessary and sufficient condition for the local rotational equilibrium of the *voussoir* at joint Mm is that:

$$|\boldsymbol{M}| \leq \frac{h}{2} \cdot R_n \tag{2.3}$$

while, if μ denotes the friction coefficient, the local equilibrium with respect to sliding requires that reaction $\boldsymbol{R}$ is internal to Coulomb's cone, defined by angle φ:

$$|\boldsymbol{R}_t| \leq \mu \cdot R_n \tag{2.4}$$

where $\mu = \tan\varphi$. Relationships (2.3) and (2.4) correspond to the analysis proposed by Coulomb, who correctly considered two inequalities for the rotational equilibrium, and two inequalities with respect to sliding.

In correspondence to a generic application point f at the key, the rotational limit condition at joint θ is reached when the reaction $\boldsymbol{R}$ crosses either the intrados point M, by identifying the local minimum value $P_{min}^{r,f}(\theta)$ for P, or the extrados point m, by identifying the local maximum value $P_{max}^{r,f}(\theta)$ for P (Fig. 2.16). These extreme values of the thrust represent the local limit equilibrium conditions with respect to the mutual rotation around the intrados (Fig. 2.14d), and the mutual rotation around the extrados (Fig. 2.14e), respectively.

Analogously, the limit condition with respect to sliding along the joint, which does not depend on the application point f at the key, is reached when the line of action of $\boldsymbol{R}$ reaches one of the boundaries of Coulomb's cone by identifying the local minimum value $P_{min}^{s}(\theta)$ for P (Fig. 2.14b), or the local maximum value $P_{max}^{s}(\theta)$ for P

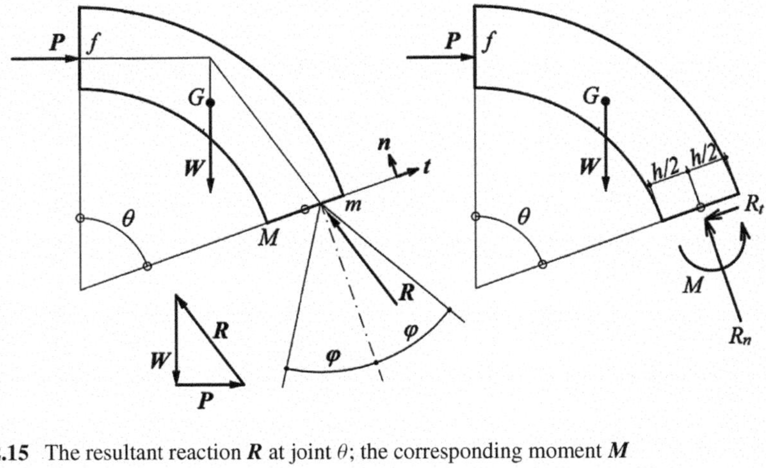

Fig. 2.15 The resultant reaction $\boldsymbol{R}$ at joint θ; the corresponding moment $\boldsymbol{M}$

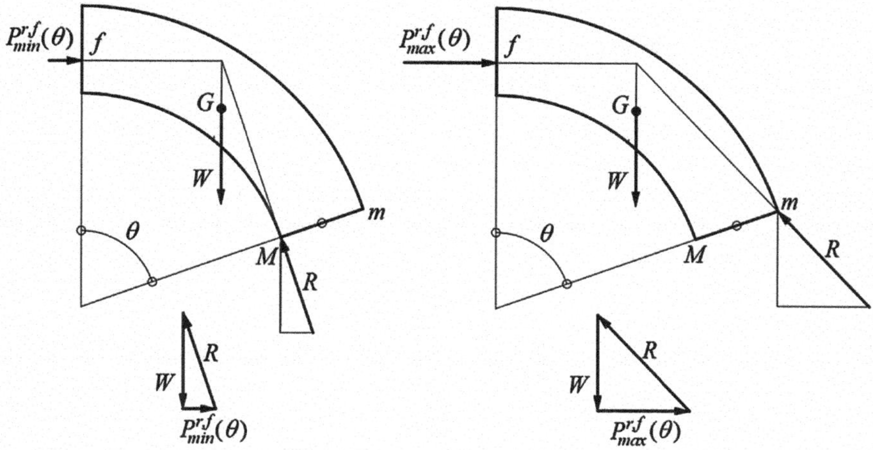

Fig. 2.16 Determining the minimum and maximum values of the crown thrust, P

(Fig. 2.14c). More details on sliding are given in Sect. 2.3, as regards Durand-Claye's method.

By extending the equilibrium conditions from the crown section to the springing, fixing the position of point f along the key joint, it is possible to determine the domain P^r of the admissible values of thrust P with respect to the rotational equilibrium:

$$max_\theta\left[P^{r,f}_{min}(\theta)\right] \leq P^r \leq min_\theta\left[P^{r,f}_{max}(\theta)\right] \tag{2.5}$$

while the domain P^s of the admissible values of thrust P with respect to sliding along the whole arch, not depending on the application point f, corresponds to:

$$max_\theta\left[P^{s}_{min}(\theta)\right] \leq P^s \leq min_\theta\left[P^{s}_{max}(\theta)\right] \tag{2.6}$$

Thus, according to Coulomb's method, the equilibrium of the whole arch is guaranteed if inequalities (2.5) and (2.6), defining the rotational and sliding equilibrium domain, respectively, are both satisfied and the intersection set between these two domains is not empty for at least one position of the application point, f, of the crown thrust.

In *Remarque* I of his *Mémoire* Coulomb states that for masonry arches the friction coefficient is generally large enough to prevent sliding so that inequality (2.6) can be disregarded. The arch stability depends therefore only on inequality (2.5), and the limit condition corresponding to the rotational collapse mode is reached when the rotational domain identifies a single value for P for a single value of the eccentricity corresponding to the application point, f (this limit condition corresponds to the only solution compatible with equilibrium).

2.2.2 *The Thrust Line Method for the Assessment of Masonry Arches*

The assessment of the equilibrium of masonry arches through the tools deriving from graphical statics was anticipated by the analogy of the inversed hanging chain, discovered by Robert Hooke (1675) and written as an anagram, *"abcccddeeeeefggi-iiiiiiillmmmmnnnnnooprrsssttttttuuuuuuuux"* (*ut pendet continuum flexile, sic stabit contiguum rigidum inversum*) (Hooke, 1675, p. 31). This analogy describes the relationship between the shape of a hanging chain under its own weight and the shape of a masonry arch.

In the nineteenth century, the concept of thrust line was first introduced by Gerstner (1831, 1834), which defined two different notions, the line of resistance and the line of pressure. More in detail, the line of resistance is the polygon joining the pressure centres on the joints of an arch; the line of pressure is the *locus* of the intersection points of the pressures acting on neighbouring joints, i.e., the envelope of the reactive forces between joint and joint.

In the case of infinite or very high friction, the thrust line method can be placed in the perspective of the 'safe theorem' (Heyman, 1966): the stability of the arch is guaranteed if there is any line of resistance contained within the arch's thickness at each joint. The line of resistance and the line of pressure are both named as 'thrust line', but it would be more correct to always remember their conceptual difference. The rotational equilibrium of an arch, with reference to the impossibility of activating rotational collapse mechanisms, is guaranteed if the line of resistance is contained anywhere within the arch's thickness. The angle according to which the pressure line intersects the joints, on the other hand, is related to sliding equilibrium: as H. Moseley (1833, 1843) will elaborate further, if it deviates too much from the right angle, mutual sliding between the *voussoirs* could arise, depending on the friction angle (Fig. 2.17).

It should be observed that for an arch in equilibrium it is possible to draw infinite admissible thrust lines (for an assigned geometric and load condition), since the statics of rigid bodies does not allow the univocal determination of the reactions and, therefore, of the thrust line (the problem is hyperstatic). Despite this, the historical contributions on the thrust line focus their attention on choosing the 'actual' thrust

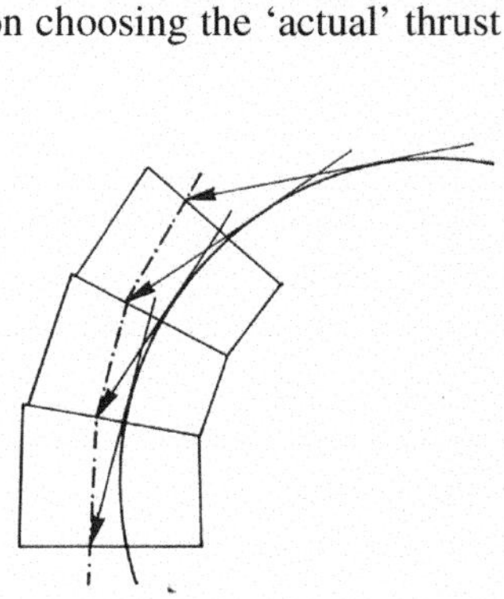

Fig. 2.17 Line of resistance and line of pressure (from Benvenuto, 1991)

line. This is the less convincing aspect of the contributions related to the thrust line method: they do not depart from the pre-elastic methodological tradition; however, they do not limit themselves at evaluating 'only' the stability of the structure, but aim at tracing the 'true' thrust line, forcibly introducing elements regarding the resistance of materials or 'metaphysical' principles (Foce & Aita, 2003).

As an example, Gerstner fails to give a correct answer to this problem, since he introduces further hypotheses, which turn out to be arbitrary.

Other scholars focus on the indeterminacy of the problem. Among these, it is interesting to cite the contribution by Barlow (1846), in which the results of some experiments on stone arches are described with the aim of showing that the same arch can admit different conditions of equilibrium. In other words, infinite solutions can be identified by requiring the thrust at the crown section to pass through a different point of the section each time.

Among the contributions aimed at searching for the 'true' thrust line, we can first mention those of Moseley (1843) and Scheffler (1857), whose solutions are based on a principle of minimum resistance. Already since 1833 Moseley had introduced a principle of minimum resistance for the solution of hyperstatic problems (Moseley, 1833); in 1839 he applied this principle to the statics of masonry arches assuming that, among all the admissible thrust lines traceable starting from a generic point of the crown section, the 'true' one minimizes the value of the horizontal thrust; to this solution, Scheffler adds the condition that the maximum compressive stress at the crown section and at the haunches must not exceed the compressive strength of the material.

It is also interesting to recall the abstract and theoretical formulations by Villarceau (1854), who proposed a method for the design of the ideal arch by conceiving the vault as an assemblage of an infinite number of *voussoirs* with joints perpendicular to the intrados; the theories of Carvallo (1853), Denfert-Rochereau (1859), Saint-Guilhelm (1859), aimed at guaranteeing the best performance of an arch with the minimum use of material, the introduction of convenient hypotheses to trace the line of pressure, formulated by Laterrade (1885). Drouets (1865) proposes a principle of minimum action, according to which the stress state that occurs inside the arch is that corresponding to the lowest value of the maximum compression stresses. For further details on this discussion (see Foce, 2002).

In this context, the method of the 'stability areas' of Durand-Claye (1867) takes on a particularly important place for its theoretical rigor. This method aims at verifying "*s'il existe des solutions d'équilibre compatibles avec un effort-limit donné*" (Durand-Claye, 1867, p. 66). Renouncing to determine the actual stress state, the graphical procedure proposed by Durand-Claye allows for defining all the values of the admissible crown thrusts with reference to the global equilibrium of the system and to the strength of the material. This method will be examined in detail in Sect. 2.3.

For further insights on the historical evolution of the thrust line method, the interested reader is referred to (Foce, 2002; Kurrer, 2008).

2.2.3 A Focus on the Méry Method

Among the various contributions concerning the thrust line method, the well-known contribution by E. Méry (1840) will be examined in detail, since it provides a simple method for determining the thrust line, by translating the Coulomb method of *maximis et minimis* into a graphical representation. The main objective of Méry's contribution is—as the author himself writes at the beginning of his work (Méry, 1840)—to present a method aimed at verifying the stability of the vaults through a geometric construction that allows to avoid any calculation and of which he, personally, has been able to appreciate simplicity across numerous applications.

The hypotheses adopted by Méry in the first part of his contribution—more precisely, up to p. 63—are the same as those adopted by Coulomb in his reknown *Essai* (1776): the subject of the study is the symmetric masonry arch, modelled as a system of rigid bodies (*voussoirs*) connected by non-smooth unilateral constraints (at the joints). The hypotheses of finite friction (characterized by a friction angle $\varphi = \arctan \mu$) and infinite compressive strength are assumed for the masonry. In the second part of his essay Méry removes this last hypothesis by examining the effects of a limited compressive strength.

With reference to Fig. 2.18, drawn from the original contribution, let *Bb* be the vertical ideal crown joint, *Cc* a generic joint of inclination θ, β the application point of the crown thrust, *P*, acting at the crown section (Fig. 2.18a), γ the application point of the resultant reaction $R(\theta)$ at generic joint *Cc*. If we determine at each joint a point analogous to γ, the locus of points γ', γ'', …, is the *courbe de pression* (thrust line).

Méry writes that, if the *courbe de pression* touches the extrados at point *B* (Fig. 2.18b), this means that *"toutes les pressions que supporte le joint Bb sont concentrées en B"* (Méry, 1840, p. 51) and, then, the joint will open at *b*, i.e., at the intrados; analogously, if the *courbe de pression* passes through point *A*, joint *Aa* will open at the intrados; if it touches point *c*, joint *cC* will open at the opposite

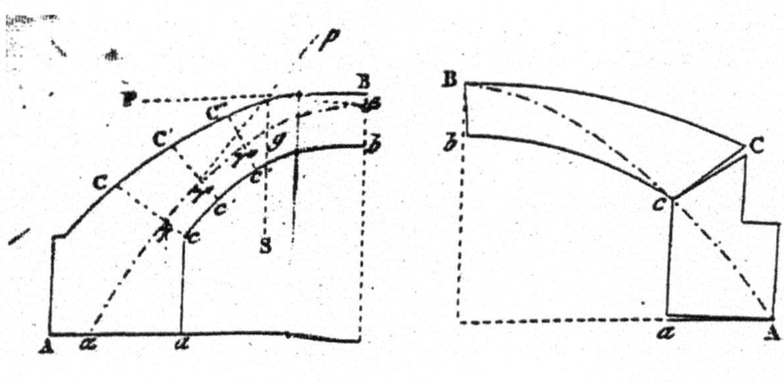

Fig. 2.18 Construction of the thrust line (**a**); *équilibre mathématique* (**b**) (from Méry, 1840)

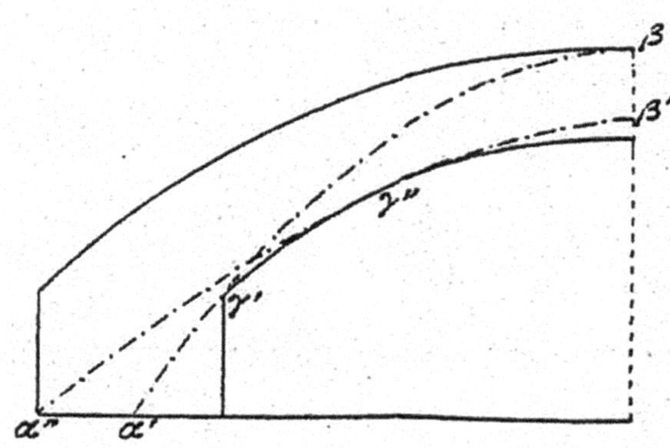

Fig. 2.19 The extreme positions of the thrust line, corresponding to the minimum and maximum crown thrust (from Méry, 1840)

side, at the extrados. In Fig. 2.18b, a case of *équilibre mathématique* is represented, corresponding to the minimum arch's thickness compatible with equilibrium and to a pure rotational collapse mode. The intrados and extrados curves are two boundaries for the thrust line. When the thrust line exits from such limits, the equilibrium is impossible for that thrust line. In this case, to have an estimate on the stability of the arch, it will be necessary to evaluate whether there is a thrust line entirely contained within the thickness of the arch (or—at most—tangent to the intrados or extrados line). Méry observes that, since the arches actually built have a thickness greater than that strictly necessary for mathematical equilibrium, infinite thrust lines can be traced, without it being possible to determine the true one.

According to Méry, however, to evaluate the stability of the arch, it is sufficient to consider "*deux positions extrêmes qui correspondent l'une au maximum de la poussée et l'autre au minimum*" (Méry, 1840, p. 52) (Fig. 2.19). This concept is of considerable importance and constitutes a significant development of Coulomb's treatment.

The domain P^r of the admissible values of thrust P with respect to the rotational equilibrium is then obtained for the whole arch:

$$min_f\left\{max_\theta\left[P^{r,f}_{min}(\theta)\right]\right\} \leq P^r \leq max_f\left\{min_\theta\left[P^{r,f}_{max}(\theta)\right]\right\},$$

where f denotes the application point of the horizontal thrust, P, at the key (see, for example, points β' and β'' in Fig. 2.19).

Before dwelling in detail on this aspect, let's analyze the geometric construction proposed by Méry to trace a *courbe de pression*. Suppose that the curve passes through two points α and β taken, respectively, along the horizontal joint, *Aa*, and the crown section, *Bb* (Fig. 2.20).

To determine the intersection points of the thrust line with the joints *Cc*, *C′c′*, …, the weights of the *voussoirs bBCc*, *bBC′c′*—denoted as $W(\theta)$, $W(\theta')$—and their respective centres of gravity are considered. Then, the horizontal βX is traced so that it passes through β which intersects at k and K, respectively, the vertical kg passing through the centre of gravity of the *voussoir bBCc* and the vertical KG passing

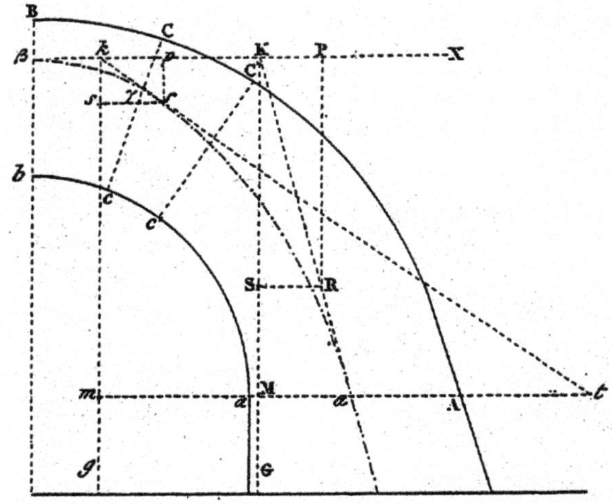

Fig. 2.20 The construction of the thrust line according to Méry (from Méry, 1840)

through the centre of gravity of the *voussoir bBAa*. Take on *kg* and *KG*, respectively, a segment *ks* equivalent to the weight of the segment *bBCc*, measured at a suitable scale, and a segment *KS* equivalent to the weight of the segment *bBAa* (denoted in the following as $W(\alpha)$, where $\alpha = 90°$ represents the inclination of the horizontal joint, *Aa*, with respect to the vertical, not to be confused with the point α belonging to the thrust line, in Méry's treatment).

By joining point α with K, the crown thrust is determined by imposing the equilibrium condition—represented by the triangle of forces *KSR*—which allows for defining its magnitude, measured by segment *SR*. Then, with reference to the generic joint *Cd*, by drawing *sr* = *SR*, the segment *kr* is obtained. This segment is proportional to the internal reaction at joint *Cc*, whose line of action will intersect the joint *Cc* at γ, point belonging to the thrust line. The same procedure can be repeated for all the other joints.

The geometrical construction is not much more complex when the points γ and α are given, while the position of the point β at the key is not known. In this case, it is necessary to trace the horizontal straight line passing through point α which intersects at points *m* and *M*, respectively, the verticals passing through the centres of gravity of the portions of the arch *bBCc* and *bBAa*.

A line segment $mt = \frac{W(\alpha)}{W(\theta)} M\alpha = \frac{KS}{ks} M\alpha$, such that

$$\frac{W(\theta)}{W(\alpha)} = \frac{ks}{KS} = \frac{M\alpha}{mt} \tag{2.7}$$

is then drawn on this horizontal line.

By joining *t* with γ, the point *k* is determined, at the intersection with the vertical passing through the centre of gravity of *bBCc*. It can be shown that this point belongs to the horizontal βX which served as the basis in the previous construction. We have, as a matter of fact:

$$\frac{ks}{sr} = \frac{km}{mt},$$

and, then,

$$sr = ks\frac{mt}{km}.$$

Furthermore, the following relation holds:

$$\frac{KS}{SR} = \frac{KM}{M\alpha} = \frac{km}{M\alpha}$$

from which

$$SR = KS\frac{M\alpha}{km}.$$

Now, by considering relation (2.7), $ks \cdot mt = KS \cdot M\alpha$, then $sr = SR$. Thus, for the equilibrium, sr corresponds to the crown thrust, P.

When the resultant reaction, $R(\theta)$, is not perpendicular to joint θ, the adjacent *voussoirs* could be subjected to mutual sliding if the friction coefficient, μ, is not sufficient. According to Méry, the inclination of the *courbe de pression* with respect to each joint, at each point γ, measures the tendency of the adjacent *voussoirs* to mutual sliding. For example, by examining the shape of the thrust line of Fig. 2.21, joint $c'C'$ is the one where sliding is most likely.

The author notes in passing that the line of action of the internal reaction at any joint is not tangent, as a rule, to the *courbe de pression*, so the previous statement is not exact. However, according to Méry, "*l'on n'exposera pas à des erreurs importantes en faisant abstraction de cette cause d'inexaxtitude*" (Méry, 1840, p. 52). This delicate issue will be rigorously clarified by Milankovitch (1907).

The following images refer to the drawing of the thrust line, proposed by Méry, in some cases of mathematical equilibrium corresponding to the minimum thickness

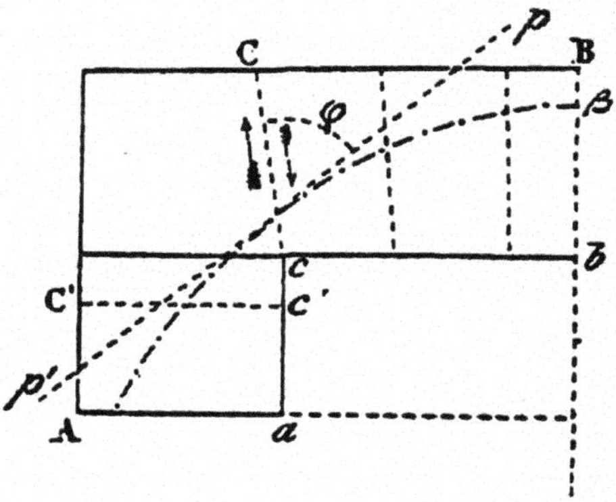

Fig. 2.21 Thrust line for a flat arch: joint $c'C'$ is the one where sliding is most likely (from Méry, 1840)

of the arch. It is interesting to note that in the applications aimed at searching for possible rotational collapse mechanisms, Méry determines the application point of the horizontal crown thrust, without imposing a priori its passage through the extrados of the key. Figure 2.22 shows some cases corresponding to the occurring of the first rotational collapse mode for symmetric masonry arches: starting from the crown section, three hinges are formed at the extrados, intrados and extrados; in Fig. 2.23 the limit thickness corresponds to the second collapse mode (the position of the hinges is opposite with respect to that of the first collapse mode: intrados, extrados, intrados); finally, in Fig. 2.24 some more complex cases are shown.

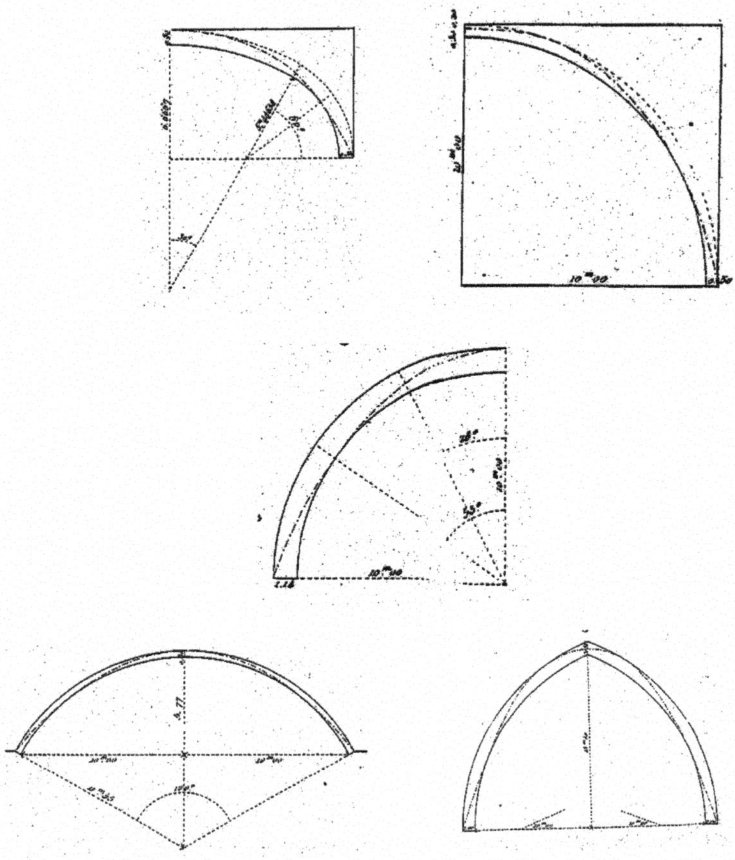

Fig. 2.22 Limit equilibrium: first rotational collapse mode (from Méry, 1840)

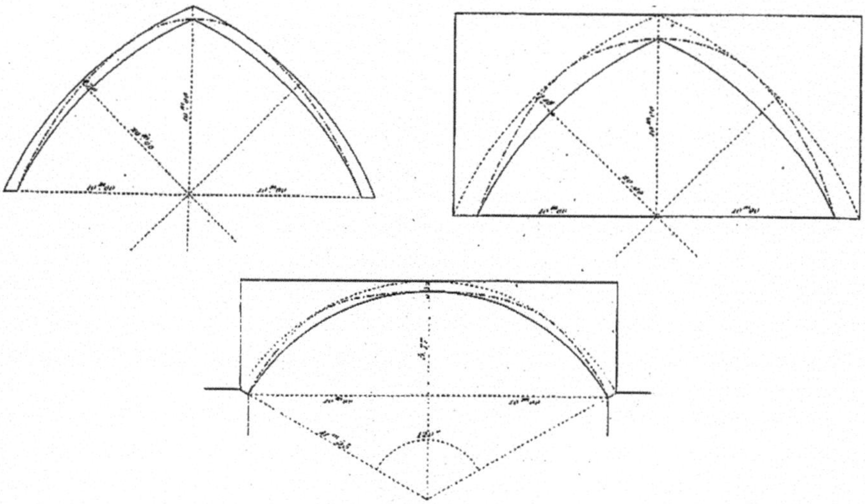

Fig. 2.23 Limit equilibrium: second rotational collapse mode (from Méry, 1840)

Fig. 2.24 Limit equilibrium: rotational collapse modes (from Méry, 1840)

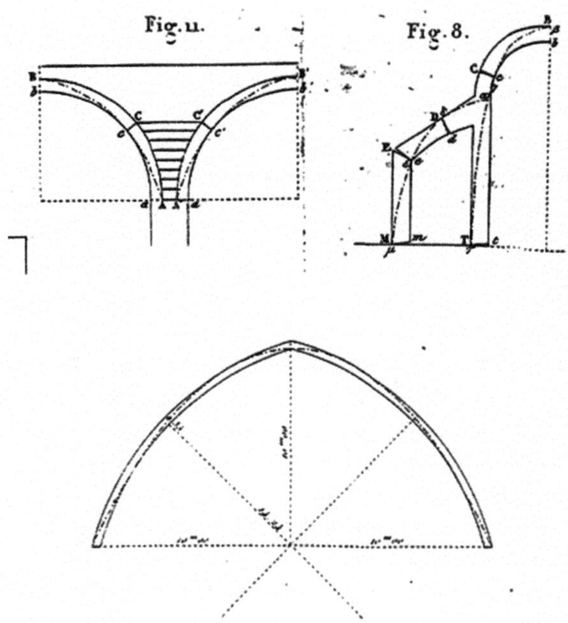

2.2.4 The True 'Method of Méry' and the So-Called 'Method of Méry'

In his contribution of 1840, p. 63, Méry writes: "*Jusqu'ici nous avons admis que l'espace où peut exister la courbe des pressions, n'avait d'autres limites que celles des voussoirs eux-mêmes. Mais nous avons fait remarquer que cette hypothèse exige des matériaux une résistance infinie; il faut donc la modifier dans la pratique*".

The author continues by observing that the thrust line will never reach the limits given by the intrados and extrados of the arch, but must remain within it: "*elle sera renfermée dans des limites plus resserrées que celles que nous avons considerées jusqu'à present*" (Méry, 1840, p. 64), since the compressive strength of the material has to be taken into account to avoid crushing.

It is interesting to observe that the search for these new limits, proposed by Méry, does not lead to the method that in literature is known as the 'Méry method', according to which the thrust line cannot go beyond the middle third at each joint. The so-called 'Méry method' is conceptually due to Navier (Sinopoli & Foce, 2001), who is not yet familiar with the concept of thrust line, introduced by Gerstner later (Gerstner, 1831, 1834). Navier writes: "*il paraît nécessaire d'avoir égard à l'élasticité de la matière des voussoirs. Cette question serait un cas particulier d'une question plus générale, qui consiste à déterminer les effets qui se produisent dans un corps élastique de figure quelconque, soumis à l'action de diverses forces*" (Navier, 1826, p. 164). However, Navier's treatment does not fully respect the purposes declared by the author. In fact, he decided to correct Coulomb's solution, obtained under the assumptions of infinite compressive strength and zero tensile strength, by considering the boundary of the middle third as the new limit for the thrust line, assuming a finite compressive strength, and a linear distribution of the stresses along each joint. Moreover, Navier excludes the possibility of partialization of the sections. In doing so, while Coulomb was concerned about the limit condition of equilibrium, Navier refers to a more severe situation, in which the critical joints are actually reactive in all their extension. The result of this procedure is ambiguous for two reasons: on the one hand, in fact, it refers to the philosophy of limit analysis, compromising its intrinsic coherence when the attention is shifted from the study of the collapse to the description of a condition of stable equilibrium; on the other hand, the hyperstatic nature of the problem makes the equations of equilibrium insufficient to solve it univocally (Foce, 2002).

The procedure proposed by Méry is different from the one described above: always assuming a linear distribution of stresses, he considers a limited compressive strength and a nil tensile stength, without excluding the possibility of partialization of the sections. By arbitrarily drawing a thrust line inside the thickness of the arch, the maximum compressive stresses to which some critical joints are subjected are roughly determined. With reference to Fig. 2.25 (*left*), related to the generic joint *dD*, Méry searches for the minimum distance of the thrust line from the end *D* "*sans que la maçonnerie s'écrase*" (Méry, 1840, p. 64). Suppose that the resultant reaction at the joint *dD*, $R(\theta)$, passes through the point *D'*. The hypothesis assumed by Méry, "*la plus conforme aux idées généralement admises sur l'élasticité des corps*" (Méry,

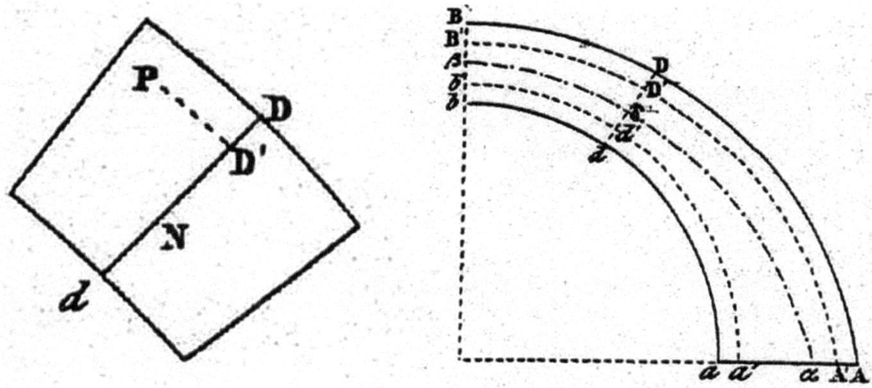

Fig. 2.25 The method proposed by Mery in order to take into account the compressive strength of the material (from Méry, 1840)

1840, p. 64), is that the compressive stresses are distributed linearly from the end D up to a point N determined so that $R(\theta)$ passes through D'. This condition requires that $ND' = 2\,D'D$.

A possible interpretation of Méry's text is set out below.

With reference to the generic joint dD of inclination θ, the maximum magnitude, σ_{max}, of the compressive stress is attained at point D (Fig. 2.25, *left*). If DD' measures the area of the joint portion of unit width between points D and D', the following relation holds:

$$R_n(\theta) = \frac{3}{2} DD' \sigma_{max},$$

where $R_n(\theta)$ denotes the normal component of $R(\theta)$ acting at joint θ. From this relation it is obtained

$$\sigma_{max} = \frac{2}{3}\frac{R_n(\theta)}{DD'}.$$

The resultant reaction at joint dD, $R(\theta)$, is then compatible with the admissible magnitude of compressive stress, σ_{amm}, if:

$$\sigma_{max} = \frac{2}{3}\frac{R_n(\theta)}{DD'} \leq \sigma_{amm}$$

i.e., according to an equivalent relation,

$$DD' \geq \frac{2}{3}\frac{R_n(\theta)}{\sigma_{amm}} \tag{2.8}$$

Méry observes that, if dD were smaller than DN, the factor 2/3 in (2.8) should be replaced by another between 1/2 and 2/3; however, for the sake of security, the author considers 2/3 in all cases, concluding that "*l'intervalle cherché DD′ doit être assez grand pour supporter les deux tiers de la pression totale*" (Méry, 1840, p. 64). By repeating the procedure for each joint, it will be possible to trace two limit curves $B'D'A'$, $b'd'a'$ (Fig. 2.25, *right*), "*destinées à marquer les limites que la courbe des pressions ne peut pas dépasser*" (Méry, 1840, p. 65). The interpretation of this last statement is not entirely clear: it would seem that, having chosen a thrust line arbitrarily, the limits found for it should be considered valid for any other curve to be traced. A more convincing reading could be, however, the following: the construction must be repeated for each thrust line, which is evaluated—case by case—as regards its compatibility with the limited compressive strength of the material. The two limit curves that result at the end of this procedure are $B'D'A'$ and $b'd'a'$.

According to Méry, through this method the analysis of the stability of masonry arches is brought back to that of their mathematical equilibrium by considering the two limit curves $B'D'A'$ and $b'd'a'$. If a limit condition occurs, it will be possible to draw only one thrust line; otherwise there will be infinite admissible thrust lines. As already observed with reference to Navier's contribution, also this procedure results to be ambiguous, since limit analysis is actually used to evaluate an operating condition. While Navier pays attention to the tensile strength of the material (preventing the partialization of the joints), Méry imposes a limit on the compressive strength. In this sense, the so-called 'Méry method'—actually due to Navier—is conceptually different from the method originally proposed by Méry.

Federico Foce (2010) investigated the matter in depth by examining the correspondence between Méry and Saint-Venant and underlining the importance of Saint-Venant's suggestions in the final draft of Méry's memoir. A last remark: as will be clarified in Sect. 2.3, the limit curves considered by Méry could be linked to the rotational domain identified by Durand-Claye (1867).

Méry's memoir is significant for its graphical approach and for the applications to masonry arches.

A mathematical formulation of the thrust line method is provided by Milankovitch (1904, 1907), who exploits differential calculus in order to determine the thrust line after assigning the law describing the load condition and the stereotomy of the joints. For the generality of the treatment, this contribution is suitable for applications on vaults of different shapes.

2.2.5 Thrust Line Analysis and Graphical Methods for the Stability of Masonry Domes

In the eighteenth century the results obtained through the thrust line method for masonry arches are applied to masonry domes. The 19th-century contributions are characterized by the attempt to assess the stability of domes exploiting the tools provided by thrust line analysis, graphical statics and membrane theory.

In this section some historical contributions will be presented, according to which the membrane solutions are translated in terms of graphical statics.

The contribution by Eddy (1877). The formulation proposed by Eddy was the first dedicated to the graphical analysis of masonry domes. In his treatise of 1877 (pp. 53–59), Eddy examines through graphical statics the stability of masonry arches and domes of revolution, by studying a dome with hemispherical shape. The author observes that the proposed construction can be applied to any dome of revolution.

The first dome under examination is a thin metallic dome, whose stability is studied through a graphical translation of the membrane theory.

In Fig. 2.26, drawn from (Eddy, 1877), the quadrant *aa* represents the part of the meridian section of a thin metallic dome, comprised between the key and the springing joint. The dome is supposed to be so thin that its thickness is not represented in the figure.

According to Eddy, by considering the membrane theory, in a thin metallic dome the thrust along a meridian section is necessarily in a tangent direction at each point of it. This consideration allows for determining the magnitude of this thrust as well as that of the hoop forces.

Eddy divides the height *ab* of the dome into a number of equal parts, $d_1u_1 = d_2u_2 = \ldots$. By conducting the horizontal planes passing through the points d_1, d_2, etc., and u_1, u_2, etc., the hemispherical surface is cut along circumferences passing through the points a_1, a_2, etc., and g_1, g_2, etc., respectively.

Suppose the thickness of the dome is constant. The vertical segment *ab* represents the weight of the portion of the dome *aa*, comprised between two meridian sections that form an infinitesimal angle between them. The weight of the portion of the lune above a_1d_1, for geometrical properties related to the area of the zone of a sphere, can be represented by segment ad_1. Similarly, au_1 is the weight of ag_1 and ad_2 the weight of aa_2. If the dome is not spherical, the weights can be determined by properly considering the actual geometry.

Now, the weight of the lune's portion aa_1 is sustained by a horizontal thrust, which is the resultant of the hoop pressures acting on the meridian planes, and by a thrust tangent to the axis line at point a_1. The procedure for determining the magnitude of the forces is the following. A horizontal straight line is drawn by starting from d_1; then, a straight line passing through *a*, parallel to the tangent at a_1, is conducted. This straight line will intersect at s_1 the horizontal straight line. The triangle ad_1s_1 is the force polygon (force triangle) which sustains the lune aa_1.

Analogously, au_1t_1 is the force triangle which guarantees the equilibrium of the lune ag_1.

If the procedure is repeated by considering all the lunes aa_1, ag_1, aa_2, ag_2, aa_3, ag_3, …, a curve *st* is obtained by joining all points s_1, t_1, s_2, t_2, s_3, t_3, … (Fig. 2.26).

Eddy writes the equation describing such a curve, with reference to a coordinate axes (*x*, *y*) system.

Finally, Eddy traces the vertical line tangent to this curve, by observing that the tangency point corresponds to a joint whose inclination is $51^\circ\ 49'$. The part of the

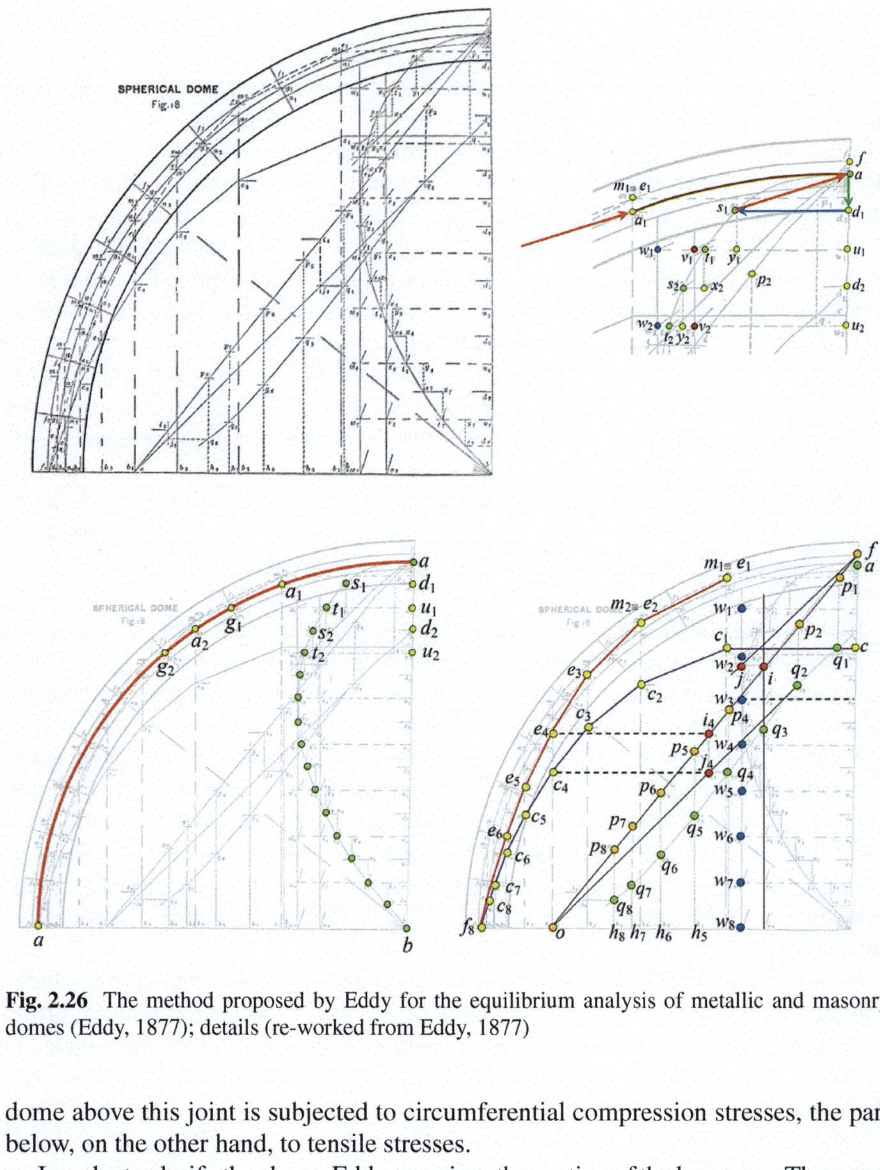

Fig. 2.26 The method proposed by Eddy for the equilibrium analysis of metallic and masonry domes (Eddy, 1877); details (re-worked from Eddy, 1877)

dome above this joint is subjected to circumferential compression stresses, the part below, on the other hand, to tensile stresses.

In order to clarify the above, Eddy examines the portion of the lune $g_1 a_2$. The upper section g_1 is subjected to a thrust whose horizontal radial component is proportional to $u_1 t_1$, while the horizontal thrust acting in correspondence with the lower section a_2 is proportional to $d_2 s_2$. The difference between these two horizontal forces is equal to $s_2 x_2$ and corresponds to a circumferential compression around the portion considered. Below the point of tangency between the vertical line and the curve, the differences of the sx (or ty) type change sign.

After exploiting the membrane theory for the assessment of a metallic dome, Eddy considers a masonry dome, whose line of axis coincides with the profile of the metallic dome, by assuming that masonry cannot withstand tensile stresses. In Fig. 2.26 this dome is represented with a uniform thickness equal to one-eighth of the intrados radius. The vertical radius of the centre line, *ab*, is ideally divided into any convenient number of equal parts (eight parts) at points u_1, u_2, etc. The straight lines of the conical joints, perpendicular to the surface of the dome, are supposed to pass through the points g_1, g_2, etc., belonging to the line of axis of the dome, at the same level with u_1, u_2, etc. Point *b* is the vertex of such conical surfaces.

Eddy considers a dome's lune, comprised between two meridian planes forming a small angle with each other. For geometrical properties (linked to the equality in area between spherical zones of the same height), the centres of gravity of the portions of the lune between the conical joints lie at points a_1, a_2, etc. obtained by tracing the horizontal corresponding to points d_1, d_2, etc. The author observes that the centres of gravity do not lie exactly on the axis line *aa*, but this simplification is acceptable if the number of the horizontal lines is large.

Then, the verticals passing through the centres of gravity a_1, a_2, etc. are drawn. The weights applied in the centres of gravity are all the same, measured by $au_1 = u_1u_2 = w_1w_2$, etc.

By choosing the point *a* as the pole and the line w_1w_8 as the weight line, the funicular polygon *c* of the weights is drawn starting from the point f_8, which corresponds to the outer point of the middle third at the base of the dome.

The graphical procedure proposed by Eddy assumes that the distance between the weight line and the pole represents the magnitude of the maximum horizontal thrust determined above, while hoop compression stresses act between the crown section and angle 51°49'. According to Eddy, below the joint corresponding to the maximum horizontal thrust, the dome portions behave as separate arches, the tensile strength being zero by hypothesis.

After determining the funicular polygon denoted as *c*, corresponding to the thrust line of the lune for the maximum value of the horizontal thrust, Eddy observes that it does not fall within the thickness of the lune. This means that the dome cannot be subjected to a thrust so great as that assumed. Eddy cites the minimum resistance principle: "no greater horizontal thrust will be called into action than is necessary to cause the dome to stand, if stability is possible" (Eddy, 1877, p. 57).

In order to guarantee the stability and to consider the nil tensile strength of masonry, Eddy imposes that the thrust line "be inscribed within the inner third of that part of the meridian section of the lune which is to act as an arch" (Eddy, 1877, p. 57). In order to find the equilibrium curve *e* so that it satisfies the conditions described above, a graphical procedure is described. The straight line *fo* is intersected at points p_1, p_2, etc. by the horizontals m_1p_1, m_2p_2 conducted from points m_1, m_2, etc., belonging to the upper border of the middle third. Then, points q_1, q_2, etc., are identified by plotting the horizontals passing through c_1, c_2, etc. (the vertex of the trial funicular polygon) and the verticals passing through p_1, p_2, etc. The ordinates of these points are of the type *qh*. Some of these ordinates must be brought to *ph*. Let oq_2 be tangent to the curve *qq*. As an example, Eddy considers the horizontal passing through c_4, which intersects oq_2 at j_4, and the vertical for j_4, which intersects *fo* at i_4, by stating that the new position of c_4 is then e_4.

Similarly, other points of curve *e* can be found. It is possible to proceed in a different way, by determining the new distance for the pole. To this aim, it is sufficient to draw $fj//oq_2$ cutting *ww* at *j*. Then, the point *i*—intersection of the horizontal through *j* and the straight line *fo*—will indicate the new position of the weights line *vv*, which will have a distance from the pole *a* diminished with respect to the first solution. The final thrust line, e_1e_8, is determined by considering the segments radiating from the pole, *a*, to the final position of the weight line, v_1v_8. Parallel segments are traced by starting from the extrados of the middle third at the springing joint.

It is interesting to observe that Eddy takes into account the nil tensile strength of masonry by imposing that the thrust surface be contained within the middle third. This choice is linked to Navier's approach (Navier, 1826), aimed at 'correcting' the limit solution proposed by Coulomb by assuming the boundary of the middle third as a more severe limit for the thrust line. As already recalled, imposing this constraint presents some critical aspects from a theoretical point of view, since on the one hand there is no reference to an equilibrium limit condition, on the other hand the structure is hyperstatic and, therefore, the solution cannot be determined by means of statics alone.

The contribution by Lévy (1888). Lévy conceives an original graphical procedure largely based on the contribution of Eddy (1877). In this section, a brief description of this method is provided, by referring the interested reader to the studies by Rapallini and Tempesta (1997) and Galassi et al. (2017).

The first purpose of the graphical procedure consists in individuating the *point neutre*, denoted by Lévy as σ_0 (Fig. 2.27, *above*, from Galassi et al., 2017). By assuming that masonry is not capable of withstanding tensile stresses, Lévy defines the *point neutre* as follows: the portion of the dome which is located above σ_0 is subjected to compressive hoop forces that are transmitted by the adjacent sections of the vault, while the one located below such a point is not subjected to hoop forces, since tensile hoop forces are not admissible. Finally, after determining the neutral point, σ_0, the funicular polygon (thrust line) related to each lune's portion below σ_0 is traced.

Here a brief description of the first step of the graphical procedure, aimed at finding σ_0, is reported.

Lévy begins by considering a lune comprised between two meridian planes. The profile of the lune is divided into an arbitrary number of parts. The weight and centre of gravity of each block are evaluated (by applying Guldin theorem).

With reference to a dome subjected to its own weight, the vertical lines of action of the vectors representing the weight of each *voussoir* (of the lune) are drawn so that each line passes through the centre of gravity of each *voussoir*.

A first trial funicular polygon is traced by assuming O_0 as the pole and imposing that this funicular polygon passes through point β_n, at the extrados of the middle third at the springing joint. This first funicular polygon is not entirely comprised within the dome's thickness.

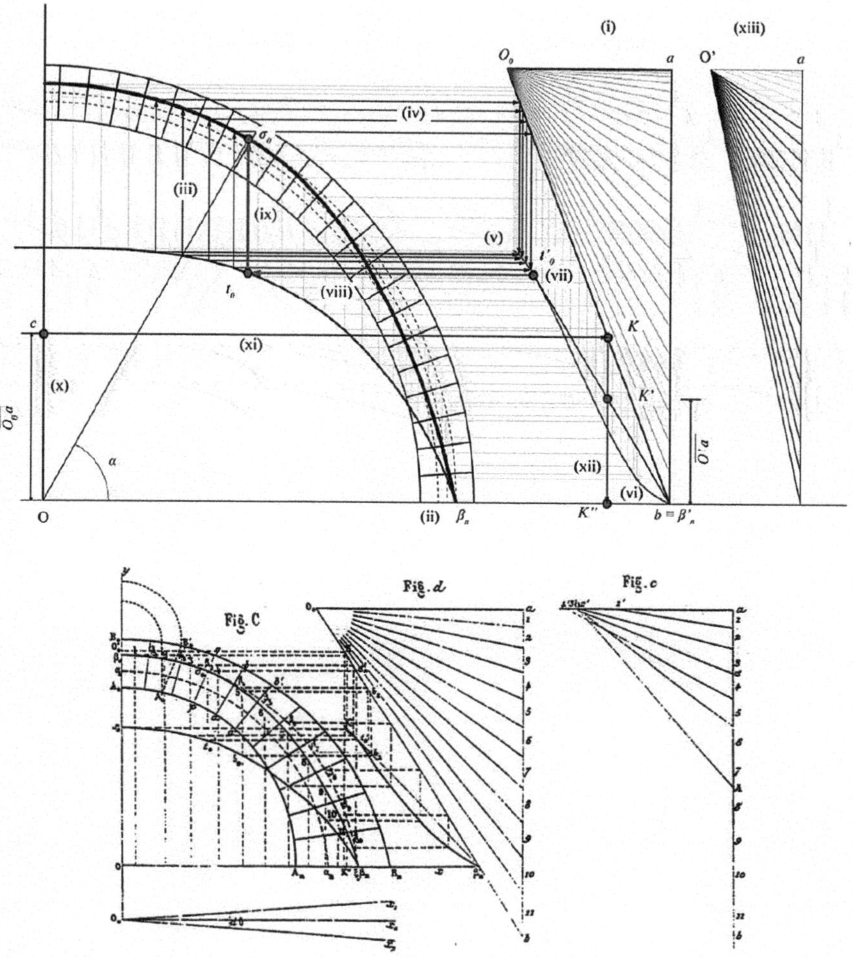

Fig. 2.27 Description of Lévy graphical method (drawn from Galassi et al., 2017), *above*; the original drawing from Lévy (1888), *below*

In order to find an admissible thrust line, a set of horizontal straight lines is drawn by starting from the intersection points between the vertical lines passing through the centres of gravity of the *voussoirs* and the curve corresponding to the outer middle third. The intersection points between these horizontal lines and an arbitrary inclined straight line, for convenience chosen coincident with the last side of the force polygon in Fig. 2.27, O_0b, are considered. From the intersection points thus identified, a set of vertical lines is traced. From the vertices of the trial funicular polygon, a set of horizontal lines is traced, which intersects the corresponding vertical lines just defined.

A set of ordered points is thus determined: starting from $b \equiv \beta'_n$ (the intersection point between the horizontal line at springing joint's level and the arbitrary sloped line chosen), a curve passing through these points is traced.

Then, the tangent to this curve from point β'_n is drawn. The tangency point is denoted as t'_0. The horizontal line passing through t'_0 intersects the trial funicular polygon in t_0. The neutral point σ_0 is identified as the intersection between the vertical line from t_0 and the curve of the outer middle third.

According to Lévy, the thrust line of the portion of the dome located below the neutral point corresponds to the funicular polygon (or curve) related to each lune, since the hoop stresses are not admissible.

The contribution by Wolfe (1921). A further development of Eddy's method (Eddy, 1877) is proposed by Wolfe in his treatise on graphical statics (Wolfe, 1921), where he presents a brief study on the stresses in domes.

After defining his contribution as an approximate study and not an exact analysis, he considers the hemispherical dome illustrated in plan in Fig. 2.28 (Fig. 497 of the original). The dome is divided into a number of small 'lunes' all identical to each other. One of them is shown in plan in Fig. 2.28 (Fig. 499 of the original).

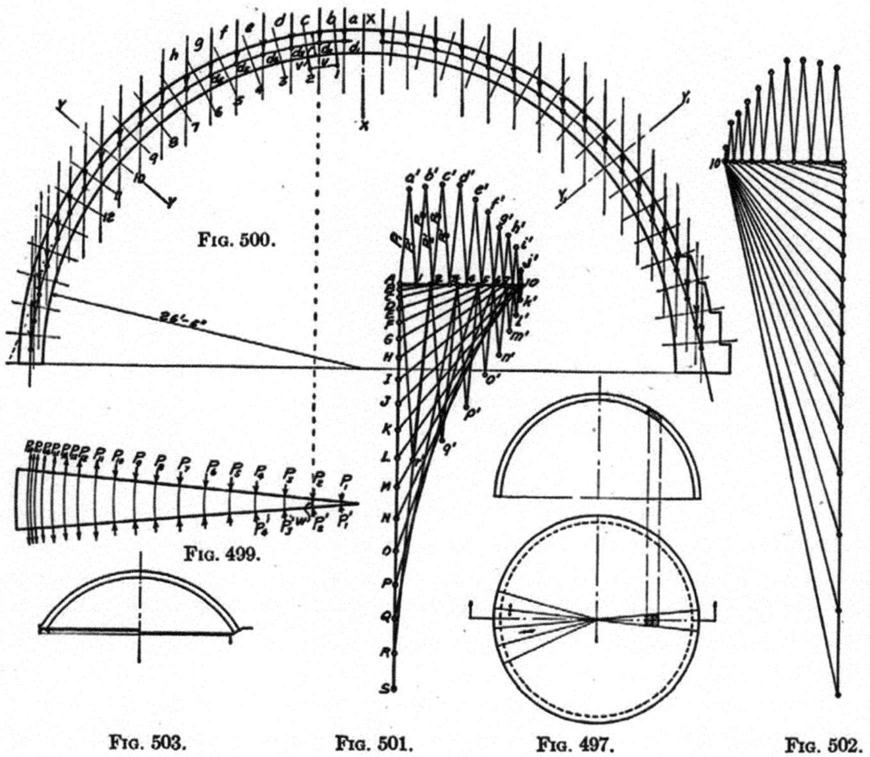

Fig. 2.28 Scheme of a hemispherical dome (from Wolfe, 1921)

Let the X-X axis be the central axis of the dome (Fig. 2.28, see Fig. 500 from the original). Suppose that each lune is subdivided into a certain number of *voussoirs* from planes 1, 2, 3 (more precisely, such surfaces are conical surface portions).

Wolfe suggests an estimate of the volume of each of these *voussoirs* and evaluates their weight, which can be represented by a concentrated force applied at the centre of gravity of each *voussoir.*

First, the *voussoir* adjacent to the vertical crown section, d_1, is considered. There are four forces acting on it: its own weight, the internal compression force at the joint separating it from the *voussoir* d_2 and the two compression forces acting on the sides of the 'lune'. If the dome is a dome of revolution of uniform thickness subject to its own weight, these two compression forces are equal and respectively orthogonal to the meridian planes. They act on the same (horizontal) plane. Their resultants, together with the other two forces under consideration, form a system of three forces in equilibrium, which therefore must be concurrent in the same point and acting on the same plane.

If a generic *voussoir* is considered, for example d_5, there are five forces acting on it: the own weight, the two forces acting on the contact surfaces with the *voussoirs* d_4 and d_6 and the two forces acting on the two sides of the 'lune'.

As shown in Fig. 2.28 (Fig. 501 from the original), the segments A-B, B-C, etc., corresponding to the weights of *voussoirs* d_1, d_2, ..., are drawn consecutively on a straight line. Then, Wolfe draws a funicular polygon that follows the axis of the 'lune' (Fig. 500 from the original). Starting from points A, B, etc. of the weights line, the straight lines parallel to the sides of this funicular polygon are drawn and extended until they intersect the horizontal line passing through point A.

The intercepts on this horizontal line, A-1, 1–2, 2–3, ..., represent the resultants of the hoop forces acting on the meridian planes of the lune, for each *voussoir*. The magnitudes of the hoop forces P_2 e P'_2, for example, are obtained by tracing, starting from point 1, a straight line parallel to the action line of P_2, and another line, starting from point 2, parallel to the action line of P'_2. Their intersection provides point b', which determines the magnitudes of P_2 and P'_2. In an analogous way also the magnitudes of P_1, P_3 e P_4, P'_1, P'_3, etc., are obtained.

It is important to observe that the forces from P_1 to P_{10} are compressive forces, while those from P_{11} to P_{18} are tensile forces (Fig. 2.28, see Fig. 499 from the original). This is clear by observing their resultant on the horizontal straight line starting from point A. Above joint Y-Y there is circumferential compression, while below this joint there is circumferential tension.

Now, if the material were unable of withstanding tensile stresses, as in the case of a masonry dome, the forces below joint Y-Y would be nil. The sides of the force polygon from point K to point R would start from point 10, by identifying a new position of the funicular polygon (see the dotted line in Fig. 500 from the original: since it exits from the arch's thickness, in this case the dome would be unsafe).

It is noteworthy that Wolfe's method is conservative, since the thrust surface related to the upper portion of the dome coincides with the median surface of the dome.

2.2.6 Modern Versions of Traditional Methods for the Equilibrium Analysis of Masonry Arches and Domes

In recent times, the thrust line method as well as procedures based on graphic statics have been re-visited by a number of scholars in order to assess the stability of masonry arches and domes. This modern re-evaluation of historic contributions on masonry vaults has been inaugurated by Jacques Heyman, who in 1966 published his work *The Stone Skeleton*, where he provides a clear explanation on how plastic theory (Drucker & Prager, 1952) can be adapted to masonry structures. Heyman stated that, if the masonry material can be characterized by certain properties, i.e., the tensile strength is negligible, the friction is sufficient to prevent sliding and the compressive strength is infinite, the classical limit analysis theorems hold. Heyman's 'safe theorem' (lower bound theorem) allows for using only equilibrium considerations in order to assess the stability of masonry structures. For this purpose, therefore, the thrust line method results to be a useful graphical tool: equilibrium is guaranteed when at least one thrust line can be identified, lying entirely within the vault's thickness (Heyman, 1966).

After Heyman's contribution, the thrust line method was exploited and numerically implemented in a variety of ways. A complete review of the modern literature on masonry structures based on the thrust line method is out of the purposes of the present chapter. Some recent works will be briefly recalled below, in order to offer an overview, necessarily incomplete, of the studies based on the thrust line method.

In the contribution by Clemente et al. (1995) the thrust line method, suitably re-elaborated, allows for performing numerical investigations on the equilibrium and collapse of masonry arches. They consider both dead and travelling loads, by determining the minimum thickness (or a load factor) corresponding to the activation of a mechanism.

Thrust lines are used also by Block et al. (2006) with the aim of visualizing the internal forces within the masonry and to predict possible collapse modes. In their contribution the thrust line method is translated in terms of graphical statics. The authors underline the drawbacks of applying classical graphic statics, in particular the trial and error process required to find acceptable solutions. For this reason, they develop computerized methods, which make graphical statics an effective tool, capable of carrying out real-time analyses on the stability of masonry structures as well as finding all of the possible funicular solutions for a certain set of loads. Thus, all possible solutions for the family of funicular polygons can be visualized. It is interesting to observe that simple two-dimensional drawing programs are exploited in order to obtain fully interactive drawings able to provide in real time information on the equilibrium conditions (Fig. 2.29).

In the frame of limit equilibrium analysis, a number of contribution have re-visited the mathematical formulation of the thrust line method proposed by Milutin Milankovitch (1907), who particularly applied it to determine the minimum thickness of a semicircular arch subjected to its own weight. Among these contributions, Makris

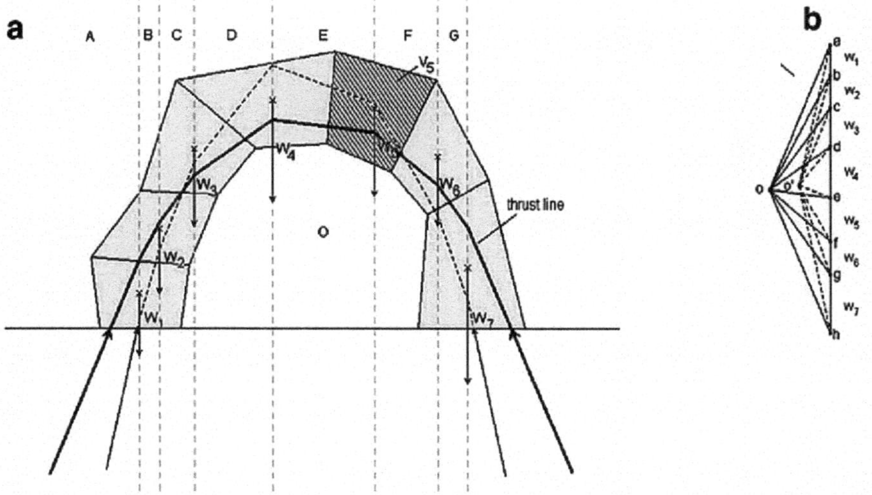

Fig. 2.29 The force polygon (**b**) is drawn to scale in order to determine admissible thrust lines (**a**) (from Block et al., 2006)

and Alexakis (2013), Alexakis and Makris (2015) and Nikolić (2017) can be cited, aimed at searching the minimum thickness of masonry arches of different shape and different stereotomies.

As regards stereotomy, the contribution by Nodargi and Bisegna (2020) applies the thrust line analysis in order to derive the minimum thickness in masonry arches whose geometry is described by means of a reference curve and a stereotomy function.

In Marmo (2021) the novel version of the thrust line method results to be a useful method for the equilibrium analysis of masonry arches: in this contribution a numerical procedure implemented in MATLAB is described in order to obtain two extreme configurations of the thrust line, corresponding to the minimum and maximum thrusts, respectively. The algorithm allows for studying arches of general shape, subjected to both vertical and horizontal loads (Fig. 2.30).

The thrust line analysis is applied also to study the equilibrium of masonry domes of revolution. As proposed by Heyman (1977) and Oppenheim et al. (1989), masonry domes can be schematized as assemblages of independent adjacent lunes, each of them considered as an arch of variable width. Under the simplifying hypothesis of nil tensile hoop forces, the equilibrium is guaranteed if the thrust line related to a single lune is contained within the dome meridian profile.

As regards the limit analysis of masonry domes, it is noteworthy to observe that the thrust line method applied to a single lune could provide a 'limit' thickness not corresponding to a kinematically admissible mechanism for the entire dome. While for the hemispherical dome the limit solution thus obtained corresponds to the collapse of the entire dome (Heyman, 1977), this does not necessarily hold for domes of different shapes.

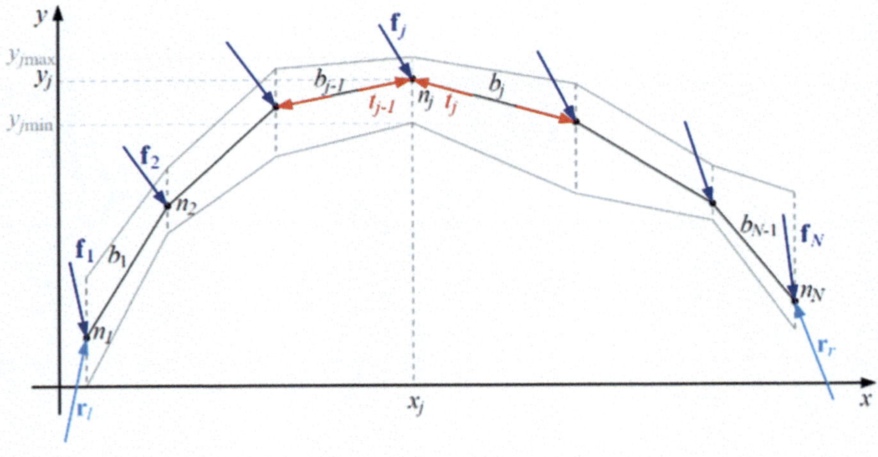

Fig. 2.30 A scheme of the thrust line (from Marmo, 2021)

The contribution of Oppenheim et al. (1989) considers a masonry ogival dome as a shell of revolution made of a rigid material with no resistance to tensile stresses. For limit state analysis of masonry shells, the admissible membrane force surface is required to lie within the shell thickness and admit only compressive forces. The analysis is also conducted by means of the thrust line method. The problem is first reduced to a two-dimensional case by dividing the dome into lunes by meridian planes (Fig. 2.31). The limit thickness is defined as the one corresponding to a kinematically admissible collapse mechanism for the entire dome. In this case, as shown in Fig. 2.31, it is identified by $\beta = 0.8462$ rad; $\gamma = 1.308$ rad; $2t/a = 0.02374$; and $H = 0.2866\ \rho a t$. Near the crown section the thrust surface is not contained within the masonry; this issue is dealt by Oppenheim et al. (1989) by considering the presence of compressive hoop forces acting in some upper part of the dome, where their influence is considered locally.

Among the contributions based on the extension of the thrust line method to domes of revolution, some recent works translate the membrane theory in terms of analytical or numerical models.

The work by Coccia et al. (2016) studies the pressure surface under the hypothesis of zero hoop stresses by extending to the three-dimensional case the concept of reversed catenary, through a re-visitation of the treatment proposed by Bossut (1778, 1779). By means of an analytical approach, the authors examine the family of the a priori unknown pressure surfaces of an axis-symmetric spherical masonry shell of constant thickness, loaded by its own weight. Among these surfaces, the authors determine the one corresponding to the minimum thrust state, and develop an accurate description of the minimum thickness state (Fig. 2.32).

Varma and Ghosh (2016) assess the stability of axi-symmetric masonry domes by combining the thrust line method with the finite element approach (Fig. 2.33). They adopt a modified axisymmetric element able to incorporate the hoop tension effects.

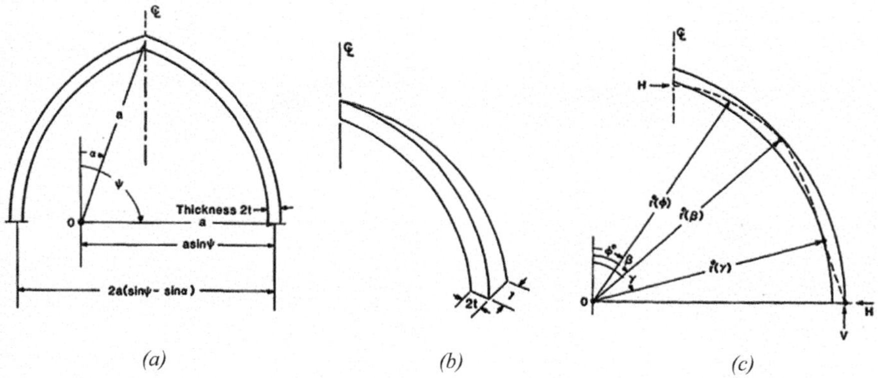

Fig. 2.31 Geometry of the ogival dome (**a**); lune (**b**); the thrust surface (**c**). Drawn from Oppenheim et al. (1989)

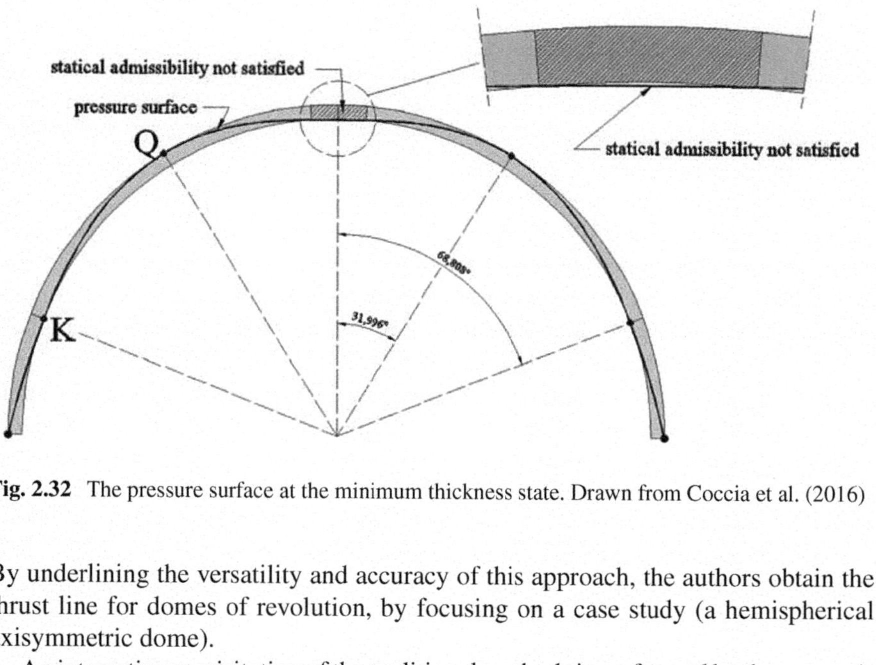

Fig. 2.32 The pressure surface at the minimum thickness state. Drawn from Coccia et al. (2016)

By underlining the versatility and accuracy of this approach, the authors obtain the thrust line for domes of revolution, by focusing on a case study (a hemispherical axisymmetric dome).

An interesting re-visitation of the traditional methods is performed by those contributions which exploit graphical methods (Eddy, 1877; Lévy, 1888; Wolfe, 1921) for assessing the stability of domes.

A brief account on the novel version of these graphical approaches is here provided.

In the already cited contribution, Galassi et al. (2017) re-visit the equilibrium analysis of masonry domes proposed by Maurice Lévy (1888), by translating the

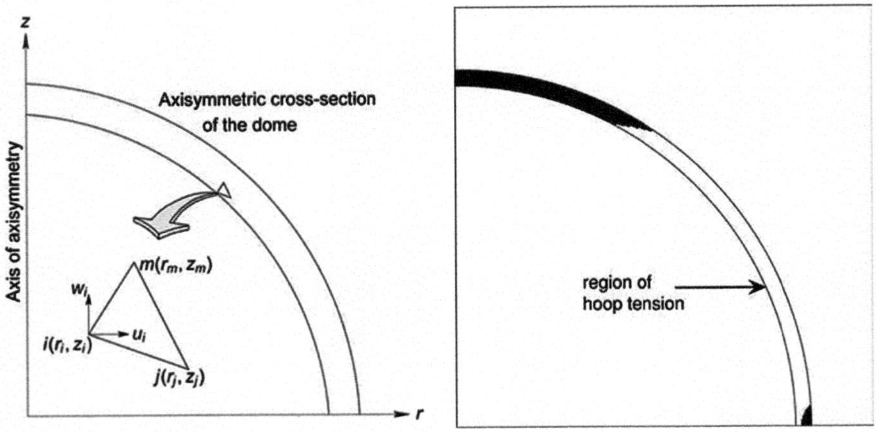

Fig. 2.33 Hemispherical dome: axisymmetric finite element (*left*); hoop compression and hoop tension regions after the final iteration (*right*) (Varma & Ghosh, 2016)

laborious graphic procedure into an analytical form. Through a computerized re-edition, this historical method turns out to be an operational tool for the structural assessment of domes. The authors study the stability of the dome of San Biagio Church in Montepulciano (Fig. 2.34) as well as some case studies (spherical and pointed dome, complete, with hole and with hole and lantern). Furthermore, the authors propose a modified version of this method in order to model the biaxial stress acting on the upper portion of the dome through the membrane theory. For each type of dome considered, the authors identify the limit thickness; then, they investigate the extension of meridian fracture by varying the thickness of the domes.

Lau (2006), Cipriani and Lau (2006) and Zessin et al. (2010), by starting from a critical review of the contributions by Eddy (1877) and Wolfe (1921), propose an approach defined as 'modified thrust line method', consisting in a computerized tool for the assessment of the stability of masonry domes essentially based on these historical methods. According to this approach, the conservative constraints imposed by Wolfe and Eddy are removed, and the line of thrust is traced within the entire thickness of the dome. If the values of the dome's external loads are assigned, both meridional and hoop forces can assume a range of values, which define different thrust line configurations. By varying the values of hoop and meridional forces, this approach allows for identifying admissible thrust line solutions; the hoop forces are determined through the membrane theory, by imposing the respect of equilibrium through an optimization algorithm (Lau, 2006).

Graphical analysis is also adopted by Cennamo et al. (2018, 2019) and Cusano et al. (2021), with reference to the case study of the dome of San Francesco di Paola in Naples, a large dome made of masonry blocks and subject to the self-weight. In (Cennamo et al., 2018) the equilibrium of this dome is studied by means of graphic statics; this approach is combined with kinematic analysis in (Cennamo et al., 2019) in order to define a range of possible displacements. Cusano et al. (2021), starting from

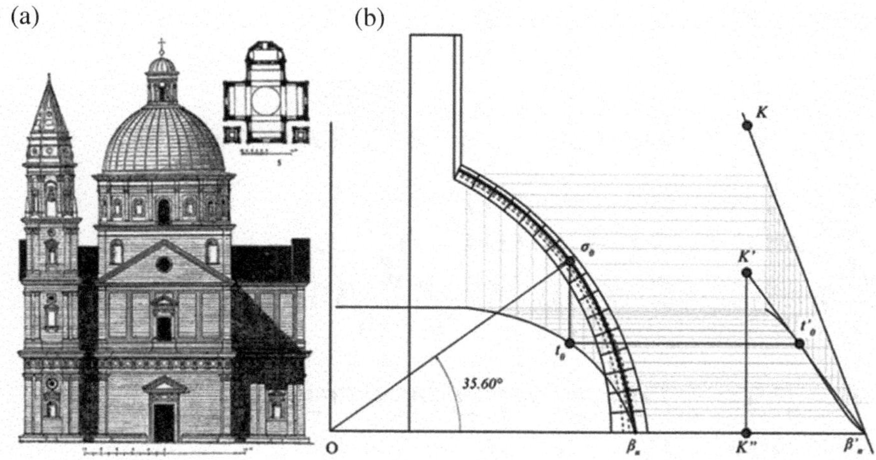

Fig. 2.34 San Biagio in Montepulciano: **a** plan and elevation; **b** thrust surface according to the Lévy graphical method (from Galassi et al., 2007)

these previous contributions, assess the equilibrium of the dome by exploiting the modified thrust line method; moreover, the results are compared with those deriving from the membrane equilibrium analysis (Fig. 2.35).

A recent contribution by Paris, Ruscic and Mirabella Roberti (2021) presents a computerized development of the modified thrust line method. The hoop forces are estimated by adopting a distribution function, denoted as *Dh*, which describes the distribution of hoop forces. This function allows for identifying a domain for the hoop forces. More in detail, by admitting only compressive forces, it is assumed that the magnitude of the hoop forces is maximum near the crown and, then, decreases. The function *Dh* is introduced in order to solve the nodal equations of equilibrium: it should be observed that several equilibrated solutions can be identified, each determined by a different distribution of hoop forces, *Dh*. Thus, the hoop forces can be determined graphically and a line of thrust traced for assessing the stability of the structure (Fig. 2.36).

2.3 A Focus on Durand-Claye's Method and its Modern Re-interpretation

2.3.1 Introduction

The historical development of the masonry arch theory is characterized by two alternative structural philosophies modernly referred to as 'limit analysis' and 'elastic analysis'.

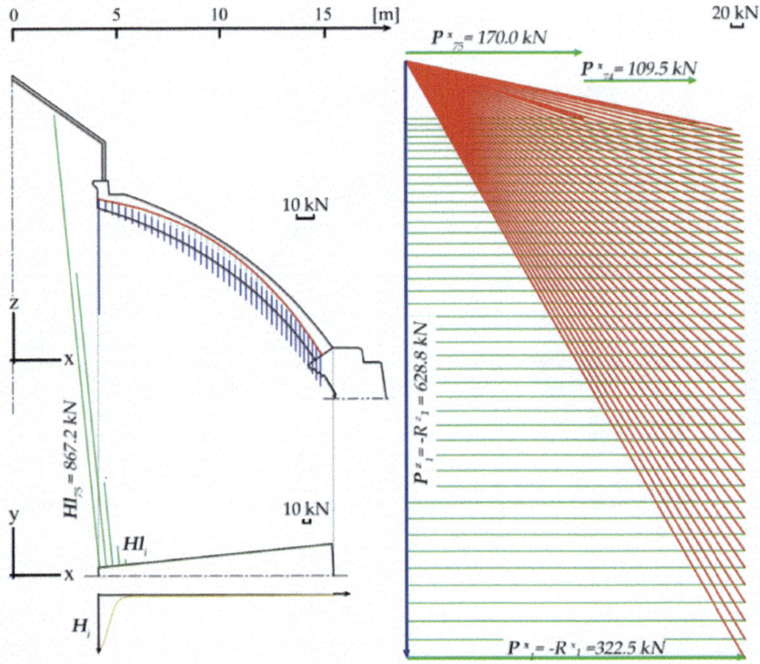

Fig. 2.35 Equilibrium analysis of the dome of San Francesco di Paola through the modified thrust line method (from Cusano et al., 2021)

According to the first approach, global stability is the object of study and the safety of the arch is guaranteed if there is any condition of equilibrium capable of making it impossible to activate collapse mechanisms. This methodology of analysis in terms of equilibrium is the one pursued since the first eighteenth-century studies on the statics of arches, vaults and domes and developed in the first half of the nineteenth century by Persy (1825), Navier (1826) and Michon (1857) by starting from the method of *maxima* and *minima* proposed by Coulomb (1776). The study of global stability is also the subject of the contribution of Mascheroni (1785), which uses the principle of virtual works to evaluate the possibility of activation of collapse mechanisms.

With the development of the thrust line method, while returning to the pre-elastic perspective, a new question opens up: the search for the 'true' thrust line among the infinite possibilities. Another important aspect is the fact that for the first time attention is also paid to the strength of the material.

From this context arises the second approach, based on the elastic analysis of the arch, developed starting from the 1920s. 'Elasticity' and 'strength' become the two fundamental concepts in the field of structural mechanics; the interest is no longer aimed at the global stability of the system, but at the local stress state: safety is considered satisfied if the actual stress at each section does not exceed a value considered admissible. Using the elastic analysis it is also possible to study deformability and,

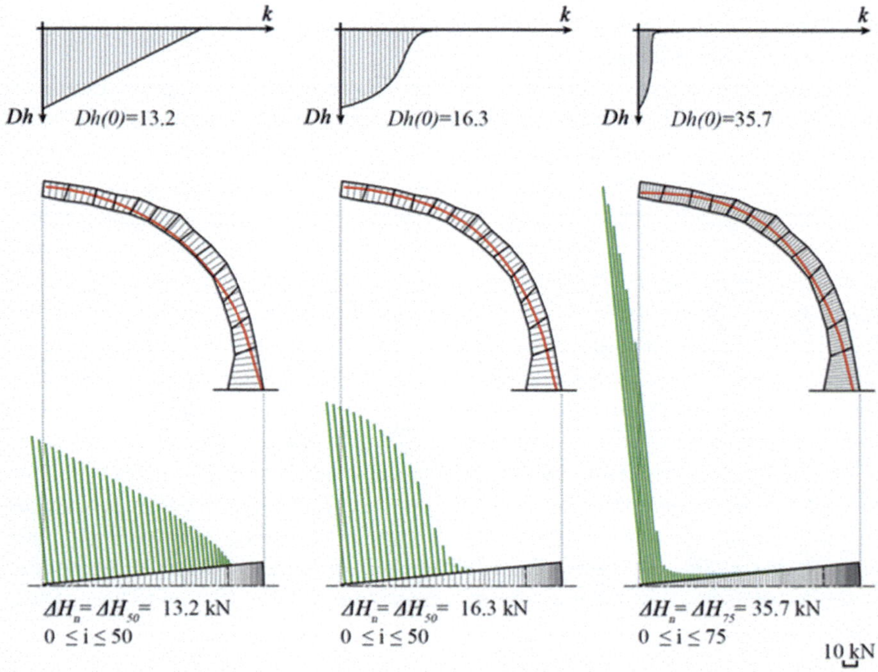

Fig. 2.36 Determining the hoop forces through the function *Dh* (from Paris et al., 2021)

therefore, to conventionally define the collapse load in terms of residual stiffness or maximum displacements. In this context, a very interesting contribution is that of Castigliano (1879–1880, 1878), who evaluates the stability of a masonry arch by using his theorems.

The approach proposed by Alfred Durand-Claye in 1867 can be considered an intermediate approach between limit and elastic analysis. As the author himself wrote, this approach aims to verify, "*s'il existe des solutions d'équilibre compatibles avec un effort-limite donné*" (Durand-Claye, 1867, p. 66). By following the limit analysis approach, such a methodology does not aim at determining the actual solution; however, it adequately takes into account a fundamental aspect related to the elastic analysis by imposing a restriction on the value of the maximum admissible stress.

From this point of view, the Durand-Claye method solves the problem that for some decades had focused the efforts of those scholars who, without abandoning the reference to the pre-elastic tradition, intended to update it by introducing aspects related to the strength of materials. According to those methods, a priori hypotheses were assumed in order to remove the static indeterminacy. As regards these hypotheses, used in order to draw the 'true' thrust line, reasonable objections arise: do those hypotheses lead to an admissible solution in terms of strength for the entire arch? Do they correspond to some boundary condition? Do they actually identify the weakest joints?

These are the questions posed by Durand-Claye when he examines the thrust lines drawn in accordance with the method of Méry (1840) or similar methods: "*Il n'échappe à personne*—the author writes—*que le tracé habituel de ces courbes présente une sorte de vague et d'incertitude. On n'arrive à trouver une courbe d'équilibre, que seule mérite le nom de courbe de pression, que par tâtonnements, en prenant arbitrairement deux points pour déterminer la courbe ou faisant varier la poussée en grandeur et en position, ce qui revient toujours à prendre au sentiment les deux conditions nécessaires pour définir la courbe. On n'a aucune idée du degré de stabilité de la voûte, puisqu'on s'arrête dès qu'une courbe d'équilibre est trouvée et que la voûte peut en comporter une infinité; l'influence de la résistance des matériaux n'est point suffisamment mise en évidence et les parties faibles du profil sont indiquées avec un certain vague*" (Durand-Claye, 1867, pp. 65–66).

The answer given by Durand-Claye to these questions is contained in the following passage: "*Nous ne déterminerons plus, un peu au hasard, une solution quelconque d'équilibre: connaissant les profils, les charges, la résistance effective ou réduite imposée aux matériaux, nous déterminerons par des constructions géométriques toutes les solutions d'équilibre que peut comporter une voûte*" (Durand-Claye, 1867, p. 66).

According to Durand-Claye, rather than searching for a single thrust line to be understood as the 'true' one, it is necessary to determine all the admissible thrust lines, among which the actual one must necessarily be found if the arch is stable. The value of this procedure lies in ascertaining the existence of admissible solutions without requiring the identification of the actual one. As Durand-Claye himself writes, "*de la possibilité de l'équilibre, nous concluons à la stabilité*" (Durand-Claye, 1867, p. 65).

This statement can be seen as an anticipation of Heyman's 'safe theorem', implicitly contained in the equilibrium conditions given by Coulomb's method (1776). Durand-Claye shrinks the range of solutions by properly taking into account the strength of the materials. In this sense, the Durand-Claye method is a rational mediation between limit analysis and elastic analysis. The interested reader is referred to (Aita, 2003; Aita et al., 2004, 2007; Foce & Aita, 2003) for further insights on this topic.

2.3.2 The Method of 'Stability Areas' as a Graphical Translation of Coulomb's Method of Maxima and Minima

This section retraces Durand-Claye's reasoning (Durand-Claye, 1867). Consider a symmetrical arch loaded symmetrically (Fig. 2.37). Let us examine the portion of the arch between the ideal vertical joint at the crown section, c_0d_0, and the generic joint, cd (whose inclination is defined by angle θ, as shown in Fig. 2.38).

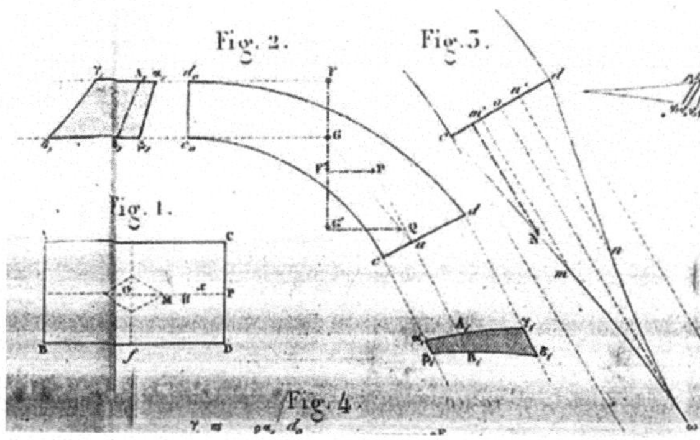

Fig. 2.37 Equilibrium of the portion of arch comprised between the crown section c_0d_0 and the generic joint *cd* (drawn from Durand-Claye, 1867)

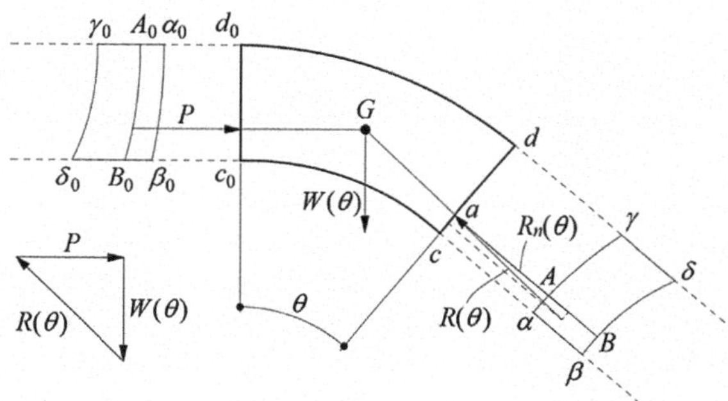

Fig. 2.38 Rotational equilibrium (nil tensile strength; infinite compressive strength of the material)

If the hypotheses implicitly considered by Coulomb are assumed (i.e., nil tensile strength, infinite compressive strength), the action line of the resultant of the horizontal thrust, P, acting at the crown section, and of the weight $W(\theta)$ of the portion of the arch c_0d_0dc, must pass inside the joint cd. This resultant force has the same line of action and magnitude as that of the reaction $R(\theta)$ at joint θ, but the opposite sense. By varying the position of the point of application of P between c_0 and d_0 and keeping fixed a generic point a belonging to the joint cd, for equilibrium the thrusts P thus identified are measured by horizontal segments conducted from the crown joint c_0d_0 to the curve A_0B_0. By making point a coincide with the extreme points of the generic joint θ, c and d, the curves $\alpha_0\beta_0$ and $\gamma_0\delta_0$. are determined, respectively. The area $\alpha_0\beta_0\gamma_0\delta_0$ therefore contains the extremes of all the segments measuring the statically admissible crown thrusts as regards the rotational equilibrium with respect to points c and d of the generic joint θ.

Now consider the normal component of the resultant reaction $R(\theta)$, denoted as $R_n(\theta)$. For the values of the crown thrust, P, corresponding to the curve A_0B_0 and, therefore, to point a, $R_n(\theta)$ is measured by segments orthogonal to joint θ between the point a and the points A and B. Similarly, the segment $\alpha\beta$ perpendicular to the joint cd and passing through c will correspond to the curve $\alpha_0\beta_0$; the segment $\gamma\delta$, always orthogonal to cd, but passing through d, will correspond to the curve $\gamma_0\delta_0$.

To evaluate the possibility of mutual sliding along the generic joint θ, it is sufficient to trace the friction cone defined by the friction angle φ, by starting from a generic point a belonging to this joint. Imagine, then, that the weight $W(\theta)$ is applied at point a. The two horizontal thrusts, $P^s_{min}(\theta)$ and $P^s_{max}(\theta)$, can be determined by imposing that, if vectorially summed with $W(\theta)$, give resultants inclined by an angle φ with respect to joint θ, corresponding to the reactions $R^+(\theta)$ and $R^-(\theta)$. These horizontal thrusts define two vertical lines at the crown section, which delimit the locus of the extremes of the horizontal segments which measure, starting from the crown joint c_0d_0, the admissible crown thrusts with respect to the sliding equilibrium along the generic joint cd (Fig. 2.39).

Obviously, in order to guarantee both rotational and sliding equilibrium of the arch's portion defined by angle θ, the thrust P must be measured by horizontal segments having the extremes contained in the intersection between the area defined by $\alpha_0\beta_0\gamma_0\delta_0$ (Fig. 2.38) and that included between the two vertical lines corresponding to $P^s_{min}(\theta)$ and $P^s_{max}(\theta)$ (Fig. 2.39).

The procedure must be extended to all joints of the arch. Thus, the area within which the extremes of the statically admissible crown thrusts must lie is obtained by considering the intersection between all the areas related to each joint. If the area shrinks to a point (or to a segment), a collapse mechanism can arise.

Under the assumptions commonly adopted in the limit analysis of the masonry arch, zero tensile strength, infinite compressive strength, finite friction, the equilibrium relations described above are sufficient to define the degree of stability of the structure and Durand-Claye's method results to be a graphical translation of Coulomb's method of *maxima* and *minima* (Coulomb, 1776).

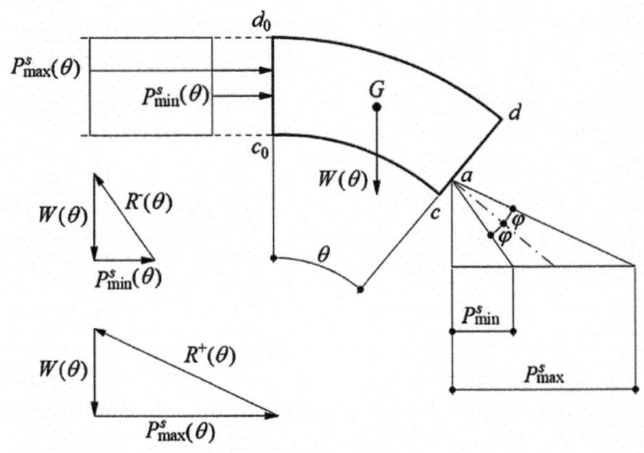

Fig. 2.39 Sliding equilibrium (nil tensile strength; infinite compressive strength; finite friction coefficient)

2.3.3 The Influence of the Limited Compressive Strength According to Durand-Claye (1867)

Durand-Claye in his contribution of 1867 describes masonry as a material not resistant to traction but with a limited compressive strength of intensity $|\sigma_c|$. Let consider an arch's section θ of thickness $h = h(\theta)$ and unit width (Fig. 2.40a). $N = N(\theta)$ is the normal force (positive if traction), $M = M(\theta)$ the bending moment, $\xi(\theta) = M(\theta)/N(\theta)$ the eccentricity of the normal force with respect to the centre of gravity of joint θ (Fig. 2.40b). A linear stress distribution along the joint is assumed, with the possibility of partialization of the section.

Durand-Claye defines the limit condition by referring to the normal force, N_{lim}, capable of producing a compressive limit stress, σ_c, in correspondence with the most compressed fibre of the section (Fig. 2.40c, d). Under these hypotheses, N_{lim} can be written as a function of the eccentricity ξ.

When the limit value σ_c is reached at the extremes of joint θ, defined by $y = \pm h/2$, Durand-Claye identifies two possible different situations, depending on the eccentricity of the axial force, ξ.

For $|\xi| \leq \frac{h}{6}$ (Fig. 2.40c), Durand-Claye uses the relations of linear elasticity corresponding to Navier's formula $\sigma_z = \frac{N}{A} + \frac{My}{J_x}$ (where $A = h$ is the area of the joint's section of unit width, $J_x = \frac{1}{12}h^3$ is the moment of inertia with respect to x-axis), which describes a linear relation between the stress, σ_z, and y and corresponds to a linear relation between σ_z and the elongation, ε. In this case, the limit value of the normal force, $N_{lim} = N_{lim}(\theta)$, is given by:

$$N_{lim} = \frac{\sigma_c h^2}{h + 6|\xi|}. \tag{2.9}$$

For $\frac{h}{6} \leq |\xi| \leq \frac{h}{2}$ (Fig. 2.40d), the limit normal force, N_{lim}, is measured by

$$N_{lim} = \frac{\sigma_c l}{2} \tag{2.10}$$

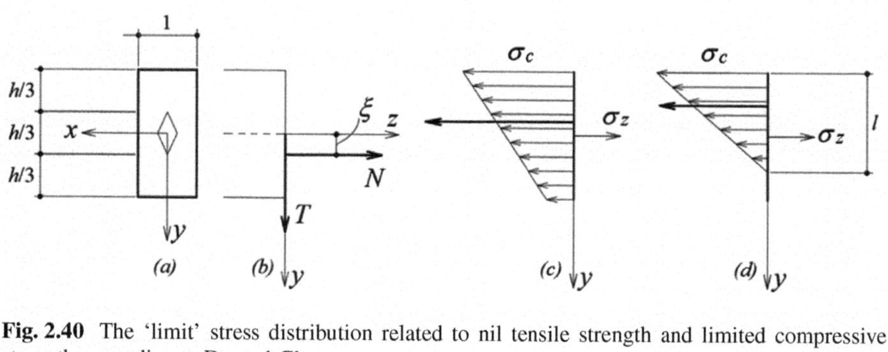

Fig. 2.40 The 'limit' stress distribution related to nil tensile strength and limited compressive strength, according to Durand-Claye

where $l = l(\theta)$ defines the region along the joint where σ_z varies linearly with y and, then, is directly proportional to elongation ε. The corresponding bending moment is given by:

$$M = -N\left(\frac{h}{2} - \frac{l}{3}\right) \quad \text{when} \quad M \geq 0 \text{ (the extrados fibre is compressed);}$$

$$M = N\left(\frac{h}{2} - \frac{l}{3}\right) \quad \text{when} \quad M < 0 \text{ (the intrados fibre is compressed).}$$

Thus, by posing $M = N\,\xi$, l can be written as follows:

$$l = \frac{3}{2}\left[\frac{Nh + 2M}{N}\right] = \frac{3}{2}[h + 2\xi] \quad \text{when } \xi \leq 0 \text{ (and } M \geq 0\text{);}$$

$$l = \frac{3}{2}[h - 2\xi] \quad \text{when } \xi > 0 \text{ (and } M < 0\text{).}$$

The two previous expressions can be summarised by:

$$l = \frac{3}{2}[h - 2|\xi|]$$

By taking the expression of l, the limit value of the normal force is derived from relation (2.10):

$$N_{lim} = \frac{3}{2}\sigma_c\left[\frac{h}{2} - |\xi|\right] \tag{2.11}$$

Equation (2.9) represents two symmetrical hyperbolas with respect to the straight line perpendicular to the joint and passing through its centre. Equation (2.11) represents two straight lines tangent to the above mentioned hyperbolas for $\xi = \pm\frac{h}{6}$ and passing through points c and d (Fig. 2.41).

With reference to Fig. 2.41, the admissibility condition related to the resistance of the material is therefore respected if the magnitude of the thrust P, acting at the crown section, and the magnitude of the normal force, $N(\theta)$, are represented, respectively, by segments orthogonal to the corresponding joints (c_0d_0 and cd) and delimited by the curves $c_0\omega_0d_0$ and $c\omega d$.

Consider again the joint θ and examine the area $\rho\sigma\varpi\chi$ determined by the intersection between the two areas $\alpha\beta\gamma\delta$ and $c\omega d$ (Fig. 2.42). This area has its reciprocal at the crown section in the area $\rho_0\sigma_0\varpi_0\chi_0$. The intersection between this last area and the domain defined by $c_0\omega_0d_0$ provides the area $r_0s_0p_0q_0$ which, in turn, has its reciprocal in the area $rspq$ at the joint θ. With reference to joint θ, it can therefore be concluded that the crown thrusts, P, compatible with both rotational equilibrium and strength of the material are represented by horizontal lines starting from the

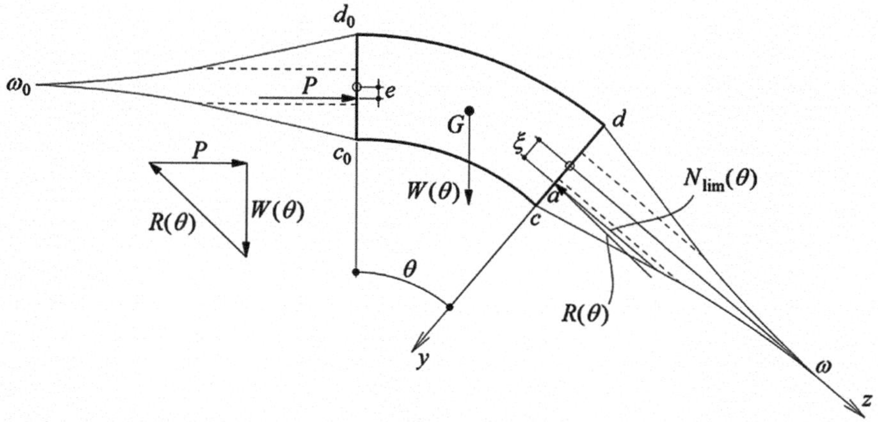

Fig. 2.41 The admissibility condition related to the resistance of the material, by assuming nil tensile strength and limited compressive strength

vertical crown joint c_0d_0, whose ends fall within the area $r_0s_0p_0q_0$. Sliding equilibrium is assessed as an additional condition for the crown thrusts by considering also the intersection between the area $r_0s_0p_0q_0$ and the rectangular domain obtained according to the procedure described in Fig. 2.39.

It can be observed that this result, presented in a somewhat convoluted way in (Durand-Claye, 1867), could have been achieved more quickly, by re-interpreting Durand-Claye's method in terms of internal forces.

Let now imagine that the procedure described above is repeated for each joint θ: this operation allows for determining, if it actually exists, the so-called '*area of stability*' (related to rotational equilibrium and strength), as the intersection between

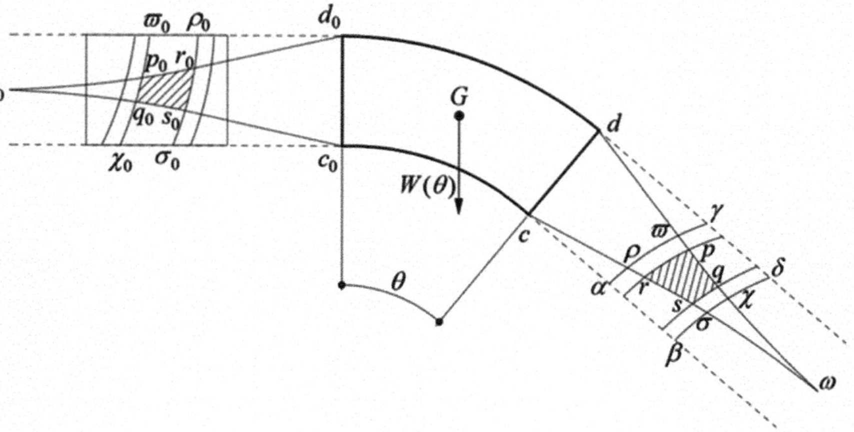

Fig. 2.42 Rotational and sliding equilibrium at joint θ, by assuming nil tensile strength, limited compressive strength and limited friction coefficient

all the areas $r_0s_0p_0q_0$ (Fig. 2.43). In order to consider sliding equilibrium, it is also necessary to take into account the intersection between the rectangular sliding domains related to all joints θ (Fig. 2.39).

The shape of the stability area and its dimension provide useful information about the safety of the arch, the position of the admissible thrust lines and the location of the most compressed joints. When the area of stability vanishes (by shrinking to a single point or a single segment), the limit equilibrium condition is attained.

If friction is supposed to be infinite, in accordance to Durand-Claye's method a safety coefficient can be defined, $\nu = \sigma_c/\sigma_{c,lim}$, as the ratio between the actual compressive strength, $|\sigma_c|$, and the limit compressive strength $|\sigma_{c,lim}|$, i.e., the ultimate value corresponding to the shrinking of the stability area to a single point.

Some critical remarks. The determination of the solutions compatible with the strength of the material is carried out by Durand-Claye (1867) for a material that is not resistant to traction, with a limited compressive strength. The author considers a linear stress distribution, with the possibility of partialization of the section, according to Fig. 2.40c, d, similarly to what Méry proposed in his contribution of 1840. However, the possibility of a piecewise linear distribution of the normal stresses induced by compression (Fig. 2.44a) is not taken into consideration. Furthermore, it should be noted that the stress distributions represented in Fig. 2.40c, d and 2.44a do not correspond to 'limit' conditions, since some of the fibres have not attained the limit value yet.

Aita (2003); Foce and Aita (2003); Aita et al., (2004, 2007, 2012; 2015; 2019a) rework Durand-Claye's method by considering, more in general, limit stress distributions compatible with the theoretical background provided by modern limit analysis.

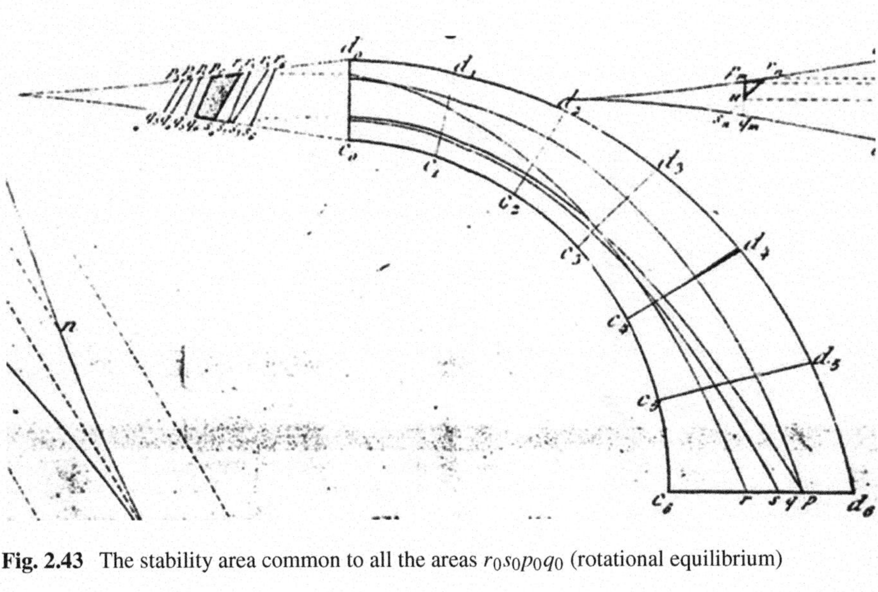

Fig. 2.43 The stability area common to all the areas $r_0s_0p_0q_0$ (rotational equilibrium)

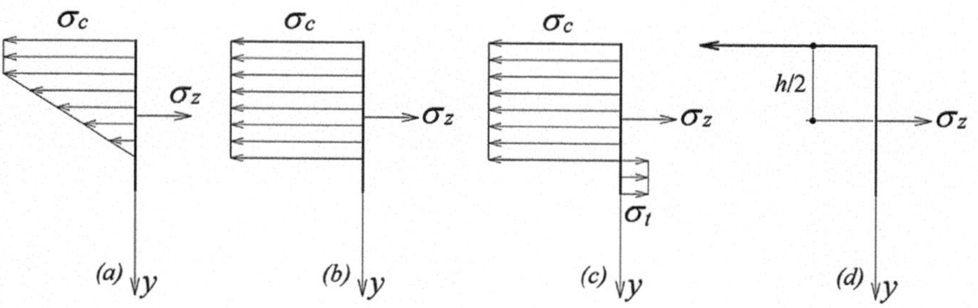

Fig. 2.44 Stress distribution along the crown section

As is known, the stress state present at each cross section of an arch, modelled as a planar beam subjected to in-plane loading, is identified by the internal forces N, T and M (Fig. 2.45), which respectively represent the normal force (positive if traction), the shear force, and the bending moment (the positive directions are shown in Fig. 2.45a).

When the tensile strength is nil, $\sigma_t = 0$, and the compressive strength, $|\sigma_c|$, is limited, the limit stress distribution is of the type shown in Fig. 2.44b and the limit bending moment is provided by

$$M_{lim} = \left| \frac{N^2 - N\sigma_c h}{2\sigma_c} \right|. \tag{2.12}$$

In this case, (N, M) and (N, ξ) planes can be subdivided into seven regions, each of them characterized by a different distribution of the normal stresses along the section. These subdivisions are plotted in Fig. 2.46a, related to the (N, M) plane, and in Fig. 2.46b, related to the (N, ξ) plane.

More in detail, region *1* of (N, M) plane (Fig. 2.46a) and region *i* of (N, ξ) plane (Fig. 2.46b) correspond to a linear stress distribution; regions 2 and 3 of (N, M) plane and regions *ii* and *iii* of (N, ξ) plane correspond to a piecewise linear distribution of stresses characterized by the partialization of the section due to the nil tensile strength; regions 4 and 5 of (N, M) plane and regions *iv* and *v* of (N, ξ) plane are characterized by a piecewise linear distribution, where the non-linearity is induced by the limited value of the compressive strength, $|\sigma_c|$; finally, regions 6 and 7 of

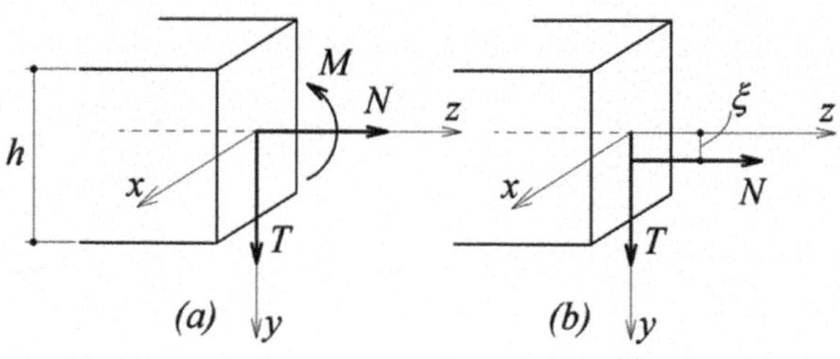

Fig. 2.45 The internal forces: N, T, and M (**a**); the eccentricity, ξ, of N with respect to the centre of gravity of the cross section (**b**)

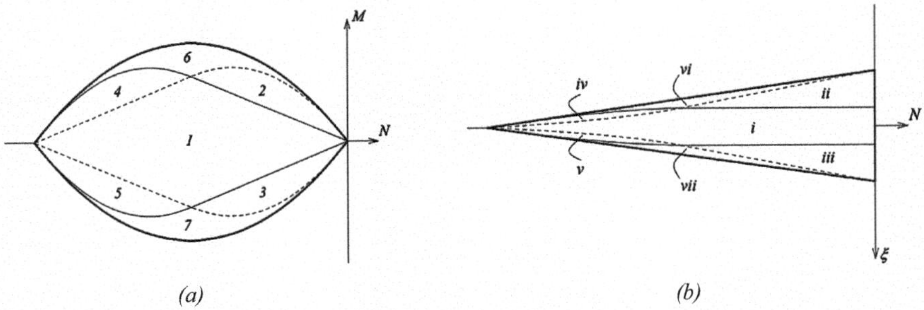

Fig. 2.46 The subdivisions of the (N, M) (**a**) and (N, ξ) (**b**) planes, when $\sigma_t = 0$ and σ_c is finite

(N, M) plane and regions *vi* and *vii* of (N, ξ) plane are characterized by a non-linear behaviour due both to zero tensile strength and the finite value of compressive strength. Regions 2, 4, 6 in Fig. 2.46a, and ii, iv, vi in Fig. 2.46b, correspond to a positive value of the bending moment, M, while regions 3, 5, 7 in Fig. 2.46a, and iii, v, vii in Fig. 2.46b, correspond to negative bending moment. Note that under the above mentioned hypotheses only negative values of the normal force, N, are admissible.

With reference to Fig. 2.46a, b, the region delimited by the dashed line identifies the domain proposed by Durand-Claye (1867); the boundaries of this region correspond to stress distributions analogous to those represented in Fig. 2.40c, d. The region delimited by the thick line represents the limit domain defined by relation (2.12), obtained under the hypotheses of a limited compressive strength, $|\sigma_c|$, and $\sigma_t = 0$, and corresponding to a rectangular distribution of the compressive stresses (Fig. 2.44b).

In the general case, under the hypotheses of limited tensile strength, σ_t, and limited compressive strength, $|\sigma_c|$, the limit stress distribution of the normal stresses corresponds to a bi-rectangular distribution (Fig. 2.44c), and the limit bending moment is given by

$$M_{lim} = \left| \frac{(N - \sigma_t h)(N - \sigma_c h)}{2(\sigma_c - \sigma_t)} \right|. \tag{2.13}$$

According to Coulomb's hypotheses, re-visited by Heyman (1966), i.e., $\sigma_t = 0$, $|\sigma_c| \rightarrow +\infty$, the limit bending moment corresponds to a thrust line touching the intrados or extrados at the joint (Fig. 2.44d), and is given by:

$$M_{lim} = \left| \frac{Nh}{2} \right|. \tag{2.14}$$

2.3.4 *Durand-Claye's treatment on the equilibrium of masonry domes*

This section, based on (Aita et al., 2017, 2019b), proposes a re-reading of an interesting contribution by Durand-Claye (1880), where the author extends his stability area method in order to assess the equilibrium of domes of revolution through a graphical approach.

According to Durand-Claye, the dome can be subdivided into single lunes. The equilibrium of one of these lunes (Fig. 2.47) is assessed in order to determine the corresponding stability area.

Durand-Claye observes that in spherical vaults or, more generally, in vaults of revolution, it is not possible to start the thrust line from the vertical section identified by the axis of symmetry as, in a large number of cases, the vault presents a cylindrical void, simple or surmounted by a lantern or a smaller dome (Fig. 2.47, *left*). Even if the dome were closed, the treatment conceived for cylindrical vaults could not be adopted, considering the hypothesis of limited compressive strength. The geometry of the *voussoirs*, due to the convergent shape of all the lunes, has to be taken into account: Durand-Claye cites a note by M. Collignon (*Cours de résistance des matériaux*), who applies the Guldin theorem in order to write the expression of the weight of the generic *voussoir* $\alpha\beta ab$ belonging to a lune comprised between two meridian planes forming an infinitely small angle between them, $d\theta$ (Fig. 2.47, *right*). As a first step, according to Durand-Claye, it is thus possible to trace the study of domes back to the general case of drawing thrust lines for the single lunes.

However, Durand-Claye is aware that the behaviour of domes differs from that exhibited by arches, and presents a specific treatment, which however is quite unsatisfactory. Suppose that the *anneau supérieur* of the dome ends at the bottom with

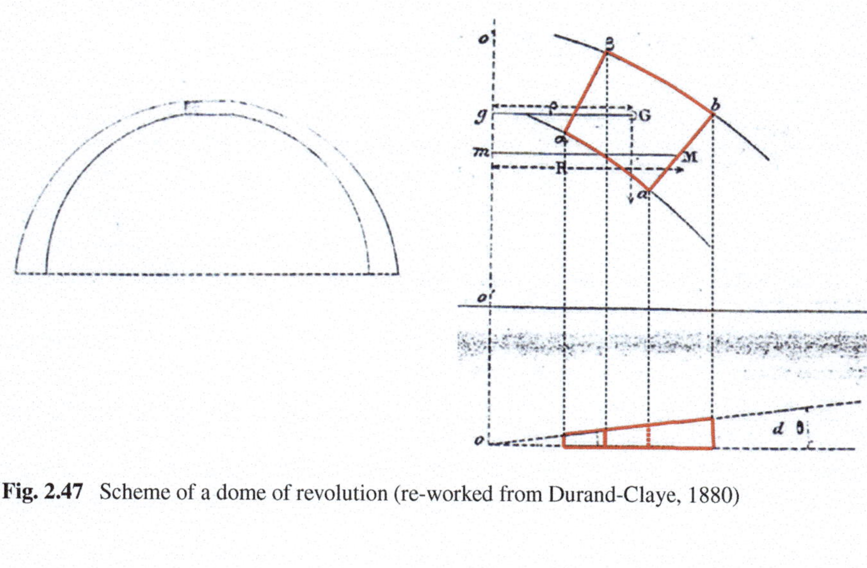

Fig. 2.47 Scheme of a dome of revolution (re-worked from Durand-Claye, 1880)

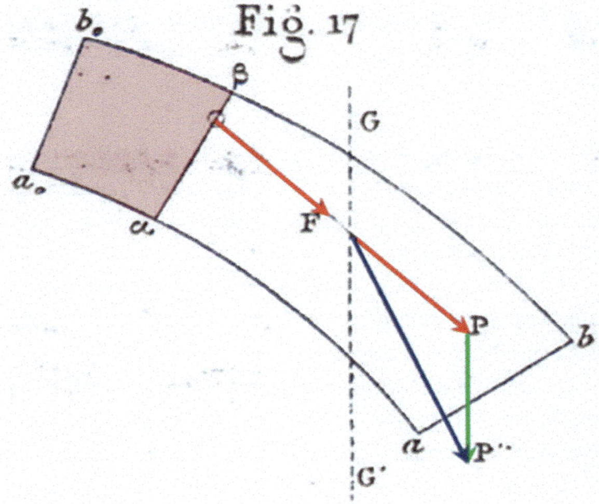

Fig. 2.48 Scheme of a portion of a dome of revolution (re-worked from Durand-Claye, 1880)

the conical surface generated by the joint a_0b_0 (Fig. 2.48). Durand-Claye denotes as Ω the area delimited by $a_0b_0\alpha\beta$ and determines the resultant F at joint $\alpha\beta$ as well as the resultant at any generic joint ab (by composing it with the weight PP' of the portion $\alpha\beta ab$).

Figure 2.49 shows a dome whose *anneau supérieur* is delimited below by the joint a_0b_0. Durand-Claye observes that, if the resultant nF of the weight of the vault portion above a_0b_0 and of the *reactions moléculaires* (hoop forces acting on the meridian planes delimiting the lune) were known, the resultant at the generic joint ab could be determined by composing $nF = O'F'$ with the weight $F'R'$ of the portion a_0b_0ab. The resultant $O'R' = NR$ is drawn at joint ab.

However, only the vertical component np of nF is known, corresponding to the weight of the lune's portion above a_0b_0. Furthermore, the vertical component of the resultant $O'P' = NP$ is the total weight of the portion above joint ab.

In order to determine the limit curve AA, locus of the extremes of the axial forces compatible with the strength of the material, Durand-Claye adopts a procedure similar to that presented in his memories on arches (Durand-Claye, 1867, 1868).

He describes the graphic procedure as follows, with reference to Fig. 2.49. Let NN' be one of the limit normal forces, so that its extreme N' lies on the limit curve AA. Now draw from point N' the parallel to joint ab; then, conduct from point P belonging to NP a horizontal line. Since the ends of the (limit) resultant must lie simultaneously on these two straight lines, the line of action of the resultant will pass through their intersection, i.e., through point R. The resultant will therefore be represented by NR. The line of action of NR intersects at O' the line GG' passing through the centre of gravity of the portion a_0b_0ab. Then, draw $O'R' = NR$ and $O'P' = NP$. Detach on the vertical $O'P'$ a length $p'P'$, representing the weight of the portion a_0b_0ab. The length $O'p'$ identifies the weight of the upper part of the dome, above the joint a_0b_0. Conduct the horizontal $p'F' = P'R' = PR$, that is the horizontal

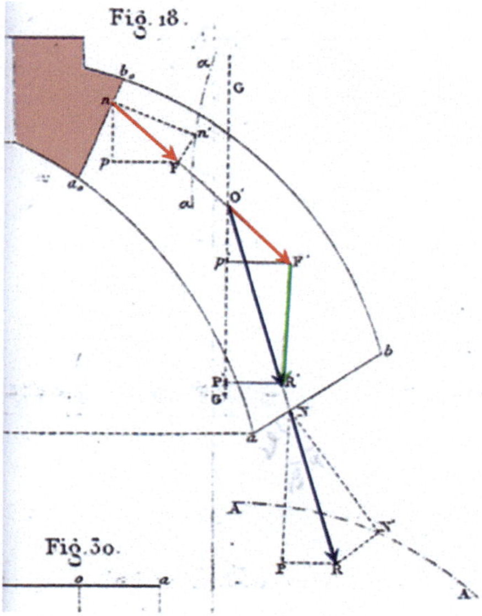

Fig. 2.49 Scheme of a portion of a dome of revolution: the effect of the limited compressive strength (re-worked from Durand-Claye, 1880)

component of the resultant on *ab*. The segment $O'F'$ represents the magnitude of the resultant on a_0b_0 (see the force nF). The force $nF = O'F'$ can in turn be decomposed into a force normal to the joint, nn', and a force tangent to it, $n'F$. The locus of the extremes n' of the normal forces thus obtained provides, at the joint a_0b_0, the curve $\alpha\alpha$ corresponding to the limit curve *AA* plotted for the joint *ab*.

The procedure described above allows for drawing the curves $\alpha\alpha$, $\beta\beta$, corresponding to curves *A* and *B* (Fig. 2.50). If it is repeated for each joint, exactly as it was done for masonry arches, the area of stability related to the lune is obtained. Sliding equilibrium is also tackled exactly as for arches. In his contribution on domes, Durand-Claye defines the area of stability as *'surface résidue'*.

Up to this point, Durand-Claye simply carries out an extension of his method by applying it to the single lunes, which are considered separately from each other. Then, Durand-Claye adds some considerations on the domes only. He observes that the resultant force related to the *anneau* above the joint a_0b_0 can be decomposed according to the horizontal and vertical directions, and that the horizontal component pF, applied at point *n*, compresses the *anneau* (Fig. 2.51). To avoid crushing, the area of the surface defined by a_0b_0xy, Ω, must be large enough. Let's consider half dome, i.e., the portion of the dome identified by a meridian plane (Fig. 2.51, *right*). According to Durand-Claye, if $|\sigma_c|$ is the compressive strength of masonry, the resistant surface offered by the two sections of the *anneau* thus identified, 2Ω, can withstand a maximum force of $2|\sigma_c|\,\Omega$. This is a limit that must not be reached by the resultant of horizontal compressions.

Now consider the force $m\varphi$ (Fig. 2.52) of components *mp* (vertical component corresponding to the weight of the portion above joint a_0b_0) and $p\varphi$, acting at joint a_0b_0. Durand-Claye conducts from point φ a straight line $\varphi'\varphi''$ parallel to joint

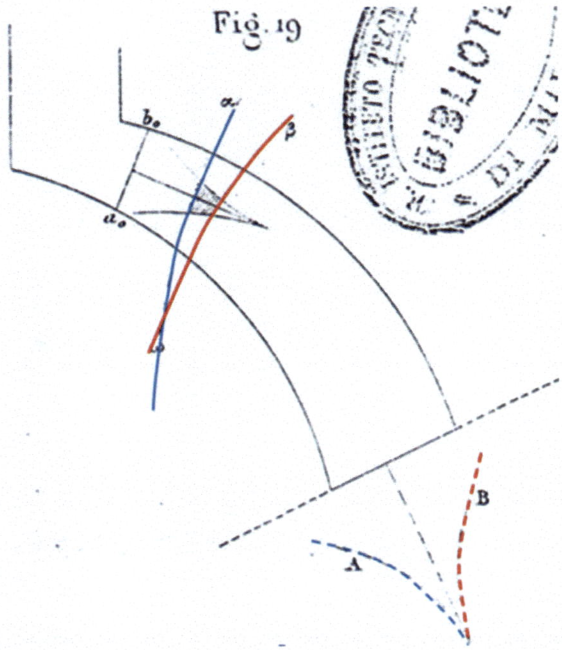

Fig. 2.50 Determination of the *surface résidue* for masonry domes (re-worked from Durand-Claye, 1880)

a_0b_0, which could further reduce the area of stability, since it defines a limit for the admissible normal forces.

As pointed out by Aita et al. (2019b), the analysis of hoop forces proposed by Durand-Claye is not convincing: only the hoop forces acting within the *anneau supérieur* (i.e. the upper part of the dome) are implicitly considered, but the beneficial effects of compressive hoop forces on the structural response are disregarded.

2.3.5 A Modern Re-interpretation of Durand-Claye's Method

A revised and enhanced version of the historical method proposed by Durand-Claye (1867) has been developed in (Aita, 2003; Foce & Aita, 2003; Aita et al., 2012, 2019a, 2020, 2021), where the graphical procedure is expressed in terms of internal forces. This newly formulated 'modern' edition of Durand-Claye's method, described here in detail, turns out to be an effective working tool for the structural analysis of masonry arches and domes.

Consider a symmetric arch and limit the analysis to the right half of the arch (for symmetry).

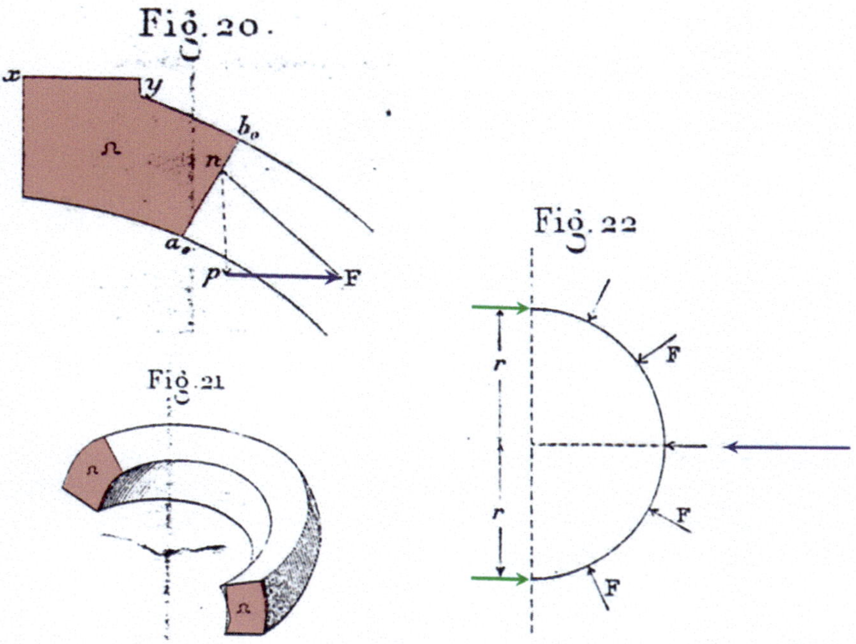

Fig. 2.51 The mechanical behaviour of the *anneau* (re-worked from Durand-Claye, 1880)

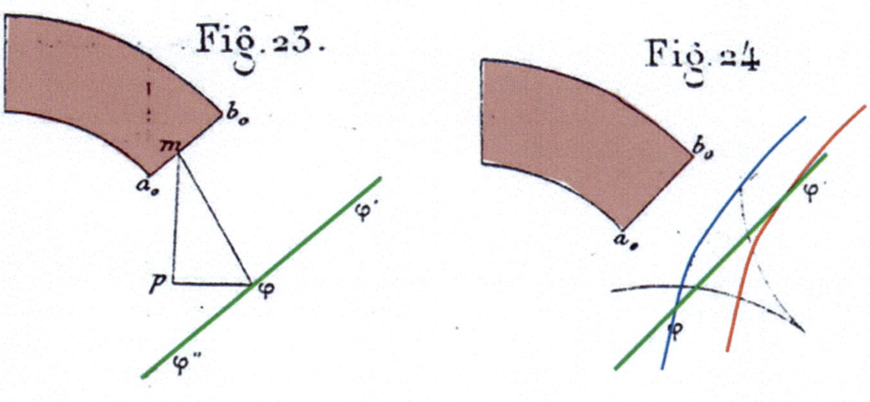

Fig. 2.52 The effects of hoop forces on the *surface résidue*, according to Durand-Claye (re-worked from Durand-Claye, 1880)

By adopting the absolute Cartesian system (O;x, y), consider now the portion of the arch c_0d_0dc, comprised between the crown section c_0d_0 and the generic joint cd (Fig. 2.53). The centre of gravity of c_0d_0dc is denoted as $G \equiv (x_G(\theta), y_G(\theta))$. The crown thrust and its eccentricity with respect to the vertical cross section's centre of

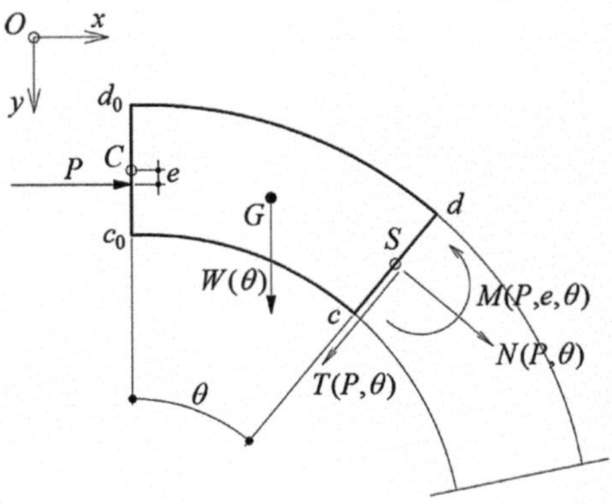

Fig. 2.53 Equilibrium of the right half of a symmetric arch: internal forces at joint θ

gravity, $C \equiv (x_C, y_C)$, are, respectively, *P* and *e*. The centre of gravity of the cross section identified by angle θ (joint *cd*) is denoted as $S \equiv (x_S(\theta), y_S(\theta))$. Equilibrium conditions enable writing the formal expressions for the normal force, $N(P, \theta)$, and the bending moment, $M(P, e, \theta)$,

$$N(P, \theta) = -P\cos\theta - W(\theta)\sin\theta \tag{2.15}$$

$$M(P, e, \theta) = P(y_S(\theta) - y_C - e) - W(\theta)(x_S(\theta) - x_G(\theta)) \tag{2.16}$$

at any given joint θ along the arch as functions of the crown thrust, *P*, and its eccentricity, *e*.

By keeping in mind the formal expressions of the limit bending moment, given by (2.12)–(2.14) and depending on the constitutive assumptions for the material, the equations which describe the attainment of the limit bending moment at any given joint θ can be written as follows:

$$M(P, e, \theta) = \pm M_{lim}(P, \theta), \tag{2.17}$$

by taking into account the expression of $N = N(P, \theta)$ given by (2.15). Relation (2.17) implicitly defines the locus of points in the (*P*, *e*) plane to which corresponds a limit condition for the bending moment at joint θ, by identifying the allowable domain that accounts for the bounded bending capacity at joint θ, denoted as $\omega_M(\theta)$.

This procedure can be repeated by scanning all joints θ along the arch and considering the intersection of all $\omega_M(\theta)$ regions in order to obtain the rotational domain, Ω_M, for the entire arch.

The problem is solved through an expressly developed numerical procedure formulated by means of the software Mathematica®. From an operational point of view, the Eqs. (2.17), solved with respect to the eccentricity, *e*, provide the explicit expression of $e^+(P, \theta)$, $e^-(P, \theta)$, corresponding to the attainment of the limit bending moment (positive or negative) at joint θ. The numerical procedure

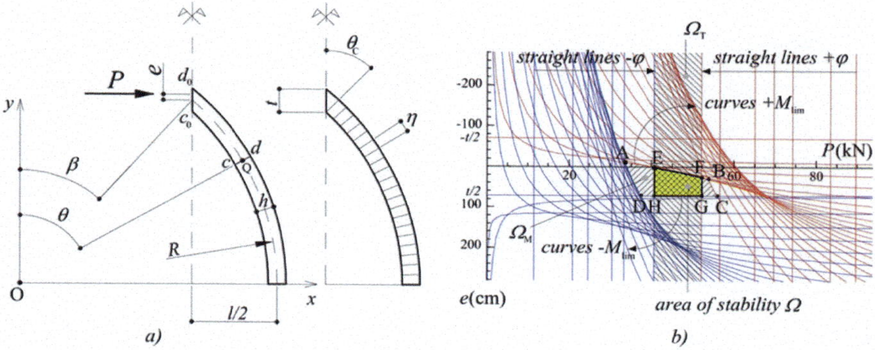

Fig. 2.54 Equilibrium analysis of a pointed arch (**a**): the area of stability Ω (**b**) (drawn from Aita et al., 2019a)

allows for plotting such functions in the (P, e) plane for each value of angle θ (Fig. 2.54b). For the sake of simplicity, these curves are denoted as $-M_{lim}$ (blue curves) and $+M_{lim}$ (red curves). Moreover, the intersection points between the $-M_{lim}$ and $+M_{lim}$ curves are determined in order to define the borders of the rotational domain, i.e., the Ω_M region. By way of example, in Fig. 2.54b the Ω_M region presents four curvilinear sides; its significant points are the vertices A, B, C, D (Aita et al., 2019a).

A second set of limitations follows from the hypothesis of limited friction coefficient, φ. For each joint θ, sliding equilibrium defines a region $\omega_T(\theta)$ in the plane (P, e) bounded by two of the vertical straight lines labelled $-\varphi$ and $+\varphi$. The numerical procedure also allows for plotting the straight lines $-\varphi$ and $+\varphi$ in the (P, e) plane as well as finding their intersections with the $-M_{lim}$ and $+M_{lim}$ curves. Thus, the borders of region $\Omega = \Omega_M \cap \Omega_T$, i.e., the area of stability for the entire arch, as well as its significant points, denoted as E, F, G, H in Fig. 2.54b, are identified.

As an example, point C in Fig. 2.54b corresponds to the intersection between the $+M$ curve at the springing joint ($\theta = \pi/2$) and the $-M$ curve at the crown joint (Fig. 2.55, *left*). Point B and point D are obtained, respectively, by searching for the maximum value of the crown thrust, P, satisfying the equations $e^+(P, \pi/2) = e^+(P, \theta)$, $e_C^-(P) = e^-(P, \theta)$, where $e^+(P, \pi/2)$ and $e_C^-(P)$ corresponds to the attainment of the limit bending moment at the springing joint (positive bending moment) and at the crown section (negative bending moment) (Fig. 2.55, *right*).

Simply observing the shape of the stability area provides a preliminary assessment of the system's stability. By decreasing—for example—the arch's thickness, h, or the friction coefficient, μ, the limit condition is attained when the stability area shrinks to a single point or a segment. The limit condition can be determined with the desired precision by evaluating the coordinates (P, e) of the significant points of the stability area.

In particular, the significant points corresponding to the maximum and minimum values of the crown thrust, P_1 and P_2, can be easily identified. In Fig. 2.54b, for

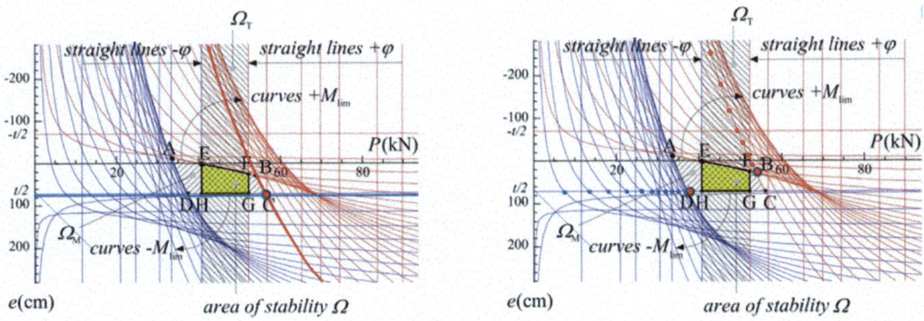

Fig. 2.55 Determining the intersection points between the $-M_{lim}$ and $+M_{lim}$ curves: point C (*left*); points B and D (*right*) (re-worked from Aita et al., 2019a)

example, points F and G correspond to the maximum value, P_1, whereas points E and H to the minimum value, P_2.

When the difference $P_1 - P_2$ is positive, the stability area is extended and the existence of statically admissible solutions is guaranteed (in Fig. 2.54b, they are identified by the points of the yellow area). On the contrary, when such a difference is negative, no statically admissible solutions are possible for the arch. In order to assess the reduction of the stability area to a single point or a segment, the values of the eccentricity, *e*, corresponding to these two significant points of the (P, e) plane also have to be checked (for further insights, the interested reader is referred to Aita et al., 2019a).

2.3.6 Some Applications of the Modern Computerized Version of Durand-Claye's Method

The modern computerized version of Durand-Claye's method lends itself to interesting applications aimed at assessing the stability of masonry arches and vaults, of which a brief review is provided.

Aita et al. (2019a) study the mechanical behaviour and collapse of symmetric circular and pointed masonry arches subject to their own weight. The influence on the arch's collapse features of its shape and thickness, as well as the friction between the arch's *voussoirs*, is analysed. Furthermore, by limiting the investigation to symmetric collapse modes characterized by one degree of freedom, the equilibrium conditions corresponding to a collapse condition are identified.

As an example, Fig. 2.56a (drawn from Aita et al., 2019a) shows the area of stability corresponding to a mixed sliding-hinging collapse mode for a lancet pointed arch. The stability area shrinks to a single point at the intersection between the vertical red straight line, r (corresponding to a limit condition related to sliding along the joint at $\theta = 45.6°$), and the two curves, $s+$ and $s-$ (corresponding to a limit condition for the bending moment at the two joints $\theta = 47.9°$and $\theta = 76°$). Figure 2.56b

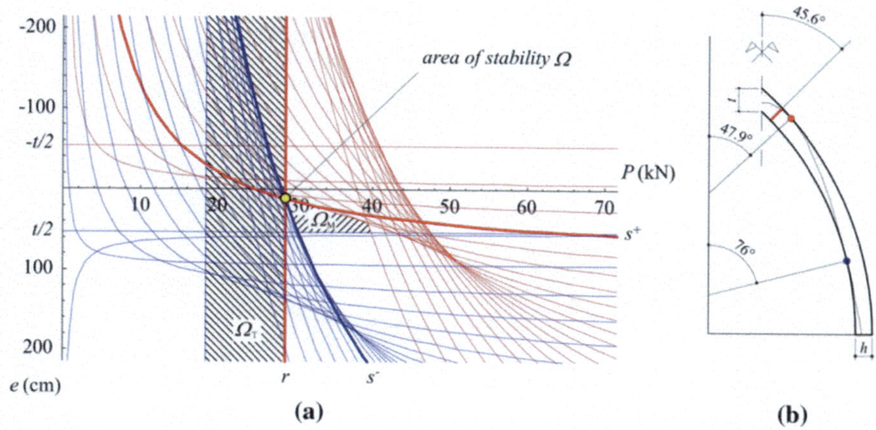

Fig. 2.56 Mixed sliding-hinging collapse mode for a lancet pointed arch; **a** stability area, **b** thrust line and critical joints (drawn from Aita et al., 2019a)

highlights the joints where a limit condition is attained: the red or blue dot shows the eccentricity at the joints where the bending moment equals a limit value, positive or negative, respectively; the red segment indicates a joint where the shear force reaches its limit value imposed by the limited friction coefficient (the upper part of the arch slides outward).

It is interesting to observe that, as the procedure also accounts for a limited friction coefficient, a *non-standard* behavior arises. As clarified in Aita et al. (2019a), the re-visitation of the Durand-Claye method allows for correctly determining the limit equilibrium conditions and the corresponding collapse mechanisms for a symmetric system. Some further remarks on *non-standard* behavior are given in Sinopoli et al. (2007), Aita and Sinopoli (2020).

In (Aita et al., 2012) a focus is performed on the influence of the masonry's limited compression strength on the arch's bearing capacity, by comparing the classical results of limit analysis under the Heyman hypotheses (1966) with those obtained through the Durand-Claye method (1867), by assuming that the tensile strength is nil, $\sigma_t = 0$, while the compressive strength is finite. The stability area method has then been applied to studying various systems composed of an arch and a superimposed wall. Figure 2.57, related to a semi-circular arch of span $L = 10$ m, shows the maximum height of the wall, H, *versus* the arch thickness, h, for different values of the masonry compressive strength: the influence of the limited compressive strength on the collapse behaviour of the arch-wall system is highlighted.

In Aita et al. (2021) a modern re-visitation of Durand-Claye's method is applied in order to study the stability and collapse modes of a system made up of a masonry arch (subject to its own weight) resting on two piers, by taking into account both the bounded bending capacity of the arch cross section and the limited friction along the joints. The system's safe domain is determined in terms of the limit conditions for arch thickness, h, pier height, H_p, and friction coefficient, μ, by extending the

Fig. 2.57 Maximum height of the wall *versus* arch thickness for different values of the compressive strength (circular arch, $L = 10$ m) (from Aita et al., 2012)

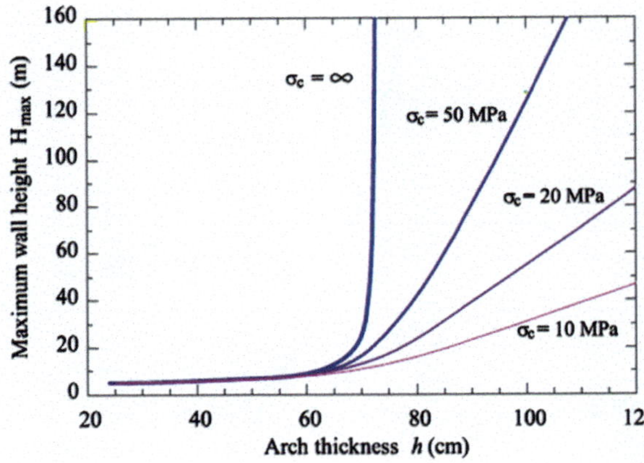

results related to stand-alone masonry arches of different shapes (Aita et al., 2019a). In Fig. 2.58, the projection of the safe domain on the (ζ, μ) plane for a low-pointed arch and piers system is shown, where the pier height and thickness are put in dimensionless form by dividing them by the span l, i.e., by setting $\eta = H_p/l$ and $\zeta = h/l$.

Other recent contributions (Aita, Barsotti & Bennati, 2017, 2019b) provide a re-visitation of Durand-Claye's method in order to study the stability of domes.

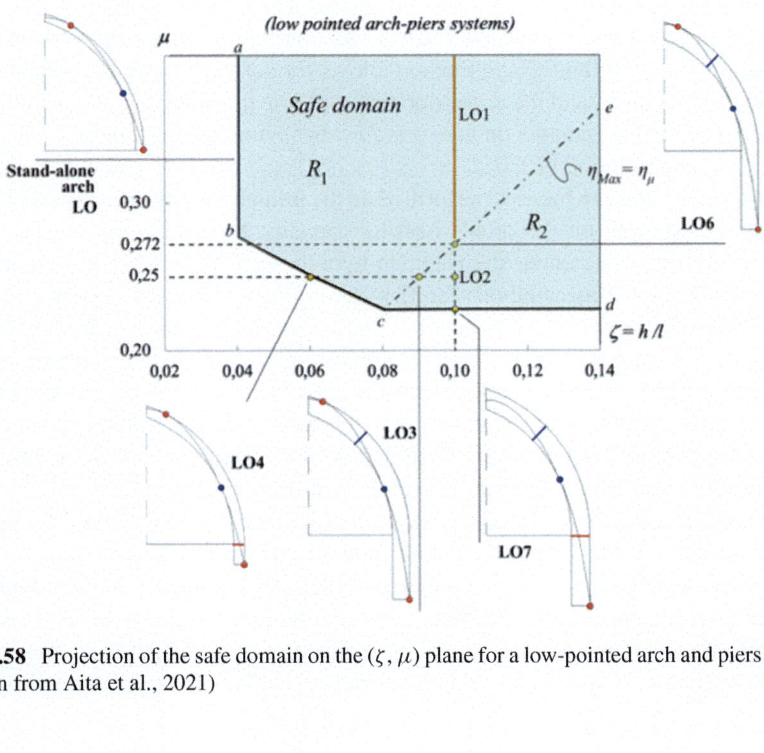

Fig. 2.58 Projection of the safe domain on the (ζ, μ) plane for a low-pointed arch and piers system (drawn from Aita et al., 2021)

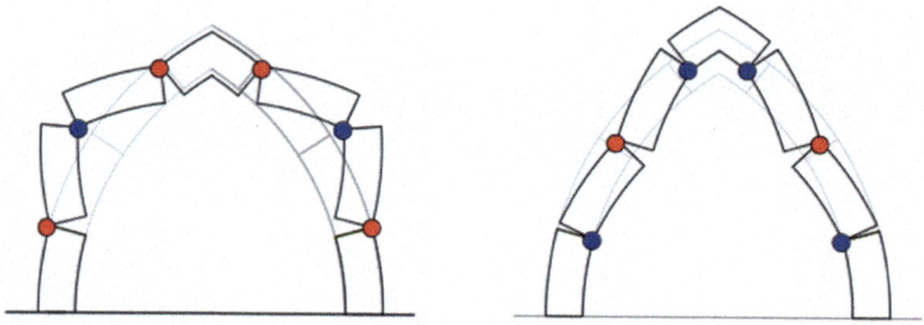

Fig. 2.59 Rotational collapse mode 1, kinematically admissible for the entire dome (*left*); rotational collapse mode 2, not admissible for the entire dome (*right*)

Aita et al. (2019b) observe that the extended version of the stability area method proposed by Durand-Claye (1880) in order to evaluate the stability of domes presents some critical issues to be carefully examined.

First, when the dome is closed, the hypothesis of limited compressive strength adopted by Durand-Claye implies that the crown thrust vanishes, since the crown's width is nil.

A second remark concerns the fact that Durand-Claye examines the equilibrium of a single lune and finds the limit condition when the stability area vanishes. It should be observed that this condition might not correspond to a collapse condition, i.e., to a kinematically admissible collapse mechanism for the entire dome. When examining the entire dome, only collapse mode 1 is admissible (Fig. 2.59, *left*). On the contrary, collapse mode 2 cannot be considered a kinematically compatible mechanism (Fig. 2.59, *right*), as in this case the lateral surfaces of each lune would act as constraints, by preventing motion towards inside.

A third critical note regards the analysis of hoop forces proposed by Durand-Claye. By examining the hoop forces acting within the *anneau supérieur*, he merely observes that they produce a further reduction of the stability area. As is evident, this approach does not take adequately into consideration the effects that compressive hoop forces have on the stability of the system.

Aita, Barsotti and Bennati (2017, 2019b) propose a re-visitation and extension of Durand-Claye's method aimed at identifying the limit thickness corresponding to the activation of a kinematically admissible mechanism for the entire dome, by assuming that the limit condition is to be investigated under the assumption that only collapse mode 1 may occur. To this purpose, an iterative method has been expressly developed by means of the Mathematica® software package. The examination of the $-M_{lim}$ and $+M_{lim}$ curves in the (P, e) plane has been performed by progressively decreasing the thickness value.

This procedure will be described more in detail with reference to the equilibrium analysis of two interesting case studies, i.e., the dome of the Pisa Cathedral and a sail vault in the Cavaniglia bastion of Fortezza Vecchia (Livorno) (Aita et al., 2019b, 2020).

Fig. 2.60 Axonometric view from the survey promoted by Università di Firenze. Istituto di restauro dei monumenti (1970)

The dome of the Pisa Cathedral. The dome of the Pisa Cathedral, dating back as early as the twelfth century, with its elliptical plan and pointed profile, represents an interesting solution from an architectural point of view (Fig. 2.60).

By starting from the laser-scanner survey, the analytical expression for the regular surface chosen to fit the intrados points coordinates was searched under the assumption that each horizontal section was an ellipse, while the two vertical sections along the major and minor axes were pointed circular arcs.

The stability of the dome has been assessed by exploiting Rankine's theorem of transformation of structures (Huerta, 2010; Rankine, 1858) and examining through a re-visitation of Durand-Claye's method two ideal domes of revolution. More in detail (Fig. 2.61), the profile of the first ideal dome of revolution corresponds to that of the actual dome that lie in the vertical meridian plane passing through the major axis of its elliptical plan. It presents a crown angle $\beta = 0.405$ rad, and an intrados span $l = 16.30$ m. The profile of the second dome corresponds to the profile along the minor axis; it presents a crown angle $\beta = 0.550$ rad, and an intrados span $l =$ 14 m. A small amplitude of the lunes ($\zeta = 1°$) and a constant thickness, $h = 0.60$ m, in the radial direction have been considered for both domes of revolution. In the numerical application—performed by means of an expressly developed, in-house algorithm implemented in Mathematica®—an infinite friction coefficient, nil tensile

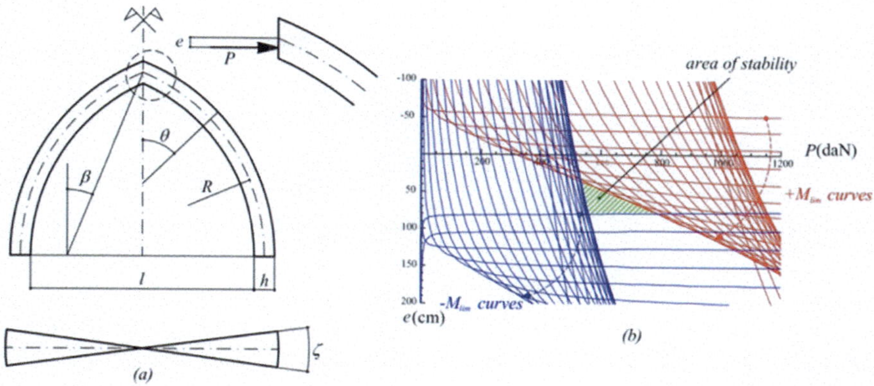

Fig. 2.61 A pointed dome profile: notation. The stability area related to a dome's lune (from Aita et al., 2019b)

strength, infinite compressive strength, and a specific weight $\gamma = 20$ kN/m^3 have been assumed.

In Fig. 2.62 (drawn from Aita et al., 2019b), the analysis of the first ideal dome of revolution is performed. The starting point consists in assuming that each lune can be considered as an independent arch. The limit curves $-M_{lim}$ and $+M_{lim}$ in the (P, e) plane are plotted for a finite number of joints in the radial direction, obtained by subdividing the intrados profile of the lune into 60 equal parts. The results thus obtained show that no point in the (P, e) plane could correspond to a statically admissible solution for the single lune (in other terms, the stability area is the empty set, shown in Fig. 2.62b). It is interesting to observe that the minimum thickness for the single lune would be twice its actual value and would correspond to the occurring of collapse mode 2 (Fig. 2.59, *right*). This result, however, cannot hold for the entire dome, since collapse mode 2 is not admissible.

The second step of analysis aims at searching for the collapse condition of the entire dome by taking into account that only collapse mode 1 (Fig. 2.59, *left*) can occur. To this purpose, the stability area method is modified, by admitting the presence of hoop forces in the upper part of the dome only. Given the above, statically admissible solutions are determined by applying Durand-Claye's method to the lower part of the lune, where hoop forces are supposed to be nil; then, in the upper part of the lune, a suitable set of hoop forces restores static admissibility.

Consider Fig. 2.62b, where the empty stability area for the single lune is shown. By progressively disregarding the $\pm M_{lim}$ curves corresponding to the joints near the crown, the procedure enables finding the inclination $\theta = \theta_a$ (Fig. 2.62a), such that the stability area related to the portion of the lune below this joint, defined by $\theta_a \leq \theta \leq \pi/2$, is the single point A in Fig. 2.62c.

An expressly developed procedure allows for determining the coordinates (P_A, e_A) of point A as well as the joints θ_a, θ_i, and θ_j (Fig. 2.62a). The solution represented by point A identifies the activation of collapse mode 2 for the lune portion under

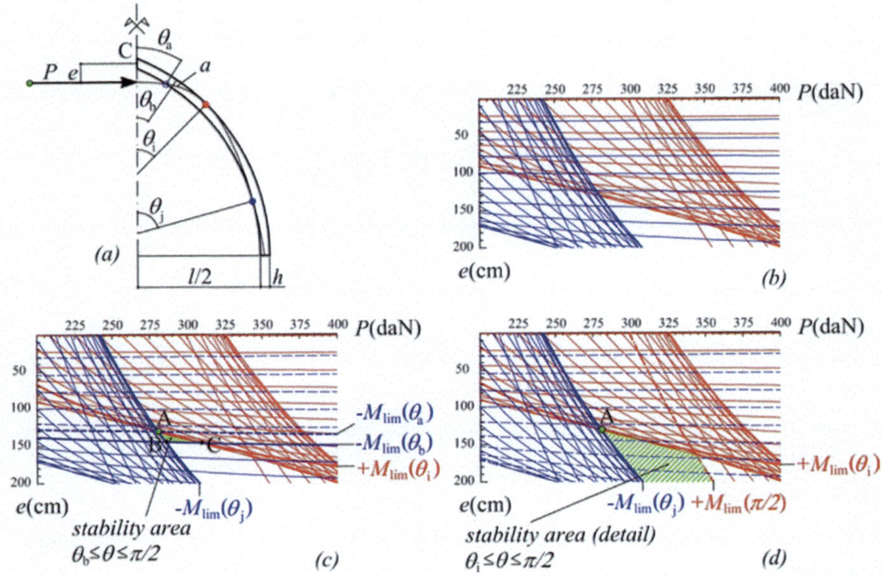

Fig. 2.62 The meridian profile of the Pisa Cathedral dome along the major axis (**a**); the limit curves $-M_{lim}$ and $+M_{lim}$ in the (P, e) plane (**b–d**) (Aita et al., 2019b)

examination. It is worth observing that the corresponding thrust line exits from the intrados in the upper portion of the lune (above joint θ_a).

In order to check if a collapse mechanism corresponding to mode 1 can occur for the entire dome, the analysis is limited to the lune portion that extends from the joint θ_i to the springing ($\theta = \pi/2$). The related stability area is the green hatched area shown in Fig. 2.62d, obtained by excluding from the analysis the dashed blue $-M_{lim}$ curves corresponding to the joints above θ_i. For the actual value of the thickness of the dome, $h = 60$ cm, such area is extended and identifies an infinite number of statically admissible solutions in the range $\theta_i \leq \theta \leq \pi/2$.

The limit value of the thickness, h^*, can be obtained by progressively decreasing h until a limit value corresponding to the shrinking of this new stability area to a single point. In this case, the limit value h^* is compatible with the occurring of collapse mode 1, kinematically admissible for the entire dome.

In order to find statically admissible solutions for the dome, the action of hoop forces in the upper part of the dome is accounted for.

The procedure allows for determining statically admissible solutions for each of the two theoretical revolution domes mentioned above (i.e., the domes, whose profiles lie along the vertical planes containing the major and minor axis of the elliptical plan, respectively). After considering the affine transformation which maps the middle surface of the larger of the two ideal domes of revolution to that of the actual dome, the stability of the dome of Pisa Cathedral is assessed: the ratio $\nu = h/h_{lim} = 2$, where $h_{lim} = h^*$, can be taken as an estimate of a geometric safety factor for the dome under its own weight (for further details, the interested reader is referred to Aita et al., 2019b).

The sail vault in the Cavaniglia bastion of Fortezza Vecchia in Livorno. In (Aita et al., 2020) a re-visitation of the stability area method has been exploited in order to study the mechanical response of another interesting masonry vault, a sail vault covering a trapezoidal room located in the Cavaniglia bastion of Fortezza Vecchia in Livorno (Fig. 2.63). Under the assumption of nil tensile hoop forces, admissible stress fields within the vault have been determined by considering the weight of the vault itself and that of the overlying soil layers.

The sail vault was modeled as an assemblage of arches obtained by slicing the vault into lunes. The results of an accurate survey allowed to define the geometry of the vault, characterized by a spherical intrados surface, with an intrados radius of 5.51 m, delimited by four vertical planes. With reference to Fig. 2.64, drawn from (Aita et al., 2020), the following geometrical and mechanical data have been assumed: $h = 0.25$ m, $a = 0.85$ m, $b = 0.35$ m, $c = 3.14$ m, and $d = 1$ m, a specific weight of 18 kN/m^3 for the vault, the 'lantern' and the overlying soil.

It is observed that the sail vault is not a dome, but its central portion, above angle $\alpha = 40.11°$ (Fig. 2.64b) is a dome of revolution.

A first equilibrium assessment was conducted by applying Durand-Claye's method to a single lune (of amplitude equal to 1°) belonging to the dome of embrace angle $\alpha = 40.11°$ by considering an infinite compressive strength. Each joint is

Fig. 2.63 Sail vault in the Cavaniglia bastion of Fortezza Vecchia: the intrados spherical surface with pendentives (from Aita et al., 2020)

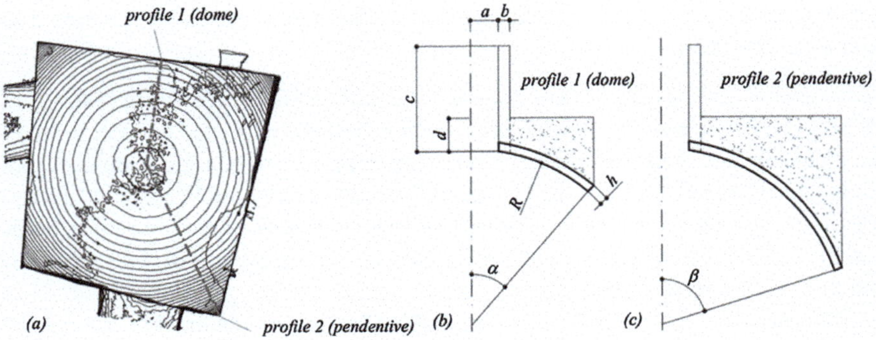

Fig. 2.64 The quadrilateral plan (**a**); profile 1, corresponding to the 'dome' above angle $\alpha = 40.11°$ (**b**); profile 2, corresponding to the pendentive of maximum embrace angle, $\beta = 70.79°$ (**c**) (from Aita et al., 2020)

assumed to be in the radial direction. In Fig. 2.65, the red and blue curves correspond to the attainment of a positive or negative limit bending moment, while the stability area is the green-hatched area. Since the stability area is extended, the lune is in equilibrium even if compressive hoop forces are assumed to be nil, as shown from the thrust lines corresponding to the maximum and minimum value of the horizontal thrust, which are contained within vault's thickness.

The stability area method can also be applied in order to obtain a first assessment of the stability of the pendentives by analyzing a lune defined by $\beta = 70.79°$ (Fig. 2.64c). Also in this case, the stability area is extended, even if noticeably reduced with respect to that in Fig. 2.65a, by showing that the lune is in equilibrium even for nil compressive hoop forces.

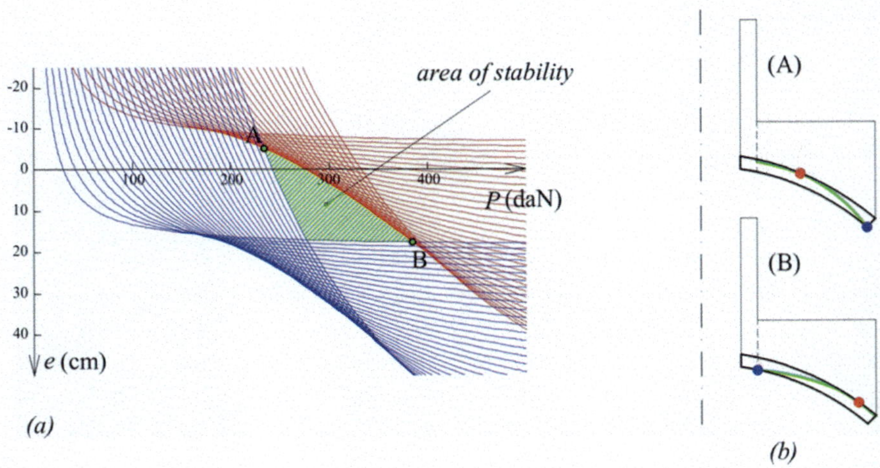

Fig. 2.65 The stability area related to the vault's portion defined by $\alpha = 40.11°$, i.e., to profile 1 (**a**); the maximum and minimum thrust lines contained within lune's thickness (**b**)

By repeating the analysis under the hypothesis of limited compressive strength, $\sigma_c = 20$ MPa, an analogous output is obtained, which confirms that the sail vault of Fortezza Vecchia is far from any collapse condition.

It is noteworthy that, even if the stability area were reduced to a single point by progressively reducing the lune's thickness, *h*, this would not lead to a limit condition for the entire vault, since this thickness would not correspond to a kinematically admissible mechanism for the entire dome. The limit thickness can be identified according to the procedure described with reference to the analysis of the dome of the Pisa Cathedral (Aita et al., 2019b), so as to check for lune stability under the restriction that only an admissible collapse mode can occur.
clearpage

References

Aita, D. (2003). *Analisi elastica non lineare e collasso di strutture ad arco in muratura.* Ph.D. Thesis, Dottorato in Ingegneria delle Strutture, XV Ciclo, Università degli Studi di Firenze; Sedi Consorziate.

Aita, D., Barsotti, R., Bennati, S., & Foce, F. (2004). The statics of pointed masonry arches between 'limit' and 'elastic' analysis. Arch Bridges IV. In P. Roca & C. Molins (Eds.), *Advances in assessment, structural design and construction* (pp. 354–362).

Aita, D., Foce, F., Barsotti, R., & Bennati, S. (2007). Collapse of masonry arches in Romanesque and Gothic constructions. In: P. B. Lourenço, D. V. Oliveira & A. Portela (Eds.), *ARCH'07, Proceedings of 5th International Conference on Arches and Bridges* (pp. 625–632). University of Minho.

Aita, D., & Sinopoli, A. (2020). Revisiting Monasterio's unpublished manuscript: A critical review of the collapse modes analysis of non-symmetric and symmetric masonry arches. *International Journal of Architectural Heritage, 14*(5), 762–793. https://doi.org/10.1080/15583058.2019.1648586.

Aita, D., Barsotti, R., & Bennati, S. (2012). Equilibrium of pointed, circular and elliptical masonry arches bearing vertical walls. *Journal of Structural Engineering, ASCE, 138*(7), 880–888. ISSN 0733-9445. https://doi.org/10.1061/(ASCE)ST.1943-541X.0000522

Aita, D., Barsotti, R., & Bennati, S. (2017). A modern reinterpretation of Durand-Claye's method for the study of equilibrium conditions of masonry domes. In *Proceedings of the XXIII conference of the italian association of theoretical and applied mechanics (AIMETA) (Salerno, Italy)*, edited by L. Ascione et al. (pp.1459–1471).

Aita, D., Barsotti, R., & Bennati, S. (2019a). Looking at the collapse modes of circular and pointed masonry arches through the lens of Durand-Claye's stability area method. *Archive of Applied Mechanics, 89*, 1537–1554. https://doi.org/10.1007/s00419-019-01526-z

Aita, D., Barsotti, R., & Bennati, S. (2019b). Studying the dome of Pisa cathedral via a modern reinterpretation of Durand-Claye's method. *JoMMS, 14*(5), 603–619. https://doi.org/10.2140/jomms.2019.14-5

Aita, D., Barsotti, R., & Bennati, S. (2020). Equilibrium analysis of a sail vault in Livorno's fortezza vecchia through a modern re-edition of the stability area method. In *Lecture notes in mechanical engineering* (pp. 989–998). https://doi.org/10.1007/978-3-030-41057-5_81

Aita, D., Barsotti, R., & Bennati, S. (2021). Arch-piers systems subjected to vertical loads: A comprehensive review of rotational, sliding and mixed collapse modes. *Archive of Applied Mechanics, 91*, 241–256. https://doi.org/10.1007/s00419-020-01766-4

Alexakis, H., & Makris, N. (2015). Limit equilibrium analysis of masonry arches. *Archive of Applied Mechanics, 85*, 1363–1381. https://doi.org/10.1007/s00419-014-0963-6

Barlow, W. H. (1846). On the existence (practically) of the line of equal horizontal thrust in arches, and the mode of determining it by geometrical constructions. *Minutes of Proceedings of the Institution of Civil Engineers, V*, 162–182.

Benvenuto, E. (1981). *La scienza delle costruzioni e il suo sviluppo storico*. Sansoni, Firenze (rpt. 2006, Edizioni di Storia e Letteratura).

Benvenuto, E. (1991). *An introduction to the history of structural mechanics*. Springer.

Block, P., Ciblac, T. & Ochsendorf, J. (2006). Real-time limit analysis of vaulted masonry buildings. *Computers & Structures, 84*(29–30), 1841–1852. https://doi.org/10.1016/j.compstruc.2006.08.002

Bossut, C. (1778). Recherches sur l'equilibre des voûtes. *Mémoires de l'Académie Royale des Sciences, année 1774*, Paris, 534–566.

Bossut, C. (1779). Nouvelles Recherches sur l'equilibre des voûtes en dôme. *Mémoires de l'Académie Royale des Sciences, année 1776*, Paris, 587–596.

Bresse, J. A. C. (1880). *Cours de mécanique appliquée, professé à l'École impériale des Ponts et Chaussées, Première partie. Résistance des matériaux et stabilité des constructions*, Gauthiers-Villars, Paris (III ed.).

Carvallo, J. (1853). Etude sur la stabilité des voûtes. *Annales Des Ponts Et Chaussées, 5*, 1–77.

Castigliano, A. (1878). *Applicazioni pratiche della teoria sui sistemi elastici*, Studi dell'Ufficio d'Arte, Strade Ferrate dell'Alta Italia, Servizio della Manutenzione e dei Lavori. Civelli, Milano.

Castigliano, A. (1879–1880). *Théorie de l'équilibre des systèmes élastiques et ses applications.* Torino: A. F. Negro.

Cennamo, C., Cusano, C., & Angelillo, M. (2018). On the statics of large domes: A static and kinematic approach for San Francesco di Paola in Naples. In *Proceedings of the British Masonry Society*, (pp. 504–517), 9–11 July 2018. Italy: Milan.

Cennamo, C., Cusano, C., & Angelillo, M. (2019). A limit analysis approach for masonry domes: The basilica of San Francesco di Paola in Naples. *International Journal of Masonary Research and Innovation, 4*, 227–242. https://doi.org/10.1504/IJMRI.2019.100568

Cipriani, B., & Lau, W. (2006). Construction Techniques in Medieval Cairo: The Domes of Mamluk Mausolea (1250 A.D.–1517 A.D.). In *Second International Congress on Construction History, Queens' College, Cambridge University* (Exeter) (pp. 695–716).

Clemente, P., Occhiuzzi, A., & Raithel, A. (1995). Limit behavior of stone arch bridges. *Journal of Structural Engineering, ASCE, 121*(7).

Coccia, S., Como, M., & Di Carlo, F. (2016). Minimum thrust and minimum thickness of hemispherical masonry domes. *Acta Mechanica, 227*, 2415–2425.

Coulomb, C. (1776). Essai sur une application de règles de maximis et minimis à quelques problèmes de Statique, relatifs à l'Architecture. *Mémoires de Mathématique et de Physique présentés à l'Académie Royale des Sciences, par divers Savans, et lûs dans les Assemblées - Année 1773.* Paris, 343–382.

Cremona, L. (1890). *Graphical statics, two treatises on the graphical calculus and reciprocal figures in graphical statics.* Trans. by Thomas Hudson Beare. Oxford: Clarendon Press.

Culmann, K. (1866) *Die graphische Statik*. Meyer & Zeller, Zürich.

Culmann, K. (1875). *Die graphische Statik*. Meyer & Zeller, Zürich (II ed.).

Culmann, K. (1880). *Traité de Statique Graphique, Traduit sur la deuxième edition en allemande par G. Glasser, J. Jacquier et A. Valat*, Paris, Dunod, Libraire des Corps Nationaux des Ponts et Chaussées, des Mines et des Télégraphes.

Cusano, C., Montanino, A., Olivieri, C., Paris, V., & Cennamo, C. (2021). Graphical and analytical quantitative comparison in the domes assessment: The case of San Francesco di Paola. *Applied Sciences, 11*, 3622. https://doi.org/10.3390/app11083622

Drouets. (1865). Sur la stabilité des voûtes. *Annales des Ponts et Chaussées, 9* (I Sem.), 179–266.

Drucker, D. C., & Prager, W. (1952). Soil mechanics and plastic analysis for limit design. *Quarterly of Applied Mathematics, 10*(2), 157–165.

Durand-Claye, A. (1867). Sur la vérification de la stabilité des voûtes en maçonnerie et sur l'emploi des courbes de pression. *Annales des Ponts et Chaussées, 13* (I Sem.), 63–93.

Durand-Claye, A. (1868). Note sur la verification de la stabilité des arcs métalliques et sur l'emploi des courbes de pression. *Annales Des Ponts Et Chaussées, 15*, 109–144.
Durand-Claye, A. (1880). Vérification de la stabilité des voûtes et des arcs. Applications aux voûtes sphériques. *Annales des Ponts et Chaussées, 19* (I Sem.), 416–440.
Eddy, H. T. (1877). *New constructions in graphical statics*. Van Nostrand.
Foce, F. (2002). Sulla teoria dell'arco murario. Una rilettura storico-critica. In A. Becchi & F. Foce (Eds.), *Degli archi e delle volte: Arte del costruire tra meccanica e stereotomia*. Venezia: Marsilio.
Foce, F., & Aita, D. (2003). The masonry arch between 'limit' and 'elastic' analysis. A critical reexamination of Durand-Claye's method. In *Proceedings of the first international congress on construction history (Madrid, 20th-24th January 2003)* Vol. II (pp. 895–908). Madrid: Instituto Juan De Herrera. ISBN 84-9728-072-5.
Foce, F. (2010). A pearl from the history of structural mechanics. Méry, Saint-Venant and the Théorie des voûtes. In A. Sinopoli (Ed.), *Mechanics and Architecture between episthéme and téchne, Edizioni di Storia e Letteratura, Roma* (pp. 125–154).
Gerstner, F. J. (1831–1834). *Handbuch der Mechanik*, Spurny (p. 405).
Galassi, S., Misseri, G., Rovero, L., & Tempesta, G. (2017). Equilibrium analysis of masonry domes. on the analytical interpretation of the Eddy-Lévy graphical method. *International Journal of Architectural Heritage, 11*(8), 1195–1211. https://doi.org/10.1080/15583058.2017.1372823
Heyman, J. (1966). The stone skeleton. *International Journal of Solids and Structures, 2*(2), 249–279.
Heyman, J. (1977). *Equilibrium of shell structures*. Oxford University Press.
Hooke, R. (1675). *A description of helioscopes, and some other instruments*. John & Martin Printer to the Royal Society.
Huerta, S. (2010). Designing by geometry: rankine's theorems of transformation of structures. In P. Cassinello, S. Huerta, J.M. de Prada Poole & R. Sanchez Lampreave (Eds.), *Geometría y proporción en las estructuras. Ensayos en honor de Ricardo Aroca, Escuela Técnica Superior de Arquitectura*. Universidad Politécnica de Madrid.
Kurrer, K.-E. (2008). *The history of the theory of structures: From arch analysis to computational mechanics*. Ernst & Sohn.
Laterrade, (1885). Note sur la stabilité des voûtes en réponse à un travail de M. Léon Durand-Claye. *Annales des Ponts et Chaussées, 10* (II Sem.), 1004–1006.
Lau, W. (2006). *Equilibrium analysis of masonry domes*, Master's thesis, Building Technology Program, Massachusetts Institute of Technology, USA.
Lévy, M. (1888). *La statique graphique et ses applications aux constructions*. Gauthier-Villars.
Makris, N., & Alexakis, H. (2013). The effect of stereotomy on the shape of the thrust-line and the minimum thickness of semicircular masonry arches. *Archive of Applied Mechanics, 83*(19), 1511–1533.
Marmo, F. (2021). ArchLab: A MATLAB tool for the thrust line analysis of masonry arches. *Curved and Layered Structures, 8*(1), 26–35.
Mascheroni, L. (1785). *Nuove ricerche sull'equilibrio delle volte*. Bergamo: Locatelli.
Maxwell, C. (1864). On reciprocal figures and diagrams of forces. *Philosophical Magazine, 26*, 250–261.
Maxwell, C. (1870). On reciprocal figures, frames and diagrams of forces. *Transactions of the Royal Society of Edinburgh, XXVI*, 1–40.
Méry, E. (1840). Sur l'équilibre des voûtes en berceau. *Annales des Ponts et Chaussées, 19* (I Sem.), 50–70.
Michon, P. F. (1857). *Instruction sur la stabilité et des murs de revêtement*. Lithographie de l'Ecole d'Application, Metz.
Milankovitch, M. (1904). *Beitrag zur Theorie der Druckkurven. Dissertation zur Erlangung der Doktorwürde*, K.K. technische Hochschule, Vienna.
Milankovitch, M. (1907). Theorie der Druckkurven. *Zeitschrift für Mathematik und Physik, 55*, 1–27.

Möbius, A. F. (1833). *Ueber eine besondere Art dualer Verhältnisse zwischen Figuren in Raume, Crelle's Journal für Mathematik.* X Bd.

Moseley, H. (1833). On a new principle in statics called the principle of least pressure. *Philosophical Magazine, 285.*

Moseley, H. (1843). On the theory of Arch. In J. Weale (Ed.), *The theory practice and architecture of bridges* (vol. 1).

Monge, G. (1794). *Géométrie descriptive.* Baudouin.

Navier, C. L. M. H. (1826). *Résumé des leçons données à l'Ecole des Ponts et Chaussées sur l'application de la mécanique à l'établissement des constructions et des machines.* F. Didot père et fils.

Nikolić, D. (2017). Thrust line analysis and the minimum thickness of pointed masonry arches. *Acta Mechanica, 228*, 2219–2236. https://doi.org/10.1007/s00707-017-1823-6

Nodargi, N. A., & Bisegna, P. (2020). Thrust line analysis revisited and applied to optimization of masonry arches. *International Journal of Mechanical Sciences, 179*, 105690. https://doi.org/10.1016/j.ijmecsci.2020.105690

Olivito, R. S. (2009). *Statica e Stabilità delle Costruzioni Murarie*, Pitagora.

Oppenheim, I. J., Gunaratnam, D. J., & Allen, R. H. (1989). Limit state analysis of masonry domes. *Journal of Structural Engineering, 115*(4), 868–882.

Paris, V., Ruscic, G., & Mirabella, R. G. (2021). Graphical modelling of hoop force distribution for equilibrium analysis of masonry domes. *Nexus Network Journal, 23*, 855–878. https://doi.org/10.1007/s00004-021-00556-x

Poncelet, J.-V. (1835). Solution graphique des principales questions sur la stabilité des voûtes. *Mémorial De L'officier Du Génie, 12*, 151–213.

Persy, N. (1825). *Cours de stabilité des constructions, à l'usage des élèves de l'Ecole Royale de l'Artillerie et du Génie.* Lithographie de l'Ecole Royale de l'Artillerie et du Génie de Metz, Metz.

Rankine, W. J. M. (1858). *A manual of applied mechanics.* Griffin.

Rankine, W. J. M. (1864). Principle of the equilibrium of polyhedral frames. *The London, Edinburgh Dublin Philosophical Magazine and Journal of Science*, *27*, 92.

Rapallini, M., & Tempesta, G. (1997). Limit analysis of masonry domes. In G.N. Pande, J. Middleton & B. Kralj (Eds.), *Computer methods in structural masonry-4: Proceedings of the fourth international symposium on computer methods in structural masonry* (pp. 3–5), September 1997, Florence, Italy, Taylor & Francis.

Ritter, W. (1888–1906). *Anwendungen der graphischen Statik nach Prof. Dr. C. Culmann*, Meyer und Zeller, Zürich.

Rocherau, D. (1859). Mémoire sur les voûtes en berceau portant une surcharge limitée à un plan horizontal. *Révue de l'Architecture et des Travaux Publics, 17*, 114–124, 158–179, 207–224, 257–268.

Roman, S. (2005). Advanced linear algebra. In *Graduate texts in mathematics* (vol. 135, 2nd ed.). Springer-Verlag, ISBN 978-0-387-24766–3.

Saint-Guilhelm, P. (1859). Mémoire sur l'établissement des arches de ponts assujettiés aux conditions du maximum de la stabilité. *Annales Des Ponts Et Chaussées, 17* (I Sem.), 83–106.

Scheffler, H. (1857). *Theorie der Gewölbe, Futtermauern und eisernen Brücken.* Brunschwick.

Sinopoli, A., Corradi, M., & Foce, F. (1997). A modern formulation for pre-elastic theories on masonry arches. *Journal of Engineering Mechanics, ASCE* 204–213.

Sinopoli, A., & Foce, F. (2001). Theoretical research on masonry and masonry arches in Italy. In C. Abundur (Ed.), *Arch '01—Third international conference on arch bridges* (pp. 323–342). Presses de l'Ecole Nationale des Ponts et Chaussées.

Sinopoli, A., Aita, D., & Foce, F. (2007). Further remarks on the collapse of masonry arches with Coulomb friction. In P.B. Lourenço, D.V. Oliveira & A. Portela (Eds.), *ARCH'07, Proceedings of 5th international conference on arch bridges (Madeira, 12–14 September 2007)* (pp. 649–657). University of Minho.

Sinopoli, A. (2012). Origine e sviluppi della Statica Grafica. In T. Levi Civita & U. Amaldi (Eds.), *Complementi alle Lezioni di Meccanica Razionale*, CompoMat (pp. 161–185).
Sinopoli, A. (2021). *Statica grafica*, Aracne, Genzano di Roma, ISBN: 1259941450.
Stevin, S. (1608). *Hypomnemata mathematica*, Leiden.
Università di Firenze. Istituto di restauro dei monumenti. (1970). *Il Duomo di Pisa, rilievo a cura dell'Istituto di Restauro dei monumenti*, Nistri-Lischi.
Varignon, P. (1725). *Nouvelle Mécanique ou Statique dont le projet fut donné en 1687, Ouvrage posthume de M. Varignon*. Paris: C. Jombert.
Varma, M. N., & Ghosh, S. (2016). Finite element thrust line analysis of axisymmetric masonry domes. *International Journal of Masonry Research and Innovation, 1*(1), 59–73.
Villarceau, A.J.F.Y. (1854). Sur l'établissement des arches de pont, *Mémoires présentés par divers savants à l'Académie des Sciences de l'Institut de France, 12*, 503–822.
Wolfe, W. S. (1921). *Graphical analysis. A text book on graphic statics*. McGrow-Hill Book Co.
Zessin, J., Lau, W., & Ochsendorf, J. (2010). Equilibrium of cracked masonry domes. *Proceedings of the Institution of Civil Engineers: Engineering and Computational Mechanics, 163*(3), 135–145.

Chapter 3
Discretised Continuum Approaches: From Continuum to Dis-Continuum

Maurizio Angelillo, Antonino Iannuzzo, and Andrea Montanino

Abstract Prologue. **The model of Heyman for Masonry as a solid fracturing continuum**. In Chap. 1, the need for special computational strategies to analyse masonry structures is motivated. Most commercial structural software tools neglect contact, fracture and friction. Basically, there exist two conceptually simple ways to approach the problem while including these aspects. The first one is represented by Discrete Element Method (DEM) approaches, in the present book briefly described in Sect. 1.2.4 and fully illustrated in Chap. 5. The second one funds upon Heyman's model and consists of a set of crude material restrictions catching at the macroscale the basic ingredients of masonry behaviour, allowing the use of Limit Analysis. Heyman's theory can also lead to a variational formulation of the boundary value problem for masonry-like structures, described in Part I of this Chapter. This variational formulation naturally leads to two dual numerical strategies described in Part II of the present Chapter. A specific Chapter, Chap. 4, is dedicated to TNA, a striking example of a lower-bound limit-analysis-based method to assess vaulted masonry structures. With TNA, admissible equilibrium states correspond to compressive force networks within the structural geometry.

Keywords Unilateral material · Total potential energy · Complementary energy · Singular stress fields · Settlements · Airy stress potential · Equilibrium method

---

M. Angelillo (✉)
Department of Civil Engineering, University of Salerno, Via Giovanni Paolo II, 84084 Fisciano, SA, Italy
e-mail: mangelillo@unisa.it

A. Iannuzzo
Department of Engineering, University of Sannio, Piazza Roma 21, 82100 Benevento, Italy

A. Montanino
Department of Structures for Engineering and Architecture, Via Claudio 21, 80125 Napoli, Italy

K. Bagi and M. Angelillo (eds.), *Discrete Computational Mechanics of Masonry Structures*, CISM International Centre for Mechanical Sciences 609,
https://doi.org/10.1007/978-3-031-32476-5_3

3.1 Formulation of BVP: Basic Ingredients of Solid Mechanics

In these notes, we essentially adopt the notation employed by Gurtin in (1972). We restrict to the plane case and identify the structure with Ω, that is with a region of R^2, whose boundary is $\partial\Omega = \partial\Omega^D \cup \partial\Omega^N$. Called $\boldsymbol{x} \in \Omega$ the position vector with respect to the origin O of any material point of Ω, the mass density-mass per unit area-of the body is denoted ρ ThCartesian reference of origin O: $\{O; x_1, x_2\}$, and the associated unit base vectors $\{\boldsymbol{e}_1, \boldsymbol{e}_2\}$ are also introduced.

Kinematics. The displacement, velocity and acceleration of material points of Ω are denoted $\boldsymbol{u}, \dot{\boldsymbol{u}}, \ddot{\boldsymbol{u}}$, superimposed dots being the classical notation for time derivatives; these are vector fields defined over Ω. $\partial\Omega^D$, the Dirichlet part, is the constrained part of the boundary where displacements $\overline{\mathbf{u}}$ are given. Restricting to the analysis of small displacements in the neighbourhood of the known geometry Ω, we adopt the infinitesimal deformation tensor

$$\boldsymbol{E}(\boldsymbol{u}) = \frac{1}{2}(\nabla\boldsymbol{u} + \nabla\boldsymbol{u}^T), \tag{3.1}$$

as the strain measure. By allowing for the presence of given eigenstrains $\overline{\mathbf{E}}$, the above definition is modified as follows

$$\boldsymbol{E}(\boldsymbol{u}) = \frac{1}{2}(\nabla\boldsymbol{u} + \nabla\boldsymbol{u}^T) - \overline{\mathbf{E}}. \tag{3.2}$$

Forces. The contact stress vector across any internal interface of unit normal $\boldsymbol{n}$ is denoted $\boldsymbol{t}_n$; in this 2D context, it is a force per unit length. The Cauchy stress tensor -dimensionally a force density per unit length- is denoted $\boldsymbol{T}$. If $\boldsymbol{t}_n$ is regular enough, then

$$\boldsymbol{T}\boldsymbol{n} = \boldsymbol{t}_n. \tag{3.3}$$

The Neumann part $\partial\Omega^N$ is the loaded part of the boundary where tractions are given. The body forces-dimensionally forces per unit area-are denoted $\boldsymbol{b}$, while the tractions-dimensionally forces per unit length-on $\partial\Omega^N$ are denoted $\overline{\boldsymbol{p}}$.

Balance laws. If all the required regularity of the motion and of the forces is assumed for any part of Ω, then, the following local form of the first Euler law holds at any point $\boldsymbol{x} \in \Omega$:

$$\text{div}\boldsymbol{T} + \boldsymbol{b} = \rho\,\ddot{\boldsymbol{u}} \tag{3.4}$$

From the second Euler law descends instead:

$$\boldsymbol{T} = \boldsymbol{T}^T, \tag{3.5}$$

that is $\boldsymbol{T} \in Sym$ at any point $\boldsymbol{x} \in \Omega$, being Sym the set of second-order symmetric tensors. If inertial effects are irrelevant and we can restrict to Statics, then, for equilibrium, at any point $\boldsymbol{x} \in \Omega$:

$$div\boldsymbol{T} + \boldsymbol{b} = 0. \tag{3.6}$$

Remark 1 Here Ω should be the current, deformed from the original, shape of the body, and div the divergence operator with respect to the current positions of material points. With the small strain assumption, there is no need to distinguish among current, reference and original shape. We have to point out that the strain/displacement smallness question is rather delicate, especially for masonry-like structures. Speaking on the order of magnitude of the displacements, we have to be aware that when masonry structures develop mechanisms, the overall position changes are much larger than those expected for structures made of conventional bilateral materials (say steel and concrete). For non-thin and non-slender modern structures, the limit post-elastic displacement order is typically about 1/10 of the overall element dimensions, whilst for traditional masonry structures, the order is about one. There are many explanations for this, but the main factor is that masonry structures can develop fractures. This means that when assessing the displacement capacity or the dynamical response of masonry structures, we have to be ready to leave the linear approximation and include geometrically nonlinear effects. In these cases, to retain the linear machinery, a step-by-step approximation strategy can be adopted.

Boundary conditions. The typical Boundary Value Problem (BVP) for a solid is complemented by the following displacement-type boundary conditions (BC):

$$\boldsymbol{u} = \overline{\mathbf{u}} \text{ on } \partial\Omega^D, \tag{3.7}$$

and by the force type BC

$$\boldsymbol{T}\boldsymbol{n} = \overline{\mathbf{p}} \text{ on } \partial\Omega^N. \tag{3.8}$$

3.1.1 Heyman's Material

Material restrictions. The constitutive restrictions defining Heyman's material, that is, the Normal Rigid No-Tension (NRNT) material, are represented by the following three conditions:

$$\boldsymbol{T} \in Sym^-, \tag{3.9}$$

with Sym^- the set of second-order negative semi-definite symmetric tensors;

$$\boldsymbol{E} \in Sym^+, \tag{3.10}$$

being Sym^+ the set of second-order positive semi-definite symmetric tensors;

$$\boldsymbol{T} \cdot \boldsymbol{E} = 0. \tag{3.11}$$

This type of material restrictions were first formulated by Heyman in (1966) (and fully exploited by him in (1997)) and subsequently adopted by Como in (1992) and some other Italian scientists (the reader is referred to (Angelillo, 1993) and references therein for more details).

If we look at the No-Tension assumption (3.9) as a constraint on stress, then a latent strain, that is, a deformation necessary to sustain such an assumption arises. Actually, for NRNT materials, the strain coincides with the latent strain, meaning that there is no other deformation besides the latent deformation, and such total strain represents a measure of fracture. The material restrictions (3.9)–(3.11) are perfectly symmetric, and we may think of them from an alternative point of view. Namely we may look at (3.10) as a constraint on the strain and the stress as the latent forces (reactions) produced by the constraint. Relation (3.11) is then the classical assumption defining simple internal constraints in the sense of Trusdell and Noll (2004). Most contemporary works in constrained theories of continuum mechanics follow the approach of Trusdell and Noll (2004), wherein the stress is decomposed a priori into reactive and active terms, with the reactive stress assumed to be powerless in all motions consistent with the constraints and the active stress given by a constitutive equation. In the present case, all the stress is non-constitutive, the active stress is zero (Angelillo & Fortunato, 2001).

Matrix notations. In the Euclidean space R^2 we introduce a fixed orthonormal Cartesian frame $\{0; x_1, x_2\}$ whose associated orthonormal basis is $\{\boldsymbol{e}_1, \boldsymbol{e}_2\}$, and decompose vectors and tensors in Cartesian components into this basis. We denote with brackets the component representation of vectors $\boldsymbol{v}$ and tensors $\boldsymbol{A}$ in this basis, namely:

$$\{\boldsymbol{v}\} = \{v_1, v_2\}, \tag{3.12}$$

$$\{\boldsymbol{A}\} = \begin{Bmatrix} A_{11} & A_{12} \\ A_{21} & A_{22} \end{Bmatrix}. \tag{3.13}$$

In the 3D space Sym of symmetric tensors of R^2, one can introduce an orthonormal dyadic basis $\{\boldsymbol{K}_1, \boldsymbol{K}_2, \boldsymbol{K}_3\}$ derived from $\{\boldsymbol{e}_1, \boldsymbol{e}_2\}$ as follows:

$$\begin{aligned} \boldsymbol{K}_1 &= \boldsymbol{e}_1 \otimes \boldsymbol{e}_1, \ \boldsymbol{K}_2 = \boldsymbol{e}_2 \otimes \boldsymbol{e}_2, \\ \boldsymbol{K}_3 &= \frac{\sqrt{2}}{2}(\boldsymbol{e}_1 \otimes \boldsymbol{e}_2 + \boldsymbol{e}_2 \otimes \boldsymbol{e}_1) \end{aligned} \tag{3.14}$$

Then, any symmetric tensor X can be decomposed into the two equivalent forms

$$\boldsymbol{X} = X_{11}\boldsymbol{e}_1 \otimes \boldsymbol{e}_1 + X_{22}\boldsymbol{e}_2 \otimes \boldsymbol{e}_2 + X_{12}\frac{\sqrt{2}}{2}(\boldsymbol{e}_1 \otimes \boldsymbol{e}_2 + \boldsymbol{e}_2 \otimes \boldsymbol{e}_1), \tag{3.15}$$

$$\boldsymbol{X} = X_1\boldsymbol{K}_1 + X_2\boldsymbol{K}_2 + X_3\boldsymbol{K}_3, \tag{3.16}$$

where $X_1 = X_{11}$, $X_2 = X_{22}$, $X_3 = \sqrt{2}X_{12}$. Then we have two alternative matrix representations of X:

$$\{\boldsymbol{X}\} = \begin{Bmatrix} X_{11} & X_{12} \\ X_{21} & X_{22} \end{Bmatrix}, \tag{3.17}$$

$$\left\{\hat{X}\right\} = \{X_1, X_2, X_3\} = \left\{X_{11}, X_{22}, \sqrt{2}X_{12}\right\} \tag{3.18}$$

Notice that in the 3D space spanned by $\{\boldsymbol{K}_1, \boldsymbol{K}_2, \boldsymbol{K}_3\}$, the only admissible rotations are those about the isotropic axis $\boldsymbol{I}$:

$$\boldsymbol{I} = \boldsymbol{K}_1 + \boldsymbol{K}_2 \tag{3.19}$$

In the space spanned by $\{\boldsymbol{K}_1, \boldsymbol{K}_2, \boldsymbol{K}_3\}$, the sets Sym^-, Sym^+ appear as conical surfaces as depicted in Fig. 3.1. Indeed, these two sets can be defined as those whose elements are symmetric tensors with non-positive and non-negative eigenvalues, respectively. Therefore, they are defined as:

$$Sym^- = \{\boldsymbol{X} \in Sym, s.t. tr\boldsymbol{X} \leq 0, det\boldsymbol{X} \geq 0\}, \tag{3.20}$$

$$Sym^+ = \{\boldsymbol{X} \in Sym, s.t. tr\boldsymbol{X} \geq 0, det\boldsymbol{X} \geq 0\} \tag{3.21}$$

Any member of Sym^- and Sym^+ is defined as a negative semidefinite or positive semidefinite symmetric tensor, respectively. The equation

$$det\boldsymbol{X} = 0, \tag{3.22}$$

defines the points of a conical surface of axis $\boldsymbol{I} = \boldsymbol{K}_1 + \boldsymbol{K}_2$, whose vertex is the origin, axially symmetric with respect to this axis, and whose span angle ψ with respect to the axis is 45°. The inequality $det\boldsymbol{X} > 0$ identifies points internal to the surface (that is, the internal part of the cone containing the axis), and the inequalities $\boldsymbol{X} \leq 0$, $tr\boldsymbol{X} \geq 0$ select the negative and positive parts of this internal volume (Fig. 3.1a).

Remark 2 The admissible domain for stress is the half cone Sym^-. This domain can be seen as a limit case of the Mohr-Coulomb admissible domain for cohesionless frictional materials. From the relation $\tan(\psi) = \sin(\varphi)$ relating the cone span angle

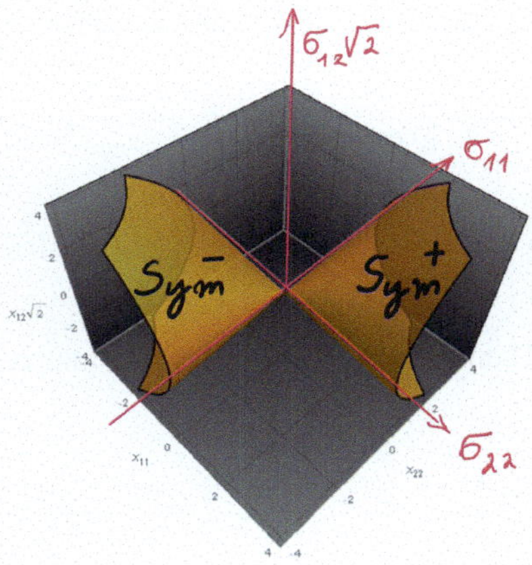

(a)

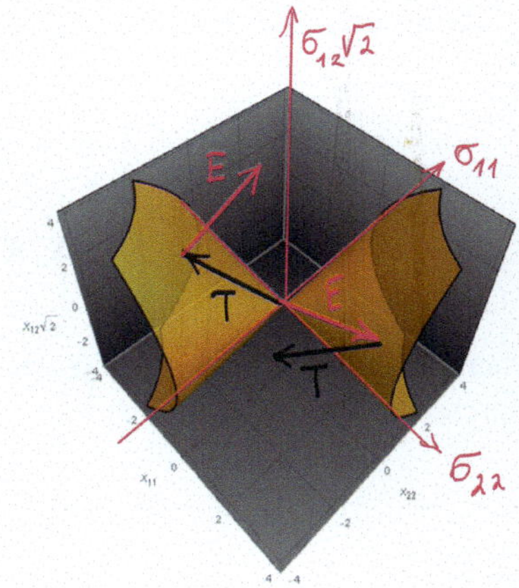

(b)

Fig. 3.1 **a** representation of the cones Sym^-, Sym^+ in an orthonormal basis of the 3D space Sym. **b** normality and dual normality conditions

ψ to the friction angle φ, we deduce that, in the case at hand, $\varphi = 90°$, that is the friction coefficient is $f = \tan(\psi) = \infty$.

Normality. Restrictions (3.9)–(3.11) are equivalent to the following condition, named "normality assumption" in Plasticity theory:

If $\boldsymbol{T} \in Sym^-$ is the stress at a point of a body composed of NRNT material, then the corresponding strain $\boldsymbol{E}$ at that point is restricted by the condition

$$\boldsymbol{E} \cdot \left(\boldsymbol{T} - \boldsymbol{T}^*\right) \geq 0, \forall \boldsymbol{T}^* \in Sym^-. \tag{3.23}$$

Dual normality. Additionally, restrictions (3.9)–(3.11) are also equivalent to the following condition, called "dual normality assumption":

If $\boldsymbol{E} \in Sym^+$ is the strain at a point of a body composed of NRNT material, then the corresponding stress $\boldsymbol{T}$ at that point is restricted by the condition

$$\boldsymbol{T} \cdot \left(\boldsymbol{E} - \boldsymbol{E}^*\right) \geq 0, \forall \boldsymbol{E}^* \in Sym^+. \tag{3.24}$$

Remark 3 Notice that, as a difference from classical Perfect Plasticity assumptions, the normality and dual normality conditions are expressed in terms of total strains rather than in rates of strain.

3.1.2 The Equilibrium BVP for NRNT Materials

The equilibrium BVP for a body Ω composed of NRNT material can be formulated as follows.

Let $\overline{\boldsymbol{p}}, \boldsymbol{b}$ be the given load, that is, the surface and body loads given on $\partial\Omega^N$ and on Ω, respectively, and let $\overline{\boldsymbol{u}}, \overline{\boldsymbol{E}}$ be the prescribed displacements (that is, the settlements) and eigenstrains given on $\partial\Omega^D$ and on Ω, respectively.

In what follows we will refer to $\overline{\boldsymbol{p}}, \boldsymbol{b}$ and $\overline{\boldsymbol{u}}, \overline{\boldsymbol{E}}$ as the load data and the distortion data.

"Look for a state $\boldsymbol{s} = \{\boldsymbol{u}, \boldsymbol{E}, \boldsymbol{T}\}$, that is a triplet of displacement, strain and stress fields defined over Ω, such that

$$\boldsymbol{E} = \frac{1}{2}(\nabla\boldsymbol{u} + \nabla\boldsymbol{u}^T) - \overline{\mathbf{E}}, \boldsymbol{E} \in Sym^+, \boldsymbol{u} = \overline{\mathbf{u}} \text{ on } \partial\Omega^D \tag{3.25}$$

$$div\boldsymbol{T} + \boldsymbol{b} = 0, \in Sym^-, \boldsymbol{Tn} = \overline{\mathbf{p}} \text{ on } \partial\Omega^N \tag{3.26}$$

$$\boldsymbol{T} \cdot \boldsymbol{E} = 0 \tag{3.27}$$

A state satisfying (3.25)–(3.27) is called a *solution* of the BVP and is denoted $\boldsymbol{s}^\circ = \{\boldsymbol{u}^\circ, \boldsymbol{E}^\circ, \boldsymbol{T}^\circ\}$.

Admissible displacement and stress fields. It is convenient to introduce the following function sets: the set of kinematically admissible displacement fields

$$K = \left\{\boldsymbol{u} \in S(\Omega), s.t.\boldsymbol{E}(\boldsymbol{u}) \in Sym^+, \boldsymbol{u} = \overline{\mathbf{u}}\, on\ \partial\Omega^D\right\} \tag{3.28}$$

where $(\boldsymbol{u}) = \frac{1}{2}(\nabla\boldsymbol{u} + \nabla\boldsymbol{u}^T) - \overline{\mathbf{E}}$, and the set of statically admissible stress fields

$$H = \left\{\boldsymbol{T} \in S'(\Omega), s.t.div\boldsymbol{T} + \boldsymbol{b} = 0, \boldsymbol{T} \in Sym^-, \boldsymbol{Tn} = \overline{\mathbf{p}}\, on\ \partial\Omega^N\right\} \tag{3.29}$$

in which $S(\Omega)$ and $S'(\Omega)$ are two function spaces of convenient regularity. The set of self-consistent admissible mechanisms

$$K^\circ = \left\{\boldsymbol{v} \in S(\Omega), s.t.\boldsymbol{E}(\boldsymbol{v}) \in Sym^+, \boldsymbol{v} = 0\, on\ \partial\Omega^D\right\} \tag{3.30}$$

where $\boldsymbol{E}(\boldsymbol{v}) = \frac{1}{2}(\nabla\boldsymbol{v} + \nabla\boldsymbol{v}^T)$, that is $\overline{\boldsymbol{E}} = 0$ is considered, and the set of admissible self-balanced stresses

$$H^\circ = \left\{\boldsymbol{\Sigma} \in S'(\Omega), s.t.div\boldsymbol{\Sigma} = 0, \boldsymbol{\Sigma} \in Sym^-, \boldsymbol{\Sigma n} = 0\, on\ \partial\Omega^N\right\} \tag{3.31}$$

The definition of the spaces $S(\Omega)$ and $S'(\Omega)$ will emerge naturally from the variational formulation of the BVP, as detailed in Sect. 3.2.3. A solution of the equilibrium BVP can be then rephrased as follows:

"Look for a state $\boldsymbol{s}^\circ = \{\boldsymbol{u}^\circ, \boldsymbol{E}^\circ, \boldsymbol{T}^\circ\}$, called a *solution* of the BVP, such that $\boldsymbol{u}^\circ \in K, \boldsymbol{T}^\circ \in H$ and $\boldsymbol{T}^\circ \cdot \boldsymbol{E}^\circ = 0$."

Remark 4 *Uniqueness*. The solution of the BVP formulated in the present Section is by no means unique. Trivial examples of non-uniqueness are the cases in which either the load data or the distortion data are homogeneous. Indeed, if the load data are zero at the interior of Ω and on the loaded boundary $\partial\Omega^N$, the vanishing stress field $\boldsymbol{T}_{null} = 0$ is a member of H. In this case, any compatible displacement $\boldsymbol{u} \in K$ is a solution of the BVP, since $\boldsymbol{T}_{null} \in H$ and $\boldsymbol{T}_{null} \cdot \boldsymbol{E}(\boldsymbol{u}) = 0$ for any $\boldsymbol{E}$.

Besides, if the distortion data are zero at the interior of Ω and on the fixed boundary $\partial\Omega^D$, the vanishing displacement field $\boldsymbol{u}_{null} = 0$ is a member of K. In this case, any statically admissible stress field $\boldsymbol{T} \in H$ is a solution of the BVP, since $\boldsymbol{u}_{null} \in K$ and $\boldsymbol{T} \cdot \boldsymbol{E}(\boldsymbol{u}_{null}) = 0$ for any $\boldsymbol{T}$.

The data somehow restrict the repertoire of admissible displacement and stress fields. In a sense, both load and displacement data contribute to restricting the set of possible solutions in the dual admissible set. If loads are given, settlements can be used to obtain a restricted class of equilibrium scenarios and get a picture of the structure's equilibrium state. These states often have no resemblance with the elastic solutions obtained by considering a bilateral behaviour of the material.

3.1.3 Variational Formulation of the BVP

The BVP problem formulated in Sect. 3.1.2, that is, the search for a state

$$\boldsymbol{s}^0 = \{\boldsymbol{u}^0, \boldsymbol{E}^0, \boldsymbol{T}^0\} \tag{3.32}$$

such that

$$\boldsymbol{u}^0 \in K, \boldsymbol{T}^0 \in H, \boldsymbol{T}^0 \cdot \boldsymbol{E}^0 = 0, \tag{3.33}$$

can be variationally formulated as the minimization of two dual functionals, the potential energy

$$E(\boldsymbol{u}) = -\int_{\partial\Omega^N} \overline{\mathbf{p}} \cdot \boldsymbol{u} ds - \int_{\Omega} \boldsymbol{b} \cdot \boldsymbol{u} da \tag{3.34}$$

defined for $\boldsymbol{u} \in K$, and the complementary energy

$$E_c(\boldsymbol{T}) = -\int_{\partial\Omega^D} \mathbf{Tn} \cdot \overline{\boldsymbol{u}} ds + \int_{\Omega} \boldsymbol{T} \cdot \overline{\boldsymbol{E}} da \tag{3.35}$$

defined for $T \in H$, in the forms

$$E(\boldsymbol{u}^0) = \min_{\boldsymbol{u}\in K} E(\boldsymbol{u}),\ E_c(\boldsymbol{T}^0) = \min_{\boldsymbol{T}\in H} E_c(\boldsymbol{T}) \tag{3.36}$$

These variational formulations of the equilibrium problem for NRNT materials are described in more detail in Sect. 3.3.

3.1.4 Basic Partitions of Ω

The content of the present Subsection follows some ideas first put forward by Salvatore Di Pasquale in (1984) and then developed in more mathematical terms by Giaquinta and Giusti in (1985).

To any given (continuous) displacement field $\boldsymbol{u}^* \in K$ and to any (continuous) stress vector field $\boldsymbol{t}_n^{**}$ corresponding to a stress field $\boldsymbol{T}^{**} \in H$, one can associate two partitions of the domain Ω. These partitions can be used to understand which stress field $\boldsymbol{T}^*$ can correspond to $\boldsymbol{E}(\boldsymbol{u}^*)$, and what strain field $\boldsymbol{E}^{**}$ can correspond to $\boldsymbol{T}^{**}$, in order to obtain a solution of the BVP for NRNT materials.

If $\boldsymbol{u}^* \in K$ then $\boldsymbol{E}(\boldsymbol{u}^*) \in Sym^+$, and, therefore, there exists an orthonormal basis

$$\{\boldsymbol{k}_1, \boldsymbol{k}_2\} \tag{3.37}$$

formed by the eigenvectors of $\boldsymbol{E}(\boldsymbol{u}^*)$, such that

$$\boldsymbol{E}^* = \lambda_1 \boldsymbol{k}_1 \otimes \boldsymbol{k}_1 + \lambda_2 \boldsymbol{k}_2 \otimes \boldsymbol{k}_2 \tag{3.38}$$

with the λ_1, λ_2 eigenvalues satisfying the restrictions

$$\lambda_1 \geq 0,\ \lambda_2 \geq 0 \tag{3.39}$$

On ordering the eigenvalues in such a way that

$$\lambda_2 \geq \lambda_1 \tag{3.40}$$

we can consider three possibilities:

(i) $\lambda_2 > 0, \lambda_1 > 0$
(ii) $\lambda_2 > 0, \lambda_1 = 0$
(iii) $\lambda_2 = 0, \lambda_1 = 0$.

Based on these three possible cases, the domain Ω can be partitioned into three disjoint regions, depending on the values taken by the eigenvalues of $\boldsymbol{E}(\boldsymbol{u}^*)$ at any point $\boldsymbol{x}$ of Ω:

$$\begin{aligned} \Sigma_1 &= \{\boldsymbol{x} \in \Omega, s.t.\lambda_2 > 0, \lambda_1 > 0\} \\ \Sigma_2 &= \{\boldsymbol{x} \in \Omega, s.t.\lambda_2 > 0, \lambda_1 = 0\} \\ \Sigma_3 &= \{\boldsymbol{x} \in \Omega, s.t.\lambda_2 = 0, \lambda_1 = 0\} \end{aligned} \tag{3.41}$$

If $\boldsymbol{T}^{**} \in H$ then $\boldsymbol{T}^{**} \in Sym^-$, and, therefore, there exists an orthonormal basis

$$\{\boldsymbol{h}_1, \boldsymbol{h}_2\} \tag{3.42}$$

formed by the eigenvectors of $\boldsymbol{T}^{**}$, such that:

$$\boldsymbol{T}^{**} = \sigma_1 \boldsymbol{h}_1 \otimes \boldsymbol{h}_1 + \sigma_2 \boldsymbol{h}_2 \otimes \boldsymbol{h}_2 \tag{3.43}$$

being σ_1, σ_2 the eigenvalues of $\boldsymbol{T}^{**}$ satisfying the restrictions

$$\sigma_1 \leq 0, \sigma_2 \leq 0 \tag{3.44}$$

On ordering the eigenvalues in such a way that

$$\sigma_2 \geq \sigma_1 \tag{3.45}$$

there are three possibilities:

(i) $\sigma_2 < 0, \sigma_1 < 0$,
(ii) $\sigma_2 = 0, \sigma_1 < 0$,
(iii) $\sigma_2 = 0, \sigma_1 = 0$.

Based on these three possibilities, the domain Ω can be partitioned into three disjoint regions, depending on the eigenvalues that the stress field $\boldsymbol{T}^{**}$ takes at any point $\boldsymbol{x}$ of Ω:

$$\begin{aligned} \Omega_1 &= \{\boldsymbol{x} \in \Omega, s.t.\sigma_2 < 0, \sigma_1 < 0\} \\ \Omega_2 &= \{\boldsymbol{x} \in \Omega, s.t.\sigma_2 < 0, \sigma_1 = 0\} \\ \Omega_3 &= \{\boldsymbol{x} \in \Omega, s.t.\sigma_2 = 0, \sigma_1 = 0\} \end{aligned} \tag{3.46}$$

Remark 5 The region Ω_3 is inert, and in there, the equilibrium problem is under-determined, meaning that body load and tractions cannot be arbitrary in order to maintain equilibrium. In particular $\boldsymbol{b} = 0$ at each internal point of Ω_3 and $\boldsymbol{Tn} = 0$ all over the region boundary $\partial\Omega_3$. Inside Ω_2, where the stress is uniaxial, the equilibrium problem is statically determined as the equilibrium problem $div\boldsymbol{T} + \boldsymbol{b} = 0$ is supplemented by the condition $det\boldsymbol{T} = 0$. Besides, the problem is parabolic, meaning that there is a family of characteristic lines along which the non-zero value of stress σ_1 propagates. In particular, if $\boldsymbol{b} = 0$, these characteristic lines are straight (Fig. 3.2) and are called compression rays (Mansfield, 1969), Fortunato (2010). Referring to Fig. 3.2 for notation, the form taken by stress along a compression ray can be expressed as a function of the length s along the compression ray:

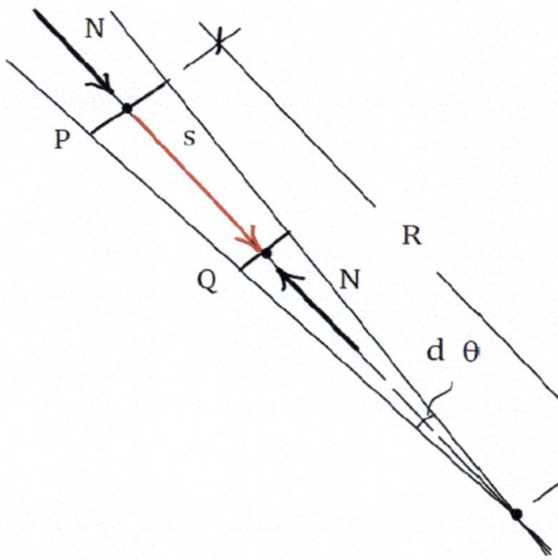

Fig. 3.2 Uniaxial stress resultants N along an infinitesimal compression wedge

$$\sigma_1 = \sigma_1^o \frac{R}{R - s} \tag{3.47}$$

in which σ_1^o is the value of σ_1 at $s = 0$, and R is the radius of curvature of the ray at $s = 0$.

3.1.5 Stress–Strain Relation

By denoting with $\{.\}'$ the Cartesian matrix representation of a tensor in the orthonormal basis $\{\boldsymbol{k}_1, \boldsymbol{k}_2\}$ corresponding to its principal reference, we can write:

$$\{\boldsymbol{E}^*\}' = \begin{Bmatrix} \lambda_1 & 0 \\ 0 & \lambda_2 \end{Bmatrix} \tag{3.48}$$

To characterize the stress $\boldsymbol{T}^*$ corresponding to $\boldsymbol{E}^*$, one must consider the dual normality condition (Fig. 3.1b), from which one deduces:

- In Σ_1

$$\boldsymbol{E}^* = \lambda_1 \boldsymbol{k}_1 \otimes \boldsymbol{k}_1 + \lambda_2 \boldsymbol{k}_2 \otimes \boldsymbol{k}_2 \text{ with } \lambda_2 > 0, \lambda_1 > 0 \tag{3.49}$$

thus

$$\boldsymbol{T}^* = 0 \tag{3.50}$$

- In Σ_2, where

$$\boldsymbol{E}^* = \lambda_2 \boldsymbol{k}_2 \otimes \boldsymbol{k}_2 \text{ with } \lambda_2 > 0 \tag{3.51}$$

thus, it results

$$\boldsymbol{T}^* = \sigma_1 \boldsymbol{k}_1 \otimes \boldsymbol{k}_1 \text{ with } \sigma_1 \leq 0 \tag{3.52}$$

And, in components:

$$\{\boldsymbol{E}^*\}' = \begin{Bmatrix} 0 & 0 \\ 0 & \lambda_2 \end{Bmatrix} \tag{3.53}$$

$$\{\boldsymbol{T}^*\}' = \begin{Bmatrix} \sigma_1 & 0 \\ 0 & 0 \end{Bmatrix} \tag{3.54}$$

- In Σ_3, where

$$\boldsymbol{E}^* = 0 \tag{3.55}$$

it follows

$$T^* = \sigma_1 \boldsymbol{k}_1 \otimes \boldsymbol{k}_1 + \sigma_2 \boldsymbol{k}_2 \otimes \boldsymbol{k}_2 \text{ with } \sigma_2 \leq 0, \sigma_1 \leq 0 \tag{3.56}$$

By denoting with $\{.\}''$ the Cartesian matrix representation of a tensor in the principal reference of $\boldsymbol{T}^{**}$ associated to the orthonormal basis $\{\boldsymbol{h}_1, \boldsymbol{h}_2\}$ formed by its eigenvectors, we can write:

$$\{\boldsymbol{T}^{**}\}'' = \begin{Bmatrix} \sigma_1 & 0 \\ 0 & \sigma_2 \end{Bmatrix} \tag{3.57}$$

To characterize the strain $\boldsymbol{E}^{**}$ corresponding to $\boldsymbol{T}^{**}$, one has to take the normality condition (3.11) into account, from which one deduces:

- In Ω_1, where

$$\boldsymbol{T}^{**} = \sigma_1 \boldsymbol{h}_1 \otimes \boldsymbol{h}_1 + \sigma_2 \boldsymbol{h}_2 \otimes \boldsymbol{h}_2 \text{ with } \sigma_2 < 0, \sigma_1 < 0 \tag{3.58}$$

it results

$$\boldsymbol{E}^{**} = 0; \tag{3.59}$$

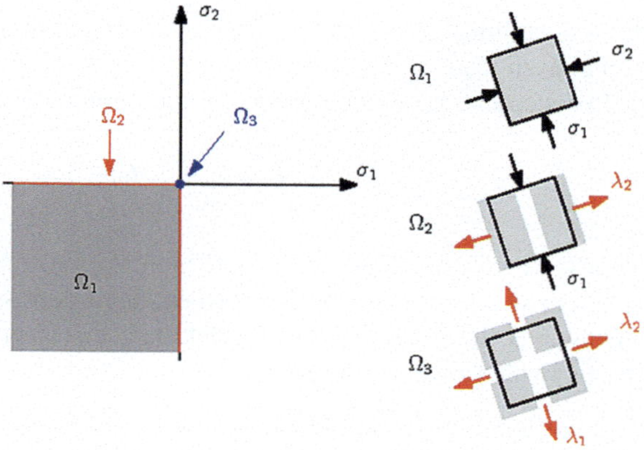

Fig. 3.3 Basic partition associated to an admissible stress and possible corresponding strain

- In Ω_2, where

$$\boldsymbol{T}^{**} = \sigma_1 \boldsymbol{h}_1 \otimes \boldsymbol{h}_1 \text{ with } \sigma_1 < 0 \tag{3.60}$$

one has

$$\boldsymbol{E}^{**} = \lambda_2 \boldsymbol{h}_2 \otimes \boldsymbol{h}_2 \text{ with } \lambda_2 \geq 0 \tag{3.61}$$

- In Ω_3, where

$$\boldsymbol{T}^{**} = 0 \tag{3.62}$$

it follows

$$\boldsymbol{E}^{**} = \lambda_1 \boldsymbol{h}_1 \otimes \boldsymbol{h}_1 + \lambda_2 \boldsymbol{h}_2 \otimes \boldsymbol{h}_2 \text{ with } \lambda_2 \geq 0,\, \lambda_1 \geq 0 \tag{3.63}$$

Remark 6 From the above derivations, we deduce that the principal directions of stress and strain corresponding to a possible BVP solution for NRNT materials do always coincide. Therefore, the NRNT material is isotropic (Fig. 3.3).

3.1.6 Compatibility of the Data

For the existence of the solution of the BVP for a body Ω composed of NRNT material, there exist non-trivial compatibility conditions on the data of both kinematical-distortion-and statical-load-types.

An example of a trivial compatibility condition on the load is the case of a traction problem (that is, a case in which the whole boundary is loaded: $\partial\Omega^D = \emptyset$) for which the compatibility condition for the existence of an equilibrium solution is the balance of the given load $\{\overline{\boldsymbol{p}}, \boldsymbol{b}\}$.

In the case at hand, the restrictions on stress and strain:

$$\boldsymbol{T} \in Sym^-, \boldsymbol{E} \in Sym^+ \tag{3.64}$$

put less trivial constraints on the data in order that statically admissible stress fields or kinematically admissible displacement fields corresponding to these data exist. In other words, the repertoire of admissible data, that is, given loads $\{\overline{\boldsymbol{p}}, \boldsymbol{b}\}$ or distorsions $\{\overline{\boldsymbol{u}}, \overline{\boldsymbol{E}}\}$, which allow for the existence of non-void sets H and K, is severely restricted.

Compatibility of loads (Theorems of Limit Analysis). The compatibility of the load data $\{\overline{\boldsymbol{p}}, \boldsymbol{b}\}$ is related to the possibility that the structure may or may not develop accelerations from the original geometry where the equilibrium is sought. Indeed, if the loads are such that the equilibrium is impossible (i.e., $H = \emptyset$), then, the balance laws can be verified only by admitting that inertia terms come into play (see Eq. (3.4)). In plasticity, this state of incipient movement is called collapse, and the question of admissibility of the loads becomes the content of the so-called Collapse Theorems, also known as Theorems of Limit Analysis.

Theorem 1 Safe Theorem. *The structure Ω cannot collapse if $H \neq \emptyset$.*

Proof If the load $\{\overline{\boldsymbol{p}}, \boldsymbol{b}\}$ is such that $H \neq \emptyset$, then, there exists at least a stress field $\tilde{\boldsymbol{T}} \in H$, and, for any virtual act of motion $\boldsymbol{v}$, that is for any displacement belonging to K°, we can express the increment of the power of the applied loads in terms of $\tilde{\boldsymbol{T}}$ as follows:

$$\int_{\partial\Omega^N} \overline{\mathbf{p}} \cdot \boldsymbol{v} ds + \int_{\Omega} \boldsymbol{b} \cdot \boldsymbol{v} da = \int_{\Omega} \tilde{\boldsymbol{T}} \cdot \boldsymbol{E}(\boldsymbol{v}) da \tag{3.65}$$

Since

$$\tilde{\boldsymbol{T}} \in Sym^-, \boldsymbol{E}(\boldsymbol{v}) \in Sym^+ \tag{3.66}$$

then $\int_{\Omega} \tilde{\boldsymbol{T}} \cdot \boldsymbol{E}(\boldsymbol{v}) da \leq 0$. If, despite the existence of $\tilde{\boldsymbol{T}}$, we admit by absurd that an actual act of motion exists and call it $\boldsymbol{v}$, on denoting $\mathbb{K}(\boldsymbol{v})$ the Kinetic Energy corresponding to $\boldsymbol{v}$, the power theorem provides

$$\int_{\partial\Omega^N} \overline{\mathbf{p}} \cdot \boldsymbol{v} ds + \int_{\Omega} \boldsymbol{b} \cdot \boldsymbol{v} da = \Delta\mathbb{K}(\boldsymbol{v}) \tag{3.67}$$

where, again, the assumed vanishing of b at the constrained part of the boundary has been exploited to reduce the surface term at the boundary.

Therefore, from (3.67) we deduce $\Delta\mathbb{K}(\boldsymbol{v}) \leq 0$, which is incompatible with any real motion starting from rest, obtaining a contradiction. ■

Remark 8 The content of the Safe Theorem can be reworded as follows: "if any equilibrated stress field can be found for which the stress is inside the limits for the material, then collapse of the structure is impossible." This way of looking at the structural stability with the equilibrium approach frees us from the need to look for the real state of stress of the structure subjected to given displacement and load data. The possibility of collapse is only dependent on the load and on the geometry, while is independent of the displacement data. This somehow counter-intuitive result depends on the assumption of small displacements, which allows linearising the kinematics and verifying approximately the equilibrium in the original configuration. The coupling of the displacement data with the equilibrium requires a nonlinear analysis. This becomes necessary when large displacements are analyzed or pathological loading cases very close to collapse are considered.

Theorem 2 Kinematic Theorem. *If there exists a self-consistent mechanism* $\boldsymbol{v} \in K^{\circ}$ *for which the power increment of the applied load is positive, then* $H = \emptyset$*. As a consequence, the structure will collapse in the sense that an act of motion can arise from the original configuration.*

Proof If, despite the existence of a self-consistent mechanism $\boldsymbol{v} \in K^{\circ}$ such that

$$\int_{\partial\Omega^N} \overline{\mathbf{p}} \cdot \boldsymbol{v} ds + \int_{\Omega} \boldsymbol{b} \cdot \boldsymbol{v} da > 0 \tag{3.68}$$

we assume by absurd that $H = \emptyset$, that is, there exists a non-zero stress field $\tilde{\boldsymbol{T}} \in H$, then, by transforming the external work into internal work, we obtain

$$\int_{\partial\Omega^N} \overline{\mathbf{p}} \cdot \boldsymbol{v} ds + \int_{\Omega} \boldsymbol{b} \cdot \boldsymbol{v} da = \int_{\Omega} \tilde{\boldsymbol{T}} \cdot \boldsymbol{E}(\boldsymbol{v}) da \tag{3.69}$$

Therefore, since $\tilde{\boldsymbol{T}} \in Sym^{-}$ and $\boldsymbol{E}(\boldsymbol{v}) \in Sym^{+}$, $\int_{\Omega} \tilde{\boldsymbol{T}} \cdot \boldsymbol{E}(\boldsymbol{v}) da \leq 0$, and we get a contradiction. ■

Compatibility of distortions. The previous paragraph was concerned with the application of the Theorems of Limit Analysis to continuous bodies composed of NRNT material. Besides some peculiarities specific to unilateral material, this treatment is classical and well-developed. What we see here is somehow new and, due to the perfect symmetry of the constitutive model defined by the conditions (3.9)–(3.11), is a sort of geometrical translation to displacement data and strain of what Limit Analysis is for load data and stress.

The incompatibility of the distortion data $\{\overline{\boldsymbol{u}}, \overline{\boldsymbol{E}}\}$, also called distortions, is related to the possibility that the structure may not be able to accommodate these data with pure elongations. Indeed, suppose the distortions are given in such a way that the compatibility is impossible. In that case, $K = \emptyset$, thus, the compatibility conditions can be verified if and only if the material either deforms by shortening or breaks by compenetration. The compatibility of the displacement data is then defined and characterized as follows.

Definition 1 The distortions $\{\overline{\boldsymbol{u}}, \overline{\boldsymbol{E}}\}$ are said to be compatible with the body Ω, if there exists at least one field of admissible displacements $\boldsymbol{u} \in K$, that is $K \neq \emptyset$.

Theorem 3 If there exists a self-balanced stress field $\boldsymbol{\Sigma} \in H^\circ$ for which the given distortions $\{\overline{\boldsymbol{u}}, \overline{\boldsymbol{E}}\}$ perform positive work, then $K = \emptyset$. Consequently, the structure will not be able to accommodate the given displacements with a pure nonnegative elongation strain (Figs. 3.4 and 3.5).

Proof If despite the existence of a self-equilibrated stress field $\boldsymbol{\Sigma} \in H^\circ$ for which

$$-\int_{\Omega} \boldsymbol{\Sigma} \cdot \overline{\boldsymbol{E}} da + \int_{\partial\Omega^D} \boldsymbol{\Sigma n} \cdot \overline{\mathbf{u}} ds > 0 \tag{3.70}$$

we assume by absurd that $K \neq \emptyset$, that is, there exists a displacement field $\tilde{u} \in K$, for which by transforming the external work into internal work and applying the assumption $\boldsymbol{\Sigma n} = 0$ on $\partial\Omega^N$, we obtain

$$-\int_{\Omega} \boldsymbol{\Sigma} \cdot \overline{\boldsymbol{E}} da + \int_{\partial\Omega} \boldsymbol{\Sigma n} \cdot \tilde{\mathbf{u}} ds = \int_{\Omega} \boldsymbol{\Sigma} \cdot \boldsymbol{E}(\tilde{\boldsymbol{u}}) da \tag{3.71}$$

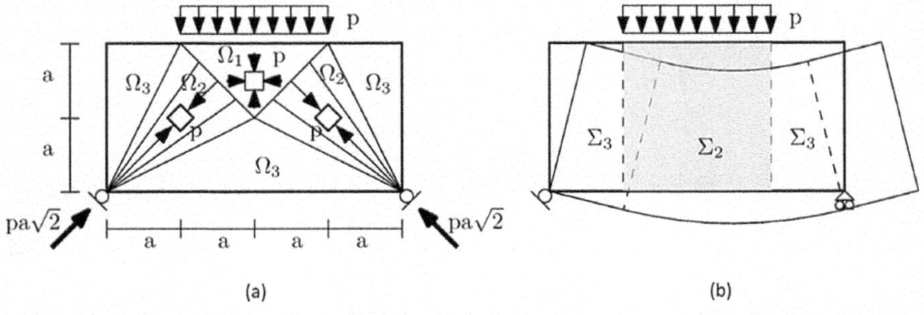

Fig. 3.4 A rectangular panel constrained at the corners and uniformly loaded on a portion of the top edge, with two different constraint settings. **a** Compatible loads. The body is divided into regions, as shown in (**a**). In Ω_1 the stress is hydrostatic of value p. In Ω_2 the stress is uniaxial along equally spaced rays passing through the bottom corners. The assumed stress field is balanced with the load and purely compressive. **b** Incompatible loads. A simple self-consistent mechanism can be constructed. The body is divided into three rectangles. The lateral rectangles Σ_3 displace rigidly and the central rectangle Σ_2 is subject to pure uniaxial bending with neutral axis the top edge of the rectangle. The load performs positive work for this displacement

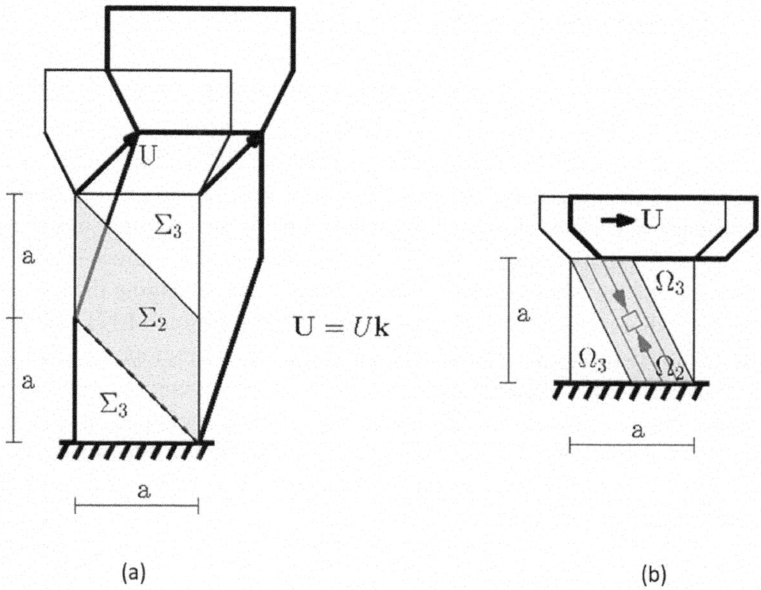

Fig. 3.5 A rectangular panel fixed at the bottom base is distorted by displacing the top base. **a** Compatible distortions. A simple strain field which is compatible with the given settlement is constructed: the body is divided into three strips. The two external strips Σ_3 displace rigidly and the central strip Σ_2 deforms uniaxially in the direction $\boldsymbol{k}$: the strain is positive semidefinite. **b** incompatible distortions. A self-equilibrated compressive uniaxial stress field is constructed. The body is divided into three strips. The lateral strips Ω_3 are inert and the central strip Ω_2 is subject to uniaxial compression. The given displacements perform positive work for this stress

Therefore, since $\boldsymbol{\Sigma} \in Sym^{-}$ and $\boldsymbol{E}(\tilde{\boldsymbol{u}}) \in Sym^{+}$, then $\int_{\Omega} \boldsymbol{\Sigma} \cdot \boldsymbol{E}(\tilde{\boldsymbol{u}}) da \leq 0$, which leads to a contradiction. ■

3.2 Admissible Strain and Stress Fields

This Section provides the reader with some geometrical insight on possible admissible strains and stresses, by using the analogy with the Tension Field Theory of Mansfield (1969) and concentrated strains and stresses.

3.2.1 Compatibility of the Data, Again

The NRNT material is unilateral in compression, meaning that normal stresses cannot be positive, and the latent elongations cannot be negative. Such assumptions are analogous to those adopted in Tension Field Theory (TFT) to model thin flat membranes, once the sign of the constraint on stress has been reverted. As a difference with NRNT masonry-like materials, elastic strains in tension are admitted, meaning that there is a constitutive part of stress. Thus, the tensile part of the stress is active, and elastic elongations are admitted. This is analogous, by just changing the signs of the constraints, to what is done for the Normal Elastic No-Tension (NENT) materials, a model first put forward by Di Pasquale in (1984) and then developed in more mathematical terms by Giaquinta and Giusti in (1985). For a thorough discussion of this material and its relations with TFT the reader can consult Fortunato (2010).

Indeed, thin membranes buckle under feeble compressive stresses by forming wrinkles, and a reasonable assumption is to consider that the material is unilateral in compression. Mutatis mutandis, regions of the type Σ_1, Σ_2, Σ_3, Ω_1, Ω_2, Ω_3, as defined in Sect. 3.1.4, can be analogously defined, and in particular, the regions Ω_1, Ω_2, Ω_3 visualized by making simple tests on thin sheets of deformable material. In TFT, the regions Ω_1, Ω_2, Ω_3 are given names: Ω_1 is the Taut region, Ω_2 is the Wrinkled region and Ω_3 is the Slack region. The result of a simple test, as shown in Fig. 3.6, can help us to visualize this partition.

Remark 6 In the test of Fig. 3.3, we may notice that the original rectangular sheet appears sensibly distorted. This is due neither to the low stiffness of the membrane (which is rather stiff) nor to the occurrence of plastic strains (the sheet actually recovers the original geometry when the loads are removed). Still, it is an interesting side effect of wrinkling, that is, of the anelastic strains predicted by TFT. Similar behaviour can be expected in the parent theory concerning NT materials. It may

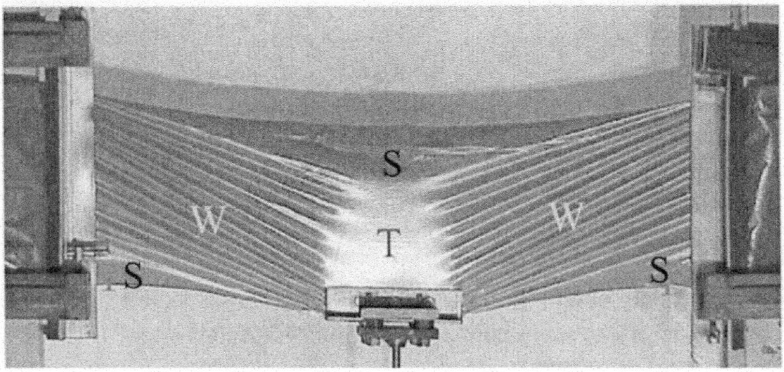

Fig. 3.6 Rectangular sheet made of a thin aluminium foil, fixed along the two lateral sides, and pulled by a force applied along a central part of the lower edge. The rest of the boundary is free (Jenkins et al., 1998)

be seen on some situations relating to walls, in which fractures appear, and some geometrical nonlinearity should be considered (Gesualdo et al., 2018).

The NT-TFT analogy gives us a way to visualize the basic features of statically admissible stress fields for NRNT materials in much the same way Hooke discovered how to visualize a possible equilibrium state of an arch made of voussoirs by using a hanging chain (Batten, 1936): take the chain in equilibrium under the given vertical load, freeze it and flip it, both the sample and the load, about the horizontal axis.

Another example is shown in Fig. 3.7, which is a simple schematization of the four-point bending for a plane beam composed of NRNT material and reinforced with a horizontal tie at the bottom.

By adopting the NENT model the material is unilateral in tension and elastic in compression. The mimisation of the potential energy, sum of the potential energy of the load and the elastic energy, is a possible way of selecting admissible stress states for NENT materials (Angelillo et al., 2010). Some examples taken from (Angelillo et al., 2010) are shown in Figs. 3.8 and 3.9.

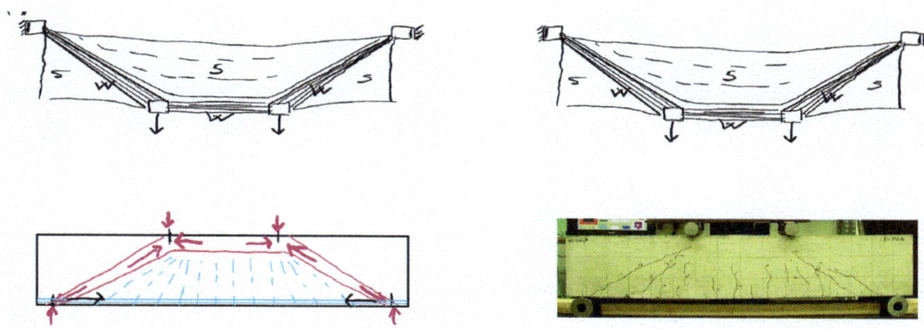

Fig. 3.7 Four-point bending test of a concrete beam reinforced with inferior longitudinal rebars only (bottom right). Membrane analogy (top). Truss analogy (bottom left)

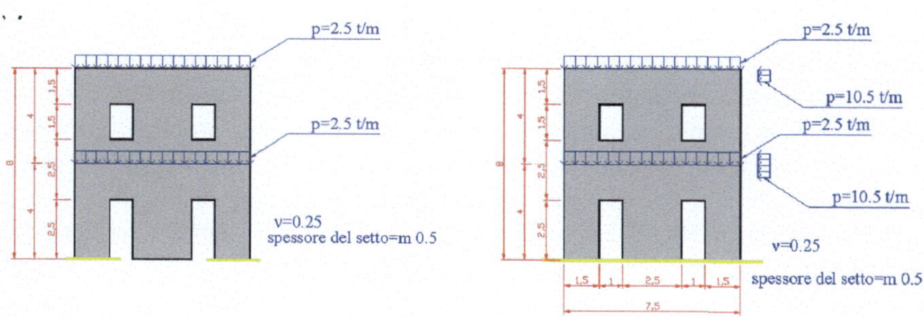

Fig. 3.8 A simple wall with openings loaded (**a**) by vertical uniform loads applied at the floor levels and subject to a uniform settlement of the central pier, and (**b**) by vertical and horizontal loads at the floor level and zero settlements

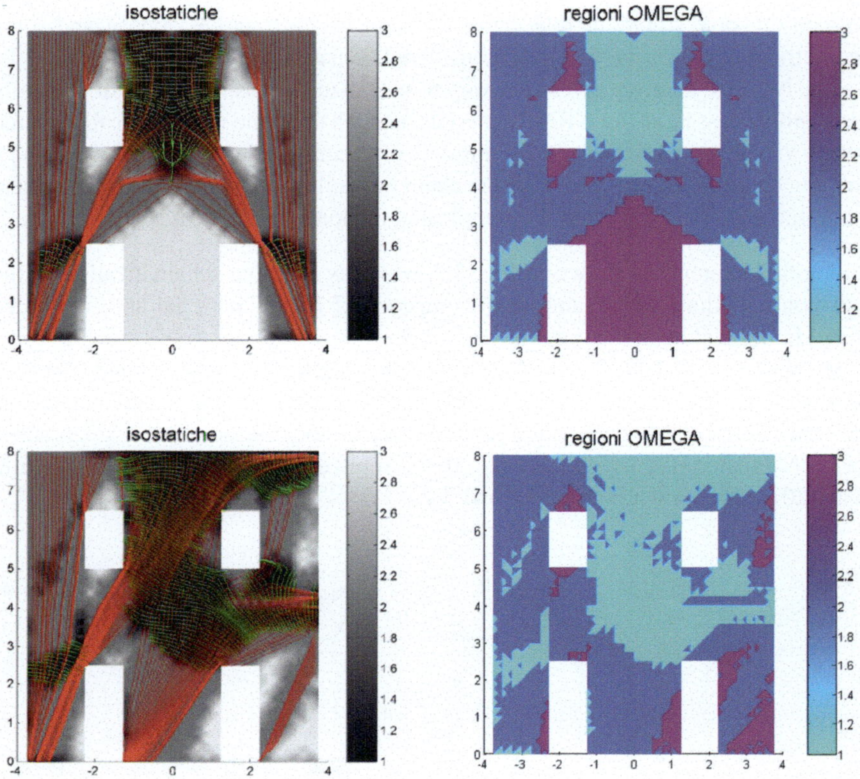

Fig. 3.9 Approximate elastic-unilateral (NENT) solutions corresponding to the schemes of Fig. 3.8. First-row shows isostatic lines and basic partition corresponding to the case of Fig.3.8a; while the second row illustrates the isostatic lines and basic partition related to Fig.3.8b

3.2.2 Strain as a Measure for Fracture

Figure 3.10 reports a a picture from "The Stone Skeleton" by Heyman (1997).

It is a circular arch subject to its self-weight and whose supports have settled laterally. This is a typical example of a mixed boundary value problem, for which the traction data at the boundary are homogeneous while the kinematical data along the boundary are not. Let us focus on the original intact arch of Fig. 3.10 a as a continuum made of NRNT material. The fractures we see can be interpreted as singular deformations in the form of line Dirac deltas applied along the fracture interfaces. The displacement and the allied deformation depicted in Fig. 3.10b are a striking example of an admissible mechanism that is of a displacement $\boldsymbol{u} \in K$. As we can see, the displacement is not continuous and this is a typical feature of perfectly plastic materials that we also accept for NRNT materials: the displacement can have finite jumps, that is, the corresponding local deformation can be either regular or singular. This possibility can be exploited in order both to establish the compatibility

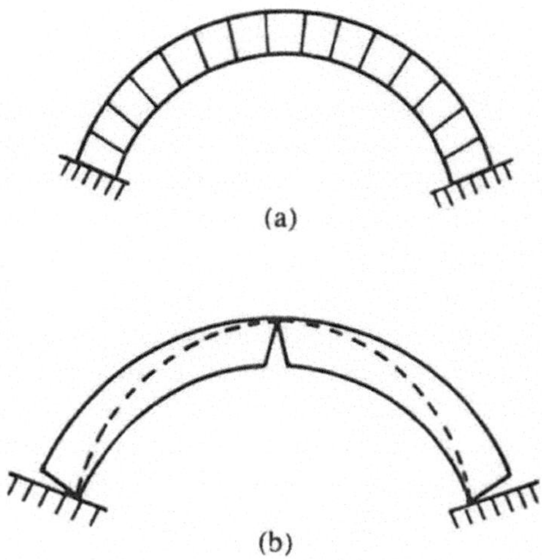

Fig. 3.10 A cracked circular arch whose supports have given way slightly: deformed configuration and corresponding thrust line (images from (Heyman, 1997))

of the data and to search for possible BVP solutions, as shown below. To establish that this is a viable BVP solution, one should find a stress field $\boldsymbol{T} \in H$ such that $\boldsymbol{T} \cdot \boldsymbol{E}(\boldsymbol{u}) = 0$

Figure 3.10b provides us with an hint of what this stress field could be: a 1d arc representing the line of thrust of the contact forces transmitted by the voussoirs is depicted with a dashed line. This can be interpreted as a 1d linear arch inside the 2d continuum and carrying a concentrated axial force that balances the applied loads. With this interpretation, we then admit that, besides the strain, also the stress can be singular, namely a line Dirac delta applied along a curve (the dashed line in Fig. 3.10b). We consider that the load consists of a distributed body load. One may admit some approximations in how this load is transmitted to the arch, allowing for some freedom in the detailed way in which the load is lumped to the line. We will return to regularity issues concerning stress and strains in Sect. 3.2.3, where the variational formulation of the BVP problem for NRNT materials is mathematically described.

Couplet and Danyzy's experiments. In 1737 Frezier published a book (1737) reporting some experiments performed by Danyzy in (1732) and suggested to him by Couplet (1729, 1730). Couplet is the first to have had the sense of what the equilibrium of arches is about from a kinematical point of view and the first to consider the effect that small movements of the supporting structures cause to the equilibrium configuration of such structural elements. When the centering of an arch is removed the voussoirs may move relatively to one another and, coming into contact, may even slide slightly among each other. Modelling these small relative displacements as settlements of the unilateral constraints among blocks, our insight is that the final equilibrium configuration minimises the potential energy of the loads under the effect of such given settlements under unilateral contact constraints among the blocks. To understand this point of view, it may be instructive to look at the experiments made by Danyzy in (1732) (Danyzy, 1732), following Couplet's theory, some of which are shown in Fig. 3.11.

As we can see from Fig. 3.11, the structures deform in a very peculiar way since the overall deformation is produced by fractures which open up among blocks and with the soil. In particular, the case in the bottom left of Fig. 3.11 is considered in Fig. 3.12. The picture shown in Fig. 3.12b is actually obtained as a result of the potential energy minimization under unilateral constraints for a multibody made of the rigid blocks shown in Fig. 3.12a.

Another example of the effect of settlements on a real arch is shown in Fig. 3.13a. This example draws our attention to the possibility that the pieces composing the structure break at their inside. The two supports of the arch, shown in a, have suffered a relative vertical Fig. 3.13 displacement due to ground settlements. In this example, we see that the arch is made of three pieces and the first piece on the left breaks into two parts, developing a crack in the middle.

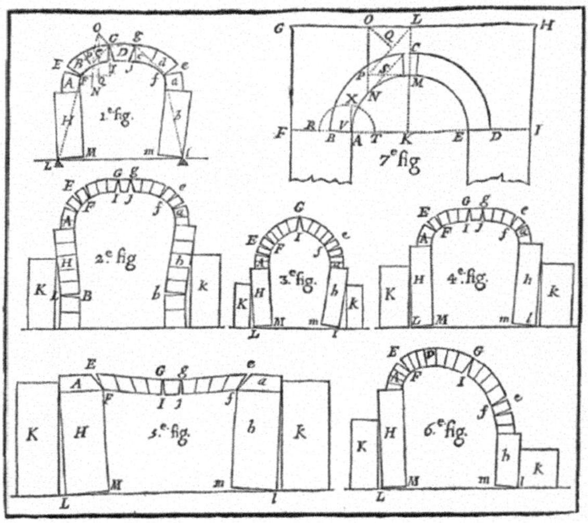

Fig. 3.11 Danyzy's experiments, from (1732)

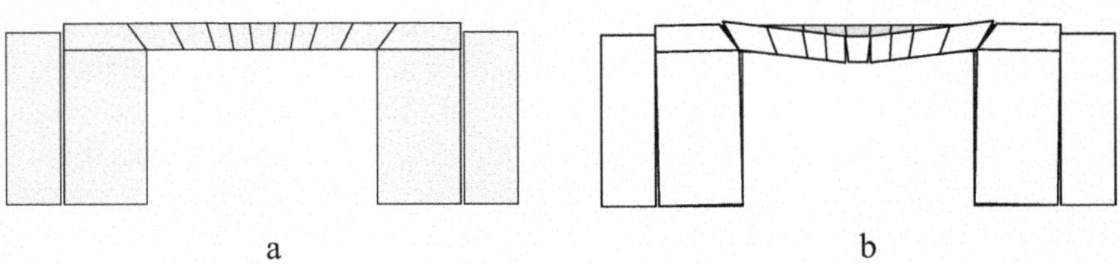

Fig. 3.12 PRD simulation (the method is described in Sect. 3.4.1.2) of the portal experiment on the bottom left of Fig. 3.11. In **a** initial geometry with the portal and two additional lateral buttresses. The portal in its ideal undeformed configuration is unstable. When it starts displacing it touches the two buttresses: this new mechanical system is stable: **b** It is worth noting that the hinges location is the same experienced by Danyzy (1732) in his tests, later reported by Frezier (1737)

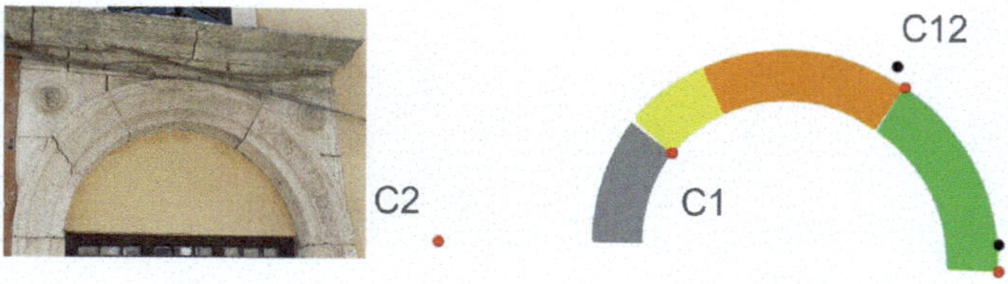

Fig. 3.13 A semicircular arch made of three large voussoirs of travertine (historic city centre of Tivoli, Italy) under the effect of settlements is shown in (**a**). The first block exhibits a fracture inside. In (**b**), a possible rigid block partition of the arch and a particular form of the settlement are proposed to interpret the effects visible in the photo (**a**)

The arch is divided into three parts articulated with three hinges (shown by dots painten black-before motion-and red-after motion-in Fig. 3.13b, and, on adopting the small displacement approximation, an elementary kinematic chain analysis is performed, considering that the grey part is fixed, and the other two (the yellow-orange and the green parts) move. The right hinge is displaced vertically, and then C1, C2 and C12 are the absolute and relative centres of rotations of the two moving parts. The displacement and the associated fracture pattern devised in Fig. 3.13b represent a possible admissible displacement for the arch structure.

Figure 3.14 shows another similar case where a hinge forms inside a voussoir.

Another real case. Figure 3.15 illustrates the damage of Deba bridge, a stone multi-span arch bridge suffering a large settlement of one of its piers. The analysis of the effects of this settlement on the deformation of the bridge, performed by adopting different modelling strategies, was presented in a recent research contribution by Malena et al. (2021).

Fig. 3.14 Example of a hinge forming inside a voussoir in a travertine arch made of three pieces

Fig. 3.15 Deba bridge distortion (Deba, Spain) under a pier settlement induced by pile shipworm infestation. This case was brought to our attention by Santiago Huerta

The structure of the arch appears severely distorted because of the settled pier. The main features of the deformation we see in Fig. 3.15 can be identified solving the BVP for NRNT materials, namely, looking for admissible displacements for which a statically admissible regime reconcilable with them in the sense of (3.11) does exist Fig. 3.16.

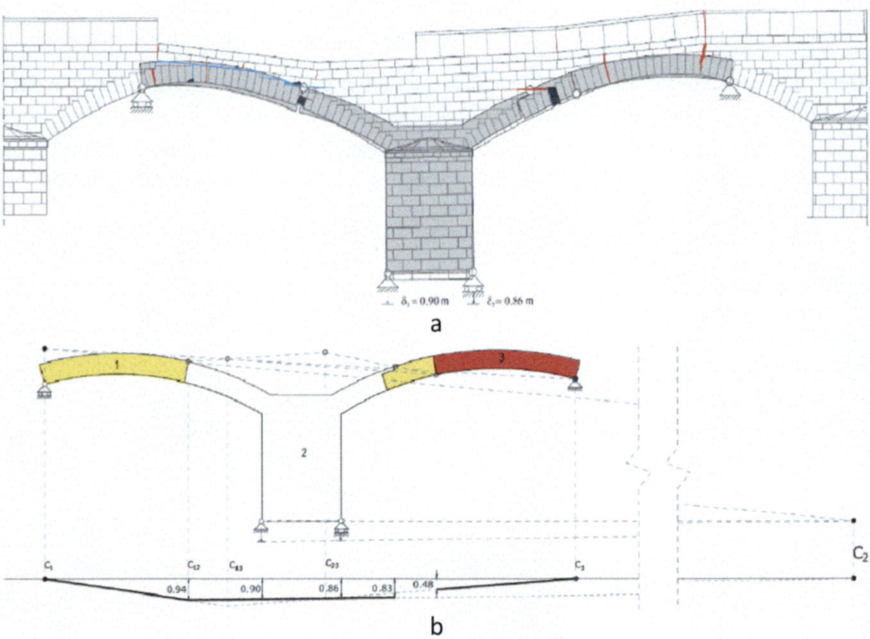

Fig. 3.16 In (**a**), *real* crack pattern superimposed on the *kinematic chain*, corresponding to the imposed displacement of the inferior external *constraints shown in the figure. In* (**b**) *partition* into *macro-rigid-blocks predicted by PRD*, showing mutual constraints and the *kinematical chain analysis*. The geometrical *result of the elementary linearized kinematic chain analysis illustrated in* (**b**) is compared with the real cracks in (**a**)

Indeed, this is the main scope of the numerical models here proposed and described in Part II of this Chapter, namely approximating the BVP solution for NRNT materials, that is, identifying classes of equilibrium stress regimes and corresponding displacements produced by given load and distortion data. This case is a perfect example to test the ability of our numerical models to do correct predictions, and this is exactly what was done in (Di Pasquale 1984). In particular, the analysis made with PRD by considering a class of possible settlements of the centre pier gave a close match with the deformations measured in situ for an optimized value of the given displacement (shown in Fig. 3.15a). With these values of vertical and horizontal displacement giving the best fit with the real deformations, the PRD method identified the formation of four macroblocks, shown with colors in Fig. 3.15b, and used in the elementary kinematic chain analysis illustrated in Fig. 3.15b. These results are compared with the real cracks in Fig. 3.15a (drawings taken from Di Pasquale (1984)).

3.2.3 Regularity of Stress and Strain: Singularities

As declared in Sect. 3.1.3, the solution sets of the equilibrium BVP for NRNT materoials, can be identified as the minimisers of convenient energy functionals that, in the case of a displacement approach is the Potential Energy, while, with a force approach, is the Complementary Energy.

Both the Potential Energy $E(\boldsymbol{u})$ and the Complementary Energy $E_c(\boldsymbol{T})$ are functionals over Ω of their arguments, which, at most, can have a linear growth. Consequently, both strain $E(\boldsymbol{u})$ nd stress $\boldsymbol{T}$ are bounded measures.

Measures are distributions that can be decomposed into absolutely continuous and singular parts. The singular part is the sum of a part supported on a null–measure set and a Cantor part supported on a fractional set. We shall restrict to measures whose Cantor part is zero. The most typical examples of such special bounded measures are point and line Dirac deltas (whose support is a 1D set), to which we shall restrict our attention.

In 3d problems, singular strains, that is deformations concentrated on lines or surfaces, are familiar in perfect plasticity. An easy way to construct kinematically admissible mechanisms, to be used as test functions for the Kinematic (upper bound) Theorem of Limit Analysis, is to consider rigid blocks moving one relative to the other and producing concentrated strains along their common interfaces.

Singular stresses are less common. Their possible use in the context of Masonry Mechanics was discovered only recently by Lucchesi et al. (2006). Still, an innovative application of concentrated stresses in elasticity was proposed by Fraternali et al. (2002), in a seminal paper where the Lumped Stress Method (LSM) is introduced. LSM is a mathematically rational way to admit singular approximations for the stress in elasticity by relaxing the energy functional. Moreover, LSM is actually one of the few existing stress-based elastic approximations that can be easily implemented on a computer (Cusano et al., 2021a, 2021b; De Chiara et al., 2019; Montanino et al., 2021; Olivieri et al., 2022).

The potential usefulness of singular stresses in Engineering can be more appreciated if we look at the support of stresses concentrated on lines or surfaces as to 1d and 2d structures that we envisage inside our 3d continuum. This is in the same spirit as the strut-and-tie method adopted to apply the safe theorem to reinforced concrete structures modelled as perfectly plastic bodies. We may think that the surface tractions and body forces applied to a 3d body are transferred to such structures through regular stresses having jumps across lines and surfaces defining them.

The use of singular stresses dramatically enlarges the set $\mathcal{H}$ of statically admissible stress fields that can be explored to apply the Safe Theorem of Limit Analysis, making it easier to check stability in a given geometry and load setting. However, the automated search of such states is far from being easy.

Actually, the BVP for the NRNT material is rather peculiar, since, in the absence of any kinematical datum, its possible solution can be any member of H. The set of possible equilibrium solutions restricts considerably once given settlements and eigenstrains have been assumed as non-zero. In general, the solution, both in terms of stress and strain, is non-unique and can be singular.

Strain. The strain tensor **E**, which for NRNT materials represents the effect of fractures solely, can be decomposed into the sum

$$\boldsymbol{E} = \boldsymbol{E}^r + \boldsymbol{E}^s \tag{3.72}$$

where $\boldsymbol{E}^r$ represents the regular part, i.e., the part which is absolutely continuous with respect to the area measure, and $\boldsymbol{E}^s$ is the singular part, here represented by line Dirac deltas, that is, a deformation concentrated on a set composed of a finite number of regular lines having finite length (each one of which is generically called Γ), and belonging to the closure of Ω. A bounded deformation concentrated on one of such lines Γ can be seen as the distributional derivative of a displacement $\boldsymbol{u}$, having a finite jump on that line Γ. We call s the arc length along Γ and $l(\Gamma)$ the length of Γ, while $\boldsymbol{t}, \boldsymbol{n}$ are the unit tangent and normal along Γ, whose orientation is the same of $\{\boldsymbol{e}_1, \boldsymbol{e}_2\}$. We consider the two parts on the two sides of Γ and define the positive part as the one where $\boldsymbol{n}$ points. Then, the jump of $\boldsymbol{u}$ along Γ is:

$$[\boldsymbol{u}] = \boldsymbol{u}^+ - \boldsymbol{u}^- \tag{3.73}$$

and

$$[\boldsymbol{u}] = w\boldsymbol{t} + v\boldsymbol{n} \tag{3.74}$$

On introducing the unit line Dirac delta on Γ as $\delta_\Gamma(\boldsymbol{x})$, that is, the operator such that

$$\int_{\Omega} \delta_{\Gamma}(\boldsymbol{x}) = l(\Gamma) \tag{3.75}$$

the strain-displacement relation for the singular part of E reads

$$\boldsymbol{E}^s = v(s)\delta_{\Gamma}(\boldsymbol{x})\boldsymbol{n}(s) \otimes \boldsymbol{n}(s) + \frac{1}{2}w(s)\delta_{\Gamma}(\boldsymbol{x})(\boldsymbol{n}(s) \otimes \boldsymbol{t}(s) + \boldsymbol{t}(s) \otimes \boldsymbol{n}(s)) \tag{3.76}$$

For NRNT materials $\boldsymbol{E} \in Sym^+$, therefore

$$\boldsymbol{E}^s = v(s)\delta_{\Gamma}(\boldsymbol{x})\boldsymbol{n}(s) \otimes \boldsymbol{n}(s) \tag{3.77}$$

and

$$[\boldsymbol{u}] = v\boldsymbol{n},\ v \geq 0 \tag{3.78}$$

Remark The decomposition $\boldsymbol{E} = \boldsymbol{E}^r + \boldsymbol{E}^s$ physically means that the fracture strain can be potentially the sum of two parts: a part $\boldsymbol{E}^r$ representing smeared fractures and a part $\boldsymbol{E}^s$ representing concentrated cracks.

Stress. For NRNT materials, the stress tensor $\boldsymbol{T}$ can be decomposed into the sum

$$\boldsymbol{T} = \boldsymbol{T}^r + \boldsymbol{T}^s \tag{3.79}$$

where $\boldsymbol{T}^r$ represents the regular part, that is, the part absolutely continuous with respect to the area measure, and $\boldsymbol{T}^s$ is the singular part. In what follows, we restrict to singular stresses represented by line Dirac deltas, that is, stresses concentrated on a set composed of a finite number of regular lines with finite length (each one generically named Γ) and belonging to the closure of Ω. A bounded stress measure concentrated on one of such lines Γ can be seen as the contact axial force transmitted inside a 1d structure having the form of Γ. If the stress is such a concentrated measure on Γ, then the stress vector $\boldsymbol{t}_n$ can have a finite jump on that line Γ. We call again s the arc length along Γ and $l(\Gamma)$ the length of Γ, and $\boldsymbol{t}, \boldsymbol{n}$ the unit tangent and normal along Γ, whose orientation is the same of $\{\boldsymbol{e}_1, \boldsymbol{e}_2\}$. We consider the two parts on the two sides of Γ and define the positive part as the one where $\boldsymbol{n}$ points. Then, the jump of $\boldsymbol{t}_n$ across Γ is:

$$[\boldsymbol{t}_n] = \boldsymbol{t}_n{}^+ - \boldsymbol{t}_n{}^- \tag{3.80}$$

and

$$[\boldsymbol{t}_n] = p\boldsymbol{t} + q\boldsymbol{n} \tag{3.81}$$

On introducing the unit line Dirac delta on Γ as $\delta_\Gamma(\boldsymbol{x})$, that is such that

$$\int_\Omega \delta_\Gamma(x) = l(\Gamma) \tag{3.82}$$

the singular stress on Γ can be expressed as

$$\boldsymbol{T}^s = P(s)\delta_\Gamma(\mathbf{x})\boldsymbol{t}(s) \otimes \boldsymbol{t}(s), \tag{3.83}$$

then the singular stress must be purely uniaxial and tangent to the line Γ.

For NRNT materials $\boldsymbol{T} \in Sym^-$, therefore

$$P(s) \leq 0 \tag{3.84}$$

Finally, in equilibrium, the relations between the stress vector jump and the regular and concentrated part of the stress are

$$\boldsymbol{T}^r\boldsymbol{n} = \boldsymbol{t}_n \tag{3.85}$$

$$\frac{d\boldsymbol{T}^s\boldsymbol{t}}{ds} + \delta_\Gamma(\boldsymbol{x})[\boldsymbol{t}_n] = 0 \tag{3.86}$$

that is

$$\frac{dP}{ds} + p = 0, \; \rho P + q = 0 \tag{3.87}$$

in which $\rho = \frac{dt}{ds} \cdot \boldsymbol{n}$ is the curvature of the line Γ.

3.2.4 Simple BVP Solutions: Examples

This Section presents two cases (Fig. 3.17) as trivial examples of BVP solutions for NRNT materials. The first is the arch already qualitatively discussed in Sect. 3.2.2, while the second concerns a similar problem for a lintel. These two examples are, in a sense, paradigmatic since exact solutions can be easily guessed, and, importantly, they contain all the peculiar ingredients of Heyman's theory, namely:

1. the compatibility of the loads (Limit Analysis);
2. the compatibility of the given displacements;
3. singular strains, that is, the appearance of fractures;
4. singular stresses, that is, the appearance of concentrated internal forces; and,

Fig. 3.17 A cracked arch: (**a**), and a cracked lintel: (**b**). These kinds of cracks are typical of such structures under the effect of spreading of their supports

5. absence of sliding.

These examples can be used, and actually, this is what is done in the subsequent Sections, in order to benchmark the numerical methods we designed to obtain approximate solutions.

The arch. A semi-circular arch, subject to its self-weight, having an aspect ratio (radius/thickness) of 4 and subject to given outer displacements of the abutments, is shown in Fig. 3.18a.

In this case, the kinematical data are compatible, as demonstrated by the kinematically admissible displacements graphically depicted in Fig. 3.18a. The loads are also compatible, since statically admissible stress states, such as that shown in Fig. 3.17b, can be constructed.

The singular strain and stress fields shown in Fig. 3.18a, b are kinematically admissible and statically admissible, respectively, and reconcilable in the sense of condition (3.11). Therefore, these represent a possible analytical BVP solution for the given vertical loads and support displacements.

The lintel. Figure 3.18 illustrates three solutions to a relevant mechanical problem: a lintel between two rigid vertical walls spreading outwardly, subject to a uniform vertical load over its upper edge. In this case, the kinematical data are compatible

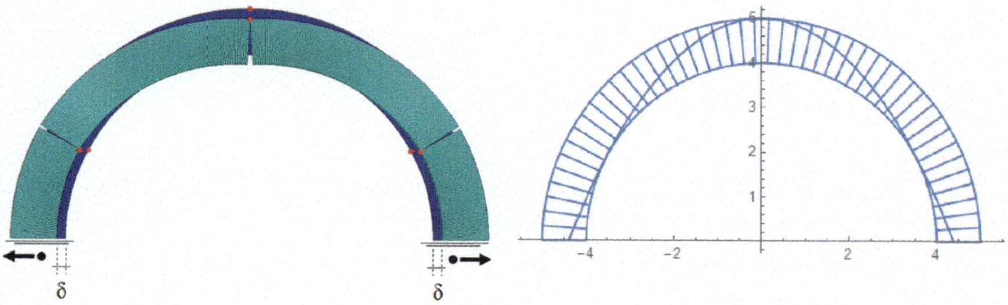

Fig. 3.18 Semi-circular arch subject to its own weight and given outward horizontal settlements of the supports

when the walls move outward and incompatible when the eternal walls move inward. Indeed, to comply with inward displacements, horizontal fibres of the body should shorten, thus, violating the material constraint (3.11) on the strain.

Another way to show the incompatibility of inward displacements of the external walls is to recall Theorem 3. Indeed, a uniform horizontal uniaxial stress field is statically admissible, but the related emerging reactions on the boundary perform positive work for inward displacement data. The compatibility of given outer displacements is demonstrated by the mechanism depicted in Fig. 3.19e. The loads are also compatible since statically admissible stress states, such as those shown in Fig. 3.19a–c, can be constructed.

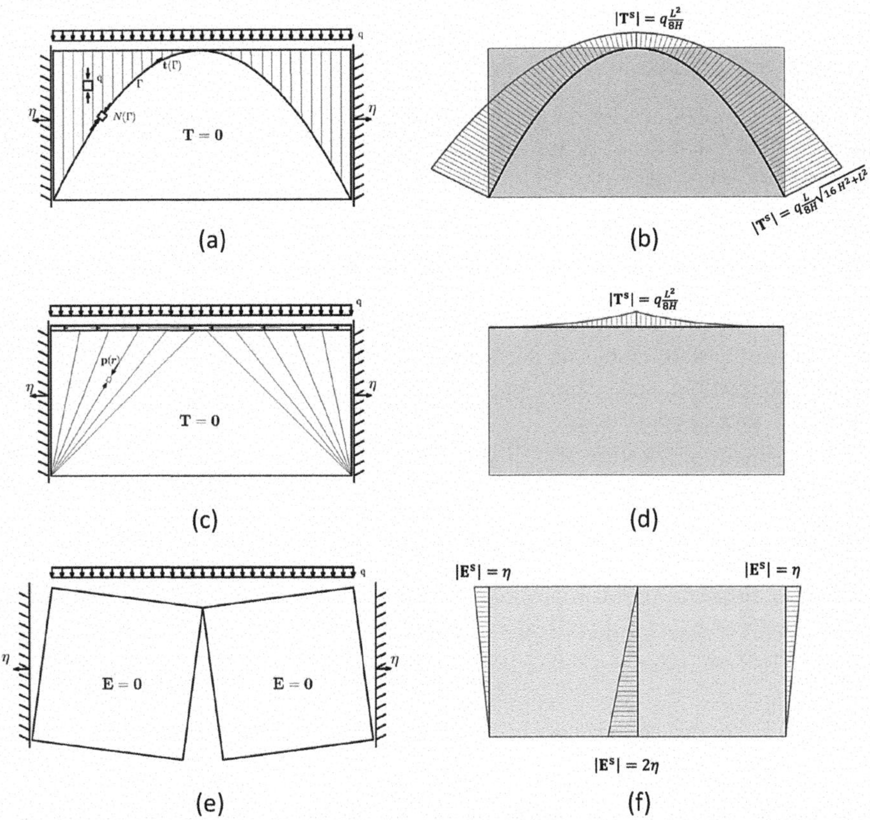

Fig. 3.19 Three solutions for a lintel subject to a uniform vertical load applied along the upper edge and to given outward horizontal displacements of the two lateral walls. (**a**) and (**c**) refer to two equilibriumsolutions in terms of stress, while (**e**) shows a displacement solution in terms of displacements. Both (**a**) and (**c**) match with (**e**) to form a solution of the BVP. In the right column the singular stress and strain corresponding to the stress and displacement fields reported in the left column are depicted

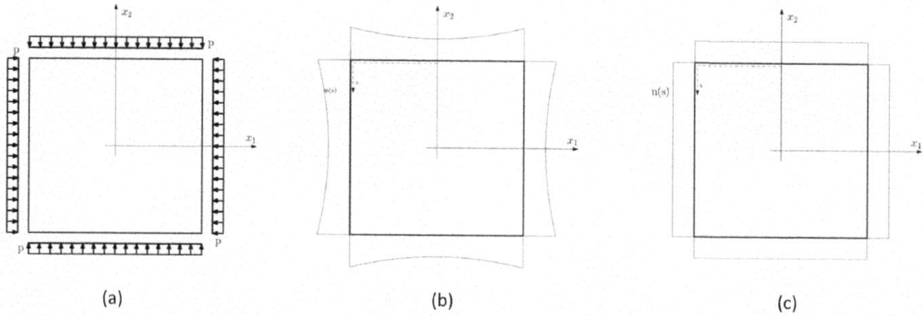

Fig. 3.20 Example of boundary value problem for an NRNT material. **a** Domain and loads; **b** boundary value $m(s)$; **c** boundary value $n(s)$

Both the displacement and stress fields represented in Fig. 3.19 exhibit singular parts, depicted with their intensity in the right column of Fig. 3.19. In particular, the strain field (Fig. 3.19f) shows three segments showing concentrated deformations with support on the vertical external edges and centre line. These three segments define the location at which cracks form (Fig. 3.19e). Figure 3.19b shows a singular stress field concentrated on a parabolic line passing through the two lower corners and tangent to the top edge of the domain at the mid point, balancing the load transmitted to it by a regular vertical uniaxial stress field. Figure 3.19c illustrates another admissible stress solution, where the singular stress field develops along a horizontal bar disposed along the top edge (Fig. 3.19d). The stress concentrated on such a bar equilibrates the unbalanced distributed forces arising from the different direction of the given vertical load and the emerging stress due to the compression rays, and shows a parabolically increasing intensity from the external borders to the mid point.

Lastly, the singular strain (Fig. 3.19e) is kinematically admissible, while the singular stress fields (Fig. 3.19a–c) are statically admissible. Both stress fields are reconcilable with the displacement solution in the sense of condition (3.11), and, together with it, represent a possible solution of the BVP. Notice that any convex combination of the two stress fields considered in Fig. 3.19a–c together with the displacement field of Fig. 3.20e, is a possible solution; therefore, this is an explicit example of an infinite class of possible non null analytical solutions of the BVP (Fig. 3.21).

3.3 Variational Formulations of the BVP

This Section provides the reader with the theoretical background concerning the two variational criteria valid for NRNT unilateral materials stated in Sect. 3.1.3, namely, the minimum of the Total Potential and Complementary Energy, in the plane case.

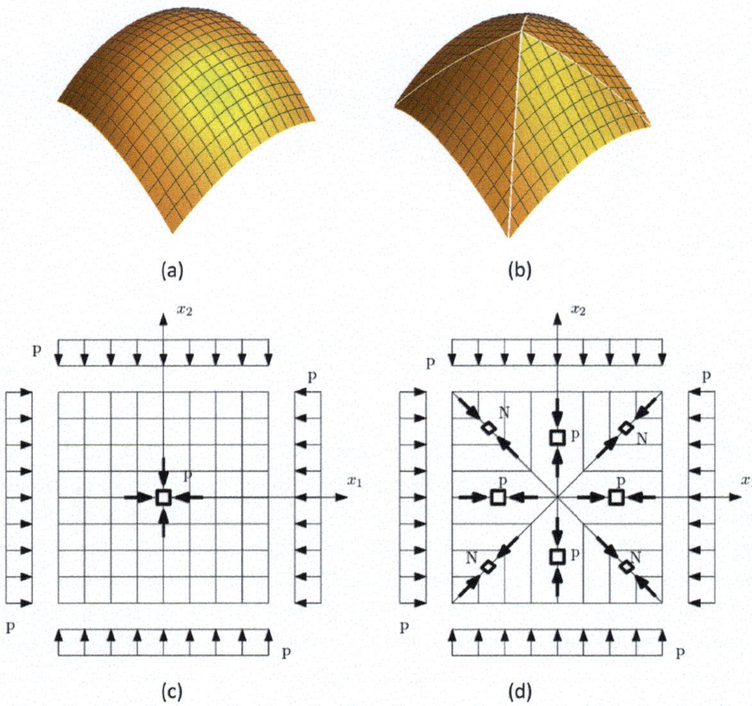

Fig. 3.21 Example of Airy stress potential meeting the boundary conditions of Fig. 3.20. **a** a smooth function; **b** a folded function; **c** stress field corresponding to the smooth Airy stress potential; **d** stress field corresponding to the folded stress potential

3.3.1 Minimum of the Total Potential Energy

A solution of the equilibrium BVP for a body Ω composed of NRNT material and subject to the load $\{\overline{\boldsymbol{p}}, \boldsymbol{b}\}$ and to the distortion data $\{\overline{\mathbf{u}}, \overline{\mathbf{E}}\}$ that is a state $\boldsymbol{s} = \{\boldsymbol{u}, \boldsymbol{E}(\boldsymbol{u}), \boldsymbol{T}\}$ for which$\boldsymbol{u} \in K$, $\boldsymbol{T} \in H$ and $\boldsymbol{T} \cdot \boldsymbol{E} = 0$, can be characterized as a minimizer of the potential energy

$$E(\mathbf{u}) = -\int_{\partial\Omega^N} \overline{\mathbf{p}} \cdot \mathbf{u} ds - \int_{\Omega}\int \mathbf{b} \cdot \mathbf{u} da, \tag{3.88}$$

which is a functional of the displacement $\boldsymbol{u}$ whose domain is K. In other words, among the infinitely many $\boldsymbol{u} \in K$ the ones, called $\boldsymbol{u}^\circ$ corresponding to the BVP solution (if any), are minimizers of $E(\boldsymbol{u})$ in K. In formulas this criterion reads:

$$E(\boldsymbol{u}^\circ) = \min_{u \in K} E(\boldsymbol{u}) \tag{3.89}$$

The question of the existence of the solution, that is, of the minimum of $E(\boldsymbol{u})$, represents a formidable mathematical task due to the complexity of the function spaces involved in this variational formulation. Indeed, these spaces are much larger and rich than those classically adopted for classical Cauchy continua. The existence of the solution remains an open problem whose discussion exceeds the scope of the present notes.

Some elementary considerations on this crucial question can be provided using only basic notions of Functional Analysis. If one wishes to prove the existence of the minimum of $E(\boldsymbol{u})$, the first trivial necessary request is that the set K be non-void, that is, the given displacement data $\{\overline{\boldsymbol{u}}, \overline{\boldsymbol{E}}\}$ be compatible. Then the compatibility of the displacement data will be assumed for granted.

By attacking the existence problem with the direct method of the Calculus of Variations, since the set K is unbounded, a growth property is needed. That is, it is necessary to make sure that $E(\boldsymbol{u})$ tends to plus infinity as some norm of the displacement $\boldsymbol{u}$ diverges. However, the functional $E(\boldsymbol{u})$ is linear in $\boldsymbol{u}$ and lacks any coercivity, that is, the simple compatibility of the loads is not sufficient to guarantee any growth property to $E(\boldsymbol{u})$. As suggested by Giaquinta and Giusti in (1985) (in the parent case on Normal Elastic No-Tension materials), to restore coercivity of the potential energy, a supplementary condition on the load, named "***safe load condition***" should be added. In short, this condition consists in adding to the given load, all over the loaded part of the boundary, a compressive pressure $-\beta n$, $\beta > 0$ being an arbitrary constant, small as one wishes.

The question of adding this small pressure all around the body is not purely formal. It has a beneficial practical effect in numerical applications, much in the same way Von Neumann and Richtmyer (1950) modified the hydrodynamics equations for compressible fluids by adding an artificial viscous term to improve the numerical solution quality of such equations in the presence of shock waves.

By introducing the safe load condition, the potential energy is endowed with a linear growth property with respect to the norm L^1 of the strain. This implies that the strain is a measure that is the sum of a regular part which is absolutely continuous with respect to the area measure and a singular part (see Sect. 3.2.3).

The complexity of this large function space for the displacements (called $BV(\Omega)$), which is a challenge for mathematicians, also represents an element of richness since larger working spaces mean a much ample repertoire of competing functions to be explored. Indeed, in the previous examples, we have seen that singular strains and stresses arise naturally and can be helpful in searching for admissible states of displacement and stress in unilateral materials. We will discuss some elementary questions concerning the nature of these functions in the following sections. The interested reader can consult (Ambrosio et al., 2000).

Existence of the infimum of $E(\boldsymbol{u})$. Another necessary condition for the existence of the minimum is that the functional $E(\boldsymbol{u})$ be bounded from below over the set K, that is $E(\boldsymbol{u})$ has an infimum in K.

Indeed, one can prove the following proposition.

Proposition 1 *Assume that the displacement data are compatible, that is $K \neq \emptyset$. Then, if $H \neq \emptyset$ there exists a finite real number $\alpha > -\infty$, such that $E(\boldsymbol{u}) \geq \alpha, \forall \boldsymbol{u} \in K$, that is the functional $E(\boldsymbol{u})$ is bounded from below over the set K.*

Proof if $H \neq \emptyset$, then there exists at least one stress field $\widetilde{\boldsymbol{T}} \in H$, and $\forall \boldsymbol{u} \in K$, we can transform the potential energy $E(\boldsymbol{u})$ as follows

$$E(\mathbf{u}) = -\int_{\partial\Omega^N} \overline{\mathbf{p}} \cdot \mathbf{u} ds - \int_{\Omega} \mathbf{b} \cdot \mathbf{u} da = \int_{\partial\Omega^D} \tilde{\mathbf{T}} n \cdot \overline{\mathbf{u}} ds - \int_{\Omega} \tilde{\mathbf{T}} \cdot \mathbf{E}(\mathbf{u}) da \tag{3.90}$$

Since

$$\tilde{\boldsymbol{T}} \in \mathrm{Sym}^-, (\boldsymbol{u}) \in \mathrm{Sym}^+, \tag{3.91}$$

Then

$$-\int_{\Omega^D} \tilde{T} \cdot \mathbf{E}(\mathbf{u}) da \geq 0, \qquad E(\mathbf{u}) \geq \int_{\partial\Omega^D} \tilde{T} \mathbf{n} \cdot \overline{\mathbf{u}} ds \tag{3.92}$$

If the displacement boundary data $\overline{\mathbf{u}}$ are continuous, and the stress $\widetilde{\boldsymbol{T}}$ is a measure over $\partial\Omega^D$, the term $\int_{\partial\Omega^D} \widetilde{\boldsymbol{T}} \boldsymbol{n} \cdot \overline{\mathbf{u}} ds$ is a finite real number, say α, and we obtain the sought result. ■

Solution as a weak minimum of $E(\boldsymbol{u})$. The BVP solution can be characterized as a weak minimum of the functional $E(\boldsymbol{u})$ over the set K.

Proposition 2 *If $s^\circ = \{\boldsymbol{u}^\circ, \boldsymbol{E}^\circ, \boldsymbol{T}^\circ\}$ is a solution of the BVP, then $E(\boldsymbol{u}^\circ)$ is a weak minimum of $E(\boldsymbol{u})$ in the set K.*

Proof For any $\boldsymbol{u} \in K$, consider the difference.

$$E(u) - E(u^\circ) = -\int_{\partial\Omega^N} \overline{\mathbf{p}} \cdot \mathbf{u} ds - \int_{\Omega} \mathbf{b} \cdot \mathbf{u} da - \left(-\int_{\partial\Omega^N} \overline{\mathbf{p}} \cdot \mathbf{u}^0 ds - \int_{\Omega} \mathbf{b} \cdot \mathbf{u}^0 da \right) \tag{3.93}$$

This energy difference can be transformed into internal work by considering the stress field $\boldsymbol{T}^{\circ} \in H$:

$$E(\mathbf{u}) - E\left(\mathbf{u}^{\circ}\right) = \int_{\partial\Omega^D} \mathbf{T}^{\circ}\mathbf{n} \cdot \bar{\mathbf{u}} ds - \int_{\Omega} \mathbf{T}^{\circ} \cdot \mathbf{E}(\mathbf{u}) da - \left(\int_{\partial\Omega^D} \mathbf{T}^{\circ}\mathbf{n} \cdot \bar{\mathbf{u}} ds - \int_{\Omega} \mathbf{T}^{\circ} \cdot \mathbf{E}\left(\mathbf{u}^{0}\right) da \right) = \int_{\Omega} T^{\circ} \cdot \left(E^{\circ} - E\right) da. \tag{3.94}$$

By recalling the dual normality inequality, this term is always non-negative, then

$$E(\boldsymbol{u}) - E(\boldsymbol{u}^{\circ}) \geq 0 \tag{3.95}$$

which proves the assertion. ■

Remark 3 In real masonry structures, the appearance of piecewise rigid mechanisms (associated with concentrated fractures) rather than continuous mechanisms (entailing diffuse fractures) is often due to mechanical characteristics (e.g. cohesion, toughness and finite friction) also depending on construction techniques. These mechanical aspects are not taken into account by the NRNT model. Indeed, for NRNT materials, it is generally not possible to prefer concentrated cracks over diffuse fractures. This circumstance is essentially due to the absence of any energy-growth property on unbounded fracture strains. An energy-growth property for displacement fields in $BD(\Omega)$ is restored by introducing the so-called **Safe Load Condition** as proved in (Giaquinta & Giusti 1985). The safe load condition consists of adding an isotropic pressure all over the loaded boundary. Without such confinement, the energy is not coercive, meaning that the displacement can grow indefinitely at zero energy price. In the proposed numerical methods, this confinement is imposed by adding, all over the loaded boundary, an artificial given uniform pressure of very small magnitude, that is, of the order of 10^{-2}–10^{-3} of the mean compressive stress σ_m. The presence of such a negligible (with respect to the given loads) value of isotropic pressure also has the effect of encouraging rigid block mechanisms over diffuse deformations. This trick is sufficient to provide the BVP with the Safe Load Condition, avoid zero-energy modes, and make rigid block deformations and concentrated cracks the preferred minimum-energy mechanism. ■

3.3.2 Minimum of the Total Complementary Energy

The solution of the equilibrium BVP for a body Ω composed of NRNT material and subject to the load data $\{\overline{\boldsymbol{p}}, \boldsymbol{b}\}$ and to the distortion data $\{\overline{\mathbf{u}}, \overline{\mathbf{E}}\}$ we recall that $\overline{\mathbf{E}}$ are the eigenstrains assigned inside Ω), that is a state $\boldsymbol{s} = \{\boldsymbol{u}, \boldsymbol{E}, \boldsymbol{T}\}$, such that $\boldsymbol{u} \in K, \boldsymbol{T} \in H$ and $\boldsymbol{T} \cdot \boldsymbol{E} = 0$, can also be characterized as the minimizer of the complementary energy

$$E_c(\mathbf{T}) = -\int_{\partial\Omega^D} \mathbf{Tn} \cdot \overline{\mathbf{u}} ds + \int_{\Omega} \mathbf{T} \cdot \overline{\mathbf{E}} da \tag{3.96}$$

that is a functional of the stress $\boldsymbol{T}$ whose domain is H. In other words, among the infinitely many $\boldsymbol{T} \in H$, the one, say $\boldsymbol{T}^{\circ}$, corresponding to a BVP solution (if any exists) is the minimizer of $E_c(\boldsymbol{T})$ in H. This concept is summarised by the following mathematical formulation:

$$E_c(\boldsymbol{T}^{\circ}) = \min_{\boldsymbol{T} \in H} E_c(\boldsymbol{T}) \tag{3.97}$$

The existence of the solution, that is, of the minimum of $E_c(\boldsymbol{T})$, represents an additional challenge as for the dual minimum problem for $E(\boldsymbol{u})$, due to the complexity of the function spaces involved in the variational formulation and remains an open problem whose discussion exceeds the scope of the present notes. However, if we wish to prove the existence of the minimum of $E_c(\boldsymbol{T})$, the first trivial necessary request is that the set H be non-void, that is, the load data $\{\overline{\boldsymbol{p}}, \boldsymbol{b}\}$ be compatible. Then we will assume the compatibility of the given loads from the start.

The functional $E_c(\boldsymbol{T})$ over H shares with the functional $E(\boldsymbol{u})$, the same problems due to the absence of any growth property: $E_c(\boldsymbol{T})$ is linear in $\boldsymbol{T}$, and lacks any coercivity, meaning that the simple compatibility of the distortions is not sufficient to guarantee that $E_c(\boldsymbol{T})$ tends to $+\infty$ as the norm of $\boldsymbol{T}$ diverges.

A supplementary condition on the given distortions, analogous to the "safe load condition", can be devised for the case at hand. For example, a uniform eigenstrain field $\overline{\boldsymbol{E}}^{\alpha} = -\alpha \boldsymbol{I}$ with $\alpha > 0$, small as one wishes, could be added to the distortion data. This has the effect of providing $E_c(\boldsymbol{T})$ of a linear growth property with respect to the L^1 norm of $\boldsymbol{T}$. This implies that even the stress is a measure. That is, the stress can be decomposed into the sum of a regular part (absolutely continuous with respect to the area measure), and a singular part. The first authors who considered measures as statically admissible stress fields for No-Tension materials were Lucchesi et al. in (2006), though singular stresses were also considered for elastic materials in (Fraternali et al., 2002).

Existence of $inf\, E_c(\boldsymbol{T})$. A necessary condition for the existence of the minimum is that the functional $E_c(\boldsymbol{T})$ be bounded from below over the set H, that is $E_c(\boldsymbol{T})$ has an infimum in H.

Indeed, one can prove the following proposition.

Proposition 1 *Assume that the load data are compatible, i.e.* $H \neq \varnothing$. *Then, if* $H \neq \emptyset$, *that is the displacement data* $\{\overline{\mathbf{u}}, \overline{\mathbf{E}}\}$ *are also compatible, there exists a finite real number* $\beta > -\infty$, *such that* $E_c(\boldsymbol{T}) \geq \beta, \forall \boldsymbol{T} \in H$, *that is the functional* $E_c(\boldsymbol{T})$ *is bounded from below over the set* H.

Proof If $K \neq \varnothing$, then there exists at least one displacement field $\tilde{\boldsymbol{u}} \in K$ and $\forall \boldsymbol{T} \in H$, we can transform the complementary energy $E_c(\boldsymbol{T})$ as follows.

$$E_c(\mathbf{T}) = -\int_{\partial\Omega^D} \mathbf{Tn}\cdot\overline{\mathbf{u}}ds + \int_{\Omega} \mathbf{T}\cdot\overline{\mathbf{E}}da = \int_{\partial\Omega^N} \overline{\mathbf{p}}\cdot\tilde{\mathbf{u}}ds - \int_{\Omega} \mathbf{T}\cdot\mathbf{E}(\tilde{\mathbf{u}})da. \tag{3.98}$$

Since

$$\boldsymbol{T} \in Sym^-, \boldsymbol{E}(\tilde{\boldsymbol{u}}) \in Sym^+ \tag{3.99}$$

Then

$$\begin{aligned} -\int_{\Omega} \mathbf{T}\cdot\mathbf{E}(\tilde{\mathbf{u}})da \geq 0, \\ E_c(\mathbf{T}) \geq \int_{\partial\Omega^N} \overline{\mathbf{p}}\cdot\tilde{\mathbf{u}}ds \end{aligned} \tag{3.100}$$

If the traction boundary data $\overline{\mathbf{p}}$ are continuous and the displacement $\tilde{\boldsymbol{u}}$ is of Bounded Variation over $\partial\Omega^N$, the term $\int_{\partial\Omega^N}\overline{\mathbf{p}}\cdot\tilde{\boldsymbol{u}}\, ds$ is a finite real number, say β, and we obtain the sought result. ■

Solution as a weak minimum of $E_c(\boldsymbol{T})$. The solution of the BVP can be characterized as a weak minimum of the functional $E_c(\boldsymbol{T})$ over the set H.

Proposition 2 *If* $s^\circ = \{\boldsymbol{u}^\circ, \boldsymbol{E}^\circ, \boldsymbol{T}^\circ\}$ *is a solution of the BVP, then* $E_c(\boldsymbol{T}^\circ)$ *is a weak minimum of* $E_c(\boldsymbol{T})$ *in the set* H.

Proof For any $\boldsymbol{T} \in H$, consider the difference.

$$\begin{aligned} E_c(\boldsymbol{T}) - E_c(\boldsymbol{T}^\circ) = -\int_{\partial\Omega^D} \boldsymbol{Tn}\cdot\overline{\boldsymbol{u}}ds + \int_{\Omega} \boldsymbol{T}\cdot\overline{\boldsymbol{E}}da \\ -\left(-\int_{\partial\Omega^D} T^\circ\boldsymbol{n}\cdot\overline{\boldsymbol{u}}ds + \int_{\Omega} T^\circ\cdot\overline{\boldsymbol{E}}da\right) \end{aligned} \tag{3.101}$$

This energy difference can be transformed into internal work by considering the displacement field $\boldsymbol{u}^\circ \in K$:

$$E_c(\boldsymbol{T}) - E_c(\boldsymbol{T}^\circ) = \int_{\partial\Omega^N} \overline{\boldsymbol{p}} \cdot \boldsymbol{u}^\circ ds - \int_\Omega \boldsymbol{T} \cdot \boldsymbol{E}(\mathbf{u}^\circ) da -$$
$$\left(\int_{\partial\Omega^N} \overline{\boldsymbol{p}} \cdot \boldsymbol{u}^\circ ds - \int_\Omega \boldsymbol{T}^0 \cdot \boldsymbol{E}(\mathbf{u}^\circ) da\right) = \tag{3.102}$$
$$= \int_\Omega (T^\circ - T) \cdot E^\circ da$$

By recalling the normality inequality, this term is always non-negative, then

$$E_c(\boldsymbol{T}) - E_c(\boldsymbol{T}^\circ) \geq 0 \tag{3.103}$$

which proves the assertion. ■

3.3.2.1 Equilibrium Solution Through the Airy Stress Potential

In the absence of body forces ($\boldsymbol{b} = 0$), the equilibrium Eq. (3.6) admit the following solution in terms of a scalar function F, called Airy stress potential:

$$T_{11} = F_{,22}, T_{22} = F_{,11}, T_{12} = -F_{,12} \tag{3.104}$$

This is the general solution to the equilibrium equations, if the loads are *self-balanced* on any closed boundary delimiting Ω (Gurtin, 1972). The b.c. $\boldsymbol{Tn} = \overline{\mathbf{p}}$ on $\partial\Omega_N$, can be reformulated in terms of F. Denoting $\boldsymbol{x}(s)$ the parametrization of $\partial\Omega_N$ with the arc length, the b.c. on F are:

$$F(s) = m(s), \frac{dF}{dv} = n(s), \quad \text{on } \partial\Omega_N, \tag{3.105}$$

in which $\frac{dF}{dv}$ is the normal derivative of F at the boundary (that is the slope of F in the direction of $\boldsymbol{n}$) and $m(s)$, $n(s)$ are the moment of contact and the axial force of contact produced by the tractions $\overline{\boldsymbol{p}}(s)$, on a beam structure having the same shape of $\partial\Omega_N$.

A simple example is shown in Fig. 3.20. Regular and singular equilibrated stress fields can be derived by any stress function meeting the prescribed boundary conditions on F and $\frac{dF}{dv}$. For this example, a regular stress field is represented by a smooth F (see Fig. 3.22a), a singular stress field by a continuous but folded F (Fig. 3.22b).

The projection of a fold of F on the plane of Ω is called a folding line and denoted Γ. On a fold of F, the second derivative of F, with respect to the normal $\boldsymbol{m}$ to the folding line Γ, is a Dirac delta with support on Γ. Therefore, along Γ the Hessian

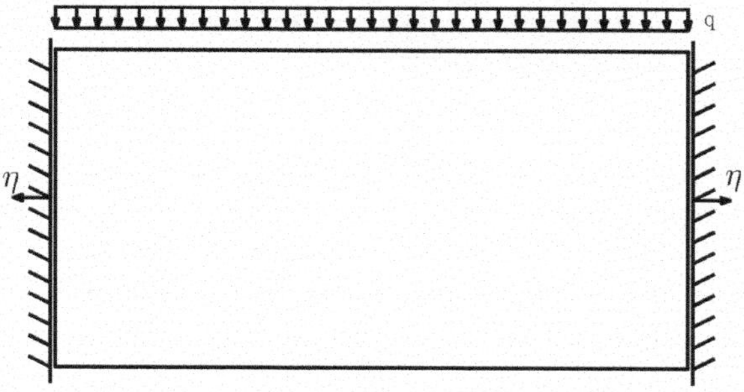

Fig. 3.22 Masonry panel under vertical load and given outward settlements of the lateral sides

$\boldsymbol{H}(F)$ of the stress function F is a dyad of the form

$$\boldsymbol{H}^s(F) = \Delta_m F \delta(\Gamma) \boldsymbol{m} \otimes \boldsymbol{m} \tag{3.106}$$

$\Delta_m F = [\nabla F] \cdot \boldsymbol{m}$ denoting the jump of the slope of ∇F in the direction of the normal $\boldsymbol{m}$ to Γ. Recalling the Airy relation, the corresponding singular part of the stress is

$$T^s = N \delta(\Gamma) \boldsymbol{t} \otimes \boldsymbol{t} \tag{3.107}$$

where N is the axial contact force, given by

$$N = \Delta_m F \tag{3.108}$$

The internal admissibility condition (3.9) on the stress field, expressed in terms of the stress potential through (3.104), are translated into conditions on the Hessian of F as

$$tr(\boldsymbol{H}(F)) \leq 0, det(\boldsymbol{H}(F)) \geq 0 \tag{3.109}$$

which imply that the Airy stress function is required to be (non-strictly) concave over the domain Ω.

In the absence of body forces $\boldsymbol{b}$, the set of the admissible stress fields H is bijectively associated with the set H_F of the admissible Airy stress functions, defined as

$$\begin{aligned} H_F = \{F \in R, \text{ s.t.tr}(\boldsymbol{H}(F)) \leq 0, \det(\boldsymbol{H}(F)) \geq \\ 0, \ F|_{\partial\Omega^N} = m(s), \ \nabla F \cdot \mathbf{n}|_{\partial\Omega^N} = n(s)\} \end{aligned} \tag{3.110}$$

Then, the variational formulation of the boundary value problem can be reformulated in terms of the Airy stress potential as

$$E_c(F^\circ) = \min_{F \in H_F} E_c(F) \tag{3.111}$$

3.4 Numerical Approaches to the BVP Variational Formulation

The mathematics of NRNT materials is non-standard, mainly because Masonry-Like materials are essentially unilateral, and then the variational forms of the mechanical problem are represented by inequalities. Indeed, inequality (unilateral) problems are governed by the inequality form of the Principle of Virtual Power, and therefore, they differ fundamentally from equality (bilateral) problems.

The inequality form of the principle of virtual power is due to Fourier; since then, until the second half of the 20th century, little work has been done on inequality problems, primarily as no sufficiently developed mathematical theory existed. In the early sixties, Fichera (1973), and Lions and Stampacchia (1967) began to study inequality problems mathematically. In the seventies, Moreau (1968) connected the theory of variational inequalities to convex analysis, especially to subdifferentiability, by introducing the notion of superpotential.

These early studies on inequality problems were all related to convex energy functions. Since the gradient (or, more generally, the subdifferential) of a convex function is a monotone mapping, it is apparent that the inequality problems studied were firmly tied to monotonicity, e.g. only monotone stress–strain laws could be treated. The stress–strain law of NRNT materials is monotone and gives rise to potential and complementary energy minimisation problems, as illustrated in Sect. 3.3.

The functional spaces in which the solutions are sought are rather large and far less regular than those typically employed in Mechanics. Panagiotopoulos lucidly exposes the reason for that in his milestone book (Panagiotopoulos, 1985):

"It is worth noting that while bilateral problems are connected with differentiable energy functions, inequality problems involve functions which are nondifferentiable. In the former case, the generalized forces are derived from a potential (by forming the gradient), and in the latter case from a convex superpotential (by forming the subdifferential). Because of the multivalued character and lack of differentiability of superpotentials, one could speak of "multivalued mechanics" or of "non-smooth mechanics."

Function spaces for variational inequality problems were introduced by Temam and Strang (1980) and called Functions of Bounded Variation (BV) to model Perfect Plasticity and subsequently developed by Suquet, who coined the term *functions of Bounded Deformation* (BD).

The function spaces BV and BD are not Sobolev spaces which are common in classical linear and non-linear elasticity and visco-elasticity. They are, actually, rather rich and interesting from a purely mathematical point of view. The most exciting aspect from an engineering point of view is that both the strain and the stress can be singular—meaning that both the displacement and the stress vector can be discontinuous—namely, in the mathematical terminology, functions of Bounded Variation. The practical use of singular strains and stresses for NRNT materials has already been introduced and discussed in Sect. 3.2.3.

The reason why the proof of the existence of the minimum of the potential energy becomes difficult resides essentially in the fact that the spaces BV and BD are not reflexive. Indeed the classical proof of existence for convex functionals in Sobolev spaces states that (see Panagiotopoulos (1985)):

"Let X be a reflexive Banach space with the norm ||. || and

$$\lim_{||x||\to\infty} f(x) = +\infty, x \in \mathbb{K} \subset X$$

Then the Minimum problem admits at least one solution. If, furthermore, f is strictly convex, the solution is unique."

The question about the lack of any growth properties both in the Potential and Complementary Energy formulation has been already anticipated and discussed in Sect. 3.3 and in Remark 3.

Concerning the numerical treatment, we must distinguish between dynamic and static inequality problems. For the former, we may employ a time discretization procedure to reduce the problem to a sequence of static problems.

For static problems, there are two main possibilities. If the problem can be formulated as a minimization problem, then optimization algorithms are used."

In the case at hand, the energy functionals are linear and then only "marginally" convex. Therefore, to solve the allied minimization problems, the well-developed machinery of Linear Programming can be used.

This Section illustrates and discusses different numerical methods used to discretize the variational BVP formulations described above. Displacement- and force-based methods are considered, and two strategies are exploited to approximate singular strains and stresses.

With the first approach, singularities in the local strain or stress are directly built in the discretization scheme; while with the second approach, these singularities are considered the weak limits of regular fields.

In particular, for what concerns the displacement approach, two strategies are presented: one looks for solutions in the space of smooth (at least continuous) functions and an 'opposite' numerical approach, which seeks solutions in terms of singular strain distributions. While the second approach, though derived from a different variational criterion, appears more aligned to standard 'rigid-block' methodologies already present in the Literature (BAggio & Trovalusci, 1993; Gilbert & Melbourne, 1994; Livesley, 1978; Orduña & Lourenço, 2005; Portioli & Cascini, 2017), the first one has been developed (Iannuzzo, 2017, 2019; Iannuzzo et al., 2021b) to overcome some typical 'rigid-block' convergence issues due to the mesh-dependency of the approximate solutions.

However, it is worth pointing out that if the normality rule applies, solving the kinematical problem through the minimum of the total potential energy is equivalent to solving the equilibrium by minimising the complementary energy. Indeed, as shown in (Gagliardo et al., 2021), the solutions are dually linked.

This theoretical result has a computational counterpart when solving variational problems through Mathematical Programming: every time a particular numerical solution is found, dual values solving the dual variational problem are automatically supplied. This aspect will be highlighted in the next sections, which are organised as follows: Sect. 3.4.1 presents two displacement-based approaches to the minimum of the Total Potential Energy (TPE), while Sect. 3.4.2 shows two force-based approaches for the Total Complementary Energy (TCE).

To illustrate and compare the proposed methodologies, the different variational formulations are illustrated referring to the simple problem of Fig. 3.19, concerning a rectangular panel uniformly loaded on the top edge and subjected to outward lateral settlements imposed on its external lateral edges. The analytical solution to this problem was discussed in Sect. 3.2.4, and will be used as a reference case to benchmark our approximate numerical solutions. It is worth emphasising that in what follows, both variational formulations are cast into suitable mathematical programming problems, that is, optimization problems with continuous variables. In particular, because of the nature of the problem at hand, it can be framed as a convex optimization problem. Indeed, the objective function, in both cases, is quadratic or linear if the elastic term is neglected; the side restrictions are either linear or conical constraints.

3.4.1 Displacement-Based Approaches: Discretisation of the Total Potential Energy

Solving the BVP through a displacement approach consists in finding a displacement field $\boldsymbol{u}^0 \in K$ for which there exists a stress field $\boldsymbol{T}^0 \in H$ such that $\boldsymbol{T}^0 \cdot \boldsymbol{E}\left(\boldsymbol{u}^0\right) = 0$. An efficient variational approach to solve the BVP with a displacement-based approach, is to minimise the potential energy $E(\boldsymbol{u})$:

$$E(\boldsymbol{u}^\circ) = \min_{u \in K} E(\boldsymbol{u}) \tag{3.112}$$

The following subsections describe two different displacement-based numerical methods: the PRD and CDF. The main objective of these displacement-based approaches is to find the effects of displacement data (e.g., settlements or eigenstrains) through the minimisation of the potential energy. The difference between these two numerical methods resides in the radically different way in which the displacement and, consequently the local strain are approximated. In particular, these two different approximations also entail different ways in which the restriction on the latent strain

$$\boldsymbol{E} \in Sym^+ \tag{3.113}$$

is imposed.

Indeed, as the strain is a measure, it can be decomposed into two additive parts, i.e. into the sum of a regular and a singular part, according to (54). More specifically, restricting to SBV displacement functions (that is, BV functions with void Cantor part (Ambrosio et al., 2000)), this decomposition, based on (3.75)–(3.77) can be written as follows:

$$\boldsymbol{E} = \boldsymbol{E}^r + v\delta(\Gamma)\boldsymbol{n} \otimes \boldsymbol{n} + \frac{1}{2} w\delta(\Gamma)(\boldsymbol{t} \otimes \boldsymbol{n} + \boldsymbol{n} \otimes \boldsymbol{t}) \tag{3.114}$$

where the singular part is expressed as a function of the displacement jumps over the set Γ (that is, the 1D jump set of $\boldsymbol{u}$), which, in what follows, will coincide with the element edges (that is, the skeleton of the mesh). More details about this modelling will be introduced in Sect. 3.4.1.2. With the PRD method, the regular part $\boldsymbol{E}^r$ of the strain is zero, and the strain consists only of its singular part: $\boldsymbol{E} = \boldsymbol{E}^s$. The domain is discretised into elements that can be thought of as rigid blocks. Consequently, the cracks are simulated by gap "openings" between adjacent blocks. As opposite to PRD, the CDF method assumes the strain to be purely regular: $\boldsymbol{E} = \boldsymbol{E}^r$, that is, cracks appear as smeared inside the elements and become mesh independent.

As a main consequence, this assumption on the latent strain leads to two different definitions of the feasible sets on which the total potential energy has to be minimised. In particular, with the PRD method, introduced to solve typical masonry mechanics problems in (Iannuzzo, 2017; Iannuzzo, et al., 2018), the set K_{PRD} of piecewise-rigid displacements is considered as the set of kinematically admissible displacements; while with the CDF method, the solutions is sought in the space K_{CDF} of continuous functions as derived from a suitable C^0 FE discretisation.

In the present notes, it will be shown how applying and integrating these two seemingly different numerical strategies, the typical rigid macroblock nucleation of masonry structures can be predicted accurately.

Once more, it is worth emphasising that the primal variables of the minimisation problem are either strains or displacements, meaning that the effect of settlements as non-homogeneous boundary conditions can be directly considered. Additionally, as the CDF method assumes only continuous displacement fields and jumps are not allowed, condition (3.73) is directly written on the strain tensor. Conversely, with the PRD method, the strain is purely singular as it is represented through displacement jumps (3.74); thus, condition (3.73) is implemented by writing conditions on the displacement jumps, i.e. on w and v directly. Finally, we point out that this difference is also reflected in the numerical way used to implement these conditions: the CDF method leads to a natural second-order cone programming, while the PRD leads to linear programming problems. In both cases, the objective function is linear, as quadratic or higher-order contributions are excluded since the NRNT model neglects any compressive elastic behaviour.

3.4.1.1 Continuous Approach: CDF Method

This Section illustrates the CDF method. The solution of the minimum problem (3.90) is approximated in the set of continuous, piecewise polynomial displacement fields generated by a quadrangular FE mesh made up of nine-node Lagrangian elements

(Bathe & Wilson, 1976). The choice of the element typology used to approximate the displacement field is not entirely free. Indeed, linear or bilinear elements cannot be selected for their inability to reproduce simple uniaxial flexure. This mechanism often appears in non-uniform cracks (i.e., due to relative rotation along a straight line). The optimal choice in shape functions, balancing accuracy and simplicity is based on a classical conforming second-order 9-node, Lagrangian quadrangular element. The numerical procedure illustrated below refer to this element and its shape functions (Bathe & Wilson, 1976). Similarly to a standard FE formulation, the displacement field is assumed to be continuous at the nodes and along the edges of the FE mesh. Therefore, the strain is regular and does not show any jump.

Let Ω be the structural domain discretised with M quadrangular elements, each one associated with a nine-node, Lagrangian element (Fig. 3.23). The total number of nodes generated by the adopted shape functions is N. The partition $(\Omega_i)_{i\in\{1,2,\ldots,M\}}$ into 9-node rectangular elements allows expressing the displacement field $\boldsymbol{u}$ as a function of the nodal displacements vector $\widehat{\boldsymbol{U}}$, that is:

$$\boldsymbol{u} = \boldsymbol{u}\left(\widehat{\boldsymbol{U}}\right) \tag{3.115}$$

With this discretisation, the minimum problem (3.10) can be approximated through the following finite-dimensional one:

$$E\left(\boldsymbol{u}^{o}_{CDF}\right) = \min_{\boldsymbol{u}\in K_C^N} E(\boldsymbol{u}) \tag{3.116}$$

where K_C^N denotes the discretised set of kinematically admissible displacements generated by the given FE mesh partition of the structural domain.

The displacement field in any finite element Ω_k can be expressed as a function of the nodal displacement $\left(U_j, V_j\right)$ as

$$u_k = u|_{\Omega_k} = f\left(U_j, V_j\right) \tag{3.117}$$

where f accounts for the adopted Lagrangian shape functions. The latent strain can thus be expressed as a function of the nodal displacements:

$$\boldsymbol{E}_k = \boldsymbol{E}|_{\Omega_k} = Sym\nabla\boldsymbol{u}_k \tag{3.118}$$

The material restrictions (3.10) are enforced directly at a specified set of internal points $(\tilde{x}, \tilde{y})$:

$$\boldsymbol{E}_k(\tilde{x}, \tilde{y}) \in Sym^{+} \tag{3.119}$$

The choice of using the Gauss points turned out to be a good numerical strategy. It is worth recalling that restriction (3.119) is equivalent to the two following inequalities:

$$trE \geq 0, detE \geq 0 \tag{3.120}$$

which in the 2D Euclidean space can be written in terms of Cartesian components as:

$$\varepsilon_{11} + \varepsilon_{22} \geq 0, \varepsilon_{11}\varepsilon_{22} - \varepsilon_{12}^2 \geq 0 \tag{3.121}$$

where ε_{ij} are the Cartesian components of the latent strain tensor $\boldsymbol{E}$.

Linear relation $(120)^1$ together with nonlinear condition $(120)^2$ define a convex set. The condition $det\mathbf{E} \geq 0$ defines a double cone in the Sym space while the additional condition $tr\mathbf{E} \geq 0$ selects one of the two parts of the cone, namely the set of positive semidefinite symmetric tensors, which is also convex. In Fig. 3.24 these restrictions are graphically represented, referring to the 3D space Sym spanned by the dyadic orthonormal basis $\{\mathbf{e}_1 \otimes \mathbf{e}_1, \mathbf{e}_2 \otimes \mathbf{e}_2, \sqrt{2}/2(\mathbf{e}_1 \otimes \mathbf{e}_2 + e_2 \otimes e_1)\}$.

From a numerical standpoint, the previous condition can be either enforced exactly through a second-order conical condition, or can be linearised, as shown in Fig. 3.24b. In the latter case, the cone is approximated through a finite number p of tangent planes, and the constraints become linear. This linearisation is produced by uniformly spacing the tangent planes along an axial cross-section of the cone, as illustrated in Fig. 3.24b. By the way, on increasing the number of points along the cross-section, a better fit of the cone Sym^+ is obtained.

By using the above-described linearisation scheme, conditions (3.121) is approximated with the following set of p inequalities:

$$\mathbf{A}_k(\tilde{x}, \tilde{y})\boldsymbol{U}_k \geq 0 \tag{3.122}$$

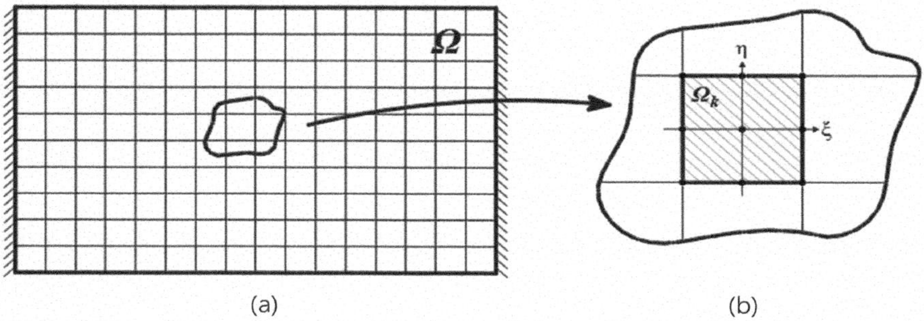

Fig. 3.23 The infinite-dimensional space K_c of continuous displacements with support in Ω is discretized considering a partition $(\Omega_i)_{i\in\{1,2,..,M\}}$ of the whole domain into quadrangular elements (**a**): the finite-dimensional approximation generated by the fixed partition is denote by K_c^N, where N is the number of nodes. In (**b**), a subdomain Ω_k and a second order Lagrangian element with a 9-node is depicted

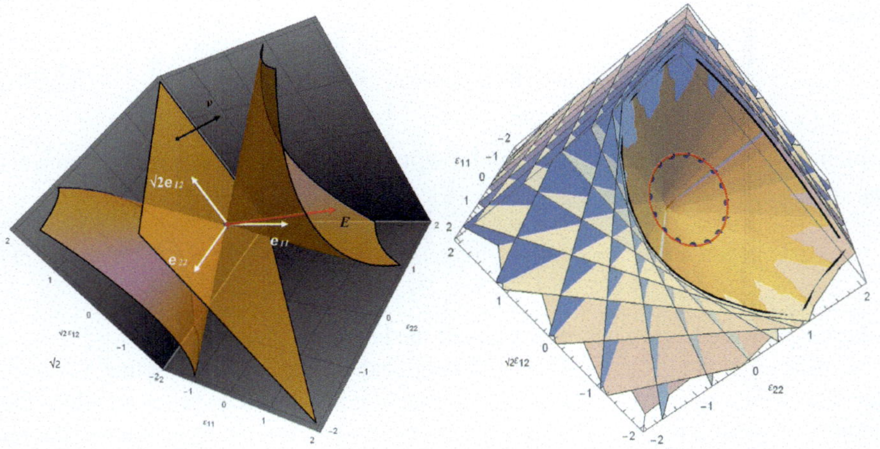

(a) (b)

Fig. 3.24 A3D graphical representation of nonlinear condition $E \in Sym^+$ in the space $(\varepsilon_{11}, \varepsilon_{22}, \sqrt{2}\varepsilon_{12})$. The subspace $E \in Sym^+$ is defined by the intersection of the subset defined by the cone $C : det\boldsymbol{E} \geq 0$ and the half-space defined by the linear inequality $tr\boldsymbol{E} \geq 0$. Specifically, the vector $\boldsymbol{v}$ is orthogonal to the plane $\pi : tr\boldsymbol{E} = 0$ and points toward the half-space $tr\boldsymbol{E} \geq 0$. An admissible latent strain $\boldsymbol{E}$ is represented as a vector. Linearisations of the nonlinear material restriction on $\boldsymbol{E}$ using or 16 tangent planes (**b**). The cross-section of the cone obtained using a plane orthogonal to the cone axis is a circumference where p equally spaced points are selected to "generate" the linear outer envelope

where $A_k(\tilde{x}, \tilde{y})$ is a matrix of p rows and U_k the n=18-vector collecting the nodal displacement components of the element k. Notice that, combining (3.117) and (3.118), condition (3.119) is expressed in terms of nodal displacements. Inequalities (3.122) are written for all Gauss points of the mesh. Once all these relations are collected, the material restriction (3.119) can be expressed on the whole domain as:

$$\boldsymbol{A}\boldsymbol{U} \geq 0 \tag{3.123}$$

The non-homogeneous boundary conditions can be easily expressed as:

$$\boldsymbol{B}_{\mathrm{e}}\boldsymbol{U} = \overline{\boldsymbol{U}}_{\mathrm{e}}, \, \boldsymbol{B}_{\mathrm{i}}\boldsymbol{U} \geq \overline{\boldsymbol{U}}_{\mathrm{i}}. \tag{3.124}$$

where $\boldsymbol{B}_{\mathrm{e}}$ and $\boldsymbol{B}_{\mathrm{i}}$ represent the extractor operators that selects the nodes on the constrained boundary subjected to equality and inequality conditions, respectively; $\overline{\boldsymbol{U}}_{\mathrm{e}}$ and $\overline{\boldsymbol{U}}_{\mathrm{i}}$ are vectors collecting boundary displacements. Looking at the boundary conditions, non-homogeneous conditions are included as non-zero elements in the vectors $\overline{\boldsymbol{U}}_{\mathrm{e}}$ and $\overline{\boldsymbol{U}}_{\mathrm{i}}$. The former looks at the equality constraints as from sliding nodal conditions, while the latter at the inequality relations as from normal unilateral boundary conditions. It is worth noting that the prescribed displacements. Inequalities (3.123) together with boundary conditions (3.124) define the set of admissible displacements:

$$K_{CDF}^{N} = \{U \in R^{2N} / \mathbf{A}U \geq \mathbf{0},\ \boldsymbol{B}_{\mathrm{e}}U = \overline{U}_{\mathrm{e}},\ \boldsymbol{B}_{\mathrm{i}}U \geq \overline{U}_{\mathrm{i}}\} \tag{3.125}$$

that represents the admissibility domain of the optimisation problem, whose objective function is represented by the total potential energy. Therefore, the minimum problem (3.112) can be discretised as:

$$\wp(U_{CDF}^{o}) = \min_{U \in K_{CDF}^{N}} \wp(U) \tag{3.126}$$

that is: "*Find a nodal displacement vector* U_{CDF}^{o} *which minimizes the potential energy* $\wp$ *in* K_{CDF}^{N}". This minimum problem is still a linear programming problem since the energy is linear and the material restrictions are reduced to be linear inequalities. In the present case, once the minimiser U_{CDF}^{o} has been obtained it is possible to construct the deformed shape of the structure, and, as we will show, the best way to graphically represent the fracture field is to plot the corresponding rotation and strain fields. For more information and examples, the reader is referred to the following contributions: Iannuzzo (2017), Iannuzzo et al. (2019, 2021b).

CDF solution of the lintel problem. Here, the lintel problem is analysed using the CDF method, on adopting a discretisation of 200 nine-node square, Lagrangian elements, as depicted in Fig. 3.25. A uniformly distributed vertical load acting along the top edge is the only load considered.

The outward spreading of the supports is modelled as inequalities enforcing Eq. (3.124) on the two vertical edges, whose nodes have vertical displacement forced to be zero through Eq. (3.124).

The solution $\mathbf{U}_{\mathrm{CDF}}^{\mathrm{o}}$ of the minimum problem (3.126) is reached using the interior-point method, and the deformed geometry is depicted in Fig. 3.25. From Fig. 3.25 it can be seen that the lintel deforms showing the formation of two, essentially rigid macroblocks, the deformation being concentrated over the two central vertical strips. This aspect is well-highlighted in Fig. 3.12a, where the contour plot of $|\mathbf{E}| = \mathrm{tr}(\mathbf{E}\mathbf{E}^{\mathrm{T}})^{1/2}$ as a measure of the intensity of the strain field is depicted.

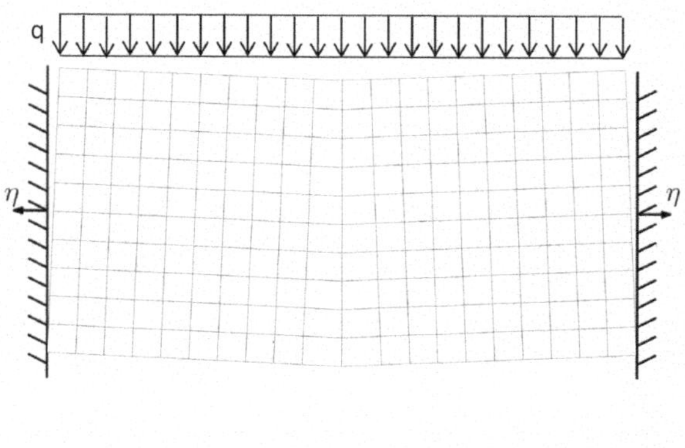

Fig. 3.25 CDF solution of the lintel discretised with 200, 9-node, Lagrangian elements. A uniform load on the top edge is considered in the objective function, while the two horizontal outward displacements are considered as non-homogeneous terms as part of the optimisation constraints

Fig. 3.26 In (**a**): contour plot of a measure of the strain field, i.e., $|\boldsymbol{E}| = tr\left(\boldsymbol{E}\boldsymbol{E}^T\right)^{1/2}$. In (**b**): contour plot of the skew-symmetric part of the displacement field, i.e., of the local rotation field

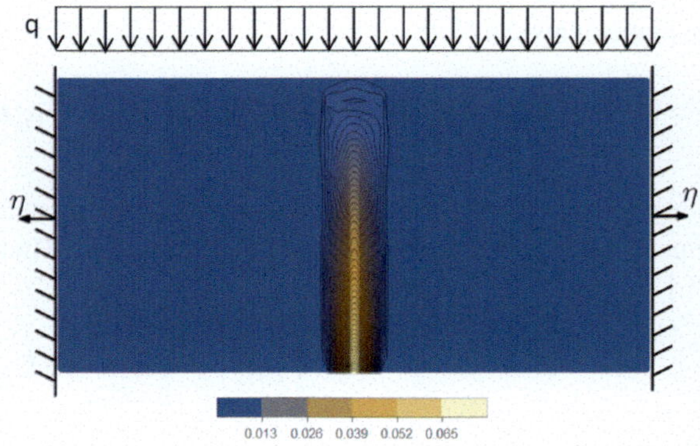

(a)

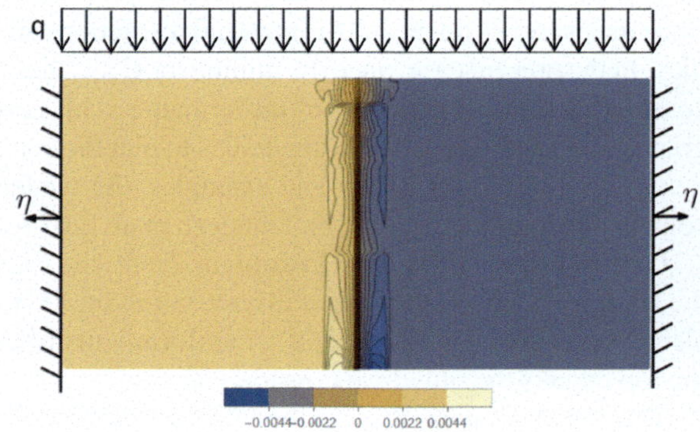

(b)

As the reader can see, the CDF analysis discloses large deformation gradients concentrated on those two central strips, obtaining a result similar to the one shown in Fig. 3.26e. The two rigid macroblocks are also highlighted in Fig. 3.26b, where the skew-symmetric part of the displacement field, i.e., the local rotation field is shown. The rotation is exactly symmetric and essentially piecewise uniform. The uniformity of the rotation field over most of the domain further indicates the smallness of the pure deformation.

3.4.1.2 Discontinuous Approach: PRD Method

This Section illustrates the PRD method. It consists of finding an approximate solution to the problem (3.112) by restricting the minimum search in the set K_{PRD} of piecewise rigid displacements. In particular, let K^M_{PRD} be the set of piecewise rigid displacements associated with a finite partition $(\Omega_i)_{i\in\{1,2,\ldots,M\}}$, of the structural domain Ω into M rigid polygonal elements Ω_i (Fig. 3.27). The boundary $\partial\Omega_i$

of the n-polygon Ω_i, is composed of n segmentsΓ, of length$l$, whose extremities are symbolically denoted 0, 1 and whose unit normal and tangent vectors are $\boldsymbol{n}$, $\boldsymbol{t}$, respectively. We call interfaces the segments Γ that are, either the common boundaries between adjacent elements or part of the constrained boundary (i.e., those Γ representing interfaces with the soil or with other interacting structures). Therefore, the goal is to solve the following discretised minimum-energy problem:

$$E\left(\boldsymbol{U}^{o}_{PRD}\right) = \min_{U \in K^{N}_{PRD}} E(\boldsymbol{U}) \tag{3.127}$$

We can represent a generic piecewise rigid displacement $\boldsymbol{u} \in K^{M}_{PRD}$ in terms of the vector $\boldsymbol{U}$ of $3M$ components representing the rigid body parameters of translation and rotation of the elements. Nevertheless, these parameters have to be constrained to fulfil the assumption that the strain must be positive semidefinite. For piecewise rigid displacements, the strain coincides with its singular part, namely:

$$\boldsymbol{E} = \boldsymbol{E}^{s} = v\delta(\Gamma)\boldsymbol{n} \otimes \boldsymbol{n} + \frac{1}{2} w\delta(\Gamma)(\boldsymbol{t} \otimes \boldsymbol{n} + \boldsymbol{n} \otimes \boldsymbol{t}) \tag{3.128}$$

$\boldsymbol{E}^{s}$ being a strain concentrated along the interfaces Γ. The material restrictions (3.113) can be translated into constraints on the displacement jumps as follows:

$$v = [\boldsymbol{u}] \cdot \boldsymbol{n} \geq 0,\, w = [\boldsymbol{u}] \cdot \boldsymbol{t} = 0 \tag{3.129}$$

Relations (3.129) ensure that tangential jumps (i.e., slidings) are not allowed while only non-negative jumps (i.e. detachments) orthogonal to Γ are admitted. Therefore, the singular strain along the interface takes the form:

$$\boldsymbol{E} = v\delta(\Gamma)\boldsymbol{n} \otimes \boldsymbol{n},\, v \geq 0 \tag{3.130}$$

We point out that with the PRD approximation, (3.130) is the only non-vanishing deformation that the body can exhibit.

Finally, conditions (3.129), stemming from the assumption of normality, represent a condition of unilateral contact with no-sliding among blocks. The static counterpart of these kinematical constraints concerns the stress vector $\boldsymbol{s}$ applied along Γ. Such a stress vector represents the reaction associated with the constraints (3.129). The stress vector $\boldsymbol{s}$ coincides with the given applied tractions $\bar{\mathbf{s}}$ where the boundary of the blocks coincides with the loaded part of the domain boundary. Denoting

$$\sigma = \bar{\boldsymbol{s}} \cdot \boldsymbol{n},\, \tau = \bar{\boldsymbol{s}} \cdot \boldsymbol{t} \tag{3.131}$$

the normal stress σ and tangential stress τ along Γ, the condition on $\boldsymbol{s}$ is

$$\sigma \leq 0 \tag{3.132}$$

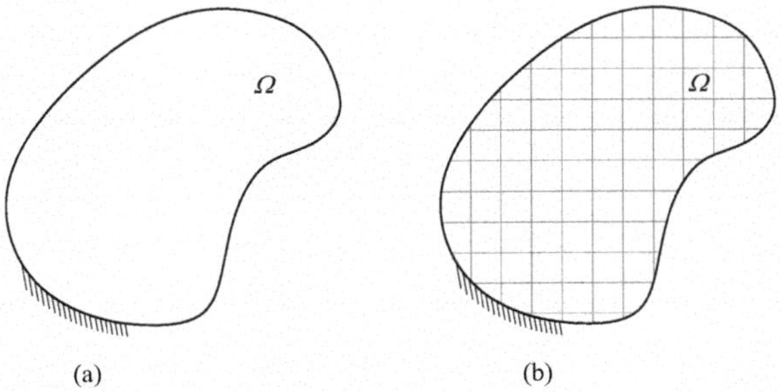

Fig. 3.27 The infinite-dimensional space K_{PRD} of piecewise rigid displacement with support in the domain Ω represented in (**a**), is discretized considering a partition of the whole domain into convex polygonal elements. In (**b**): an example of a partition of Ω with a grid of M square elements is shown. The finite-dimensional approximation generated by this partition is K^M_{PRD}

Notice that the tangential component of $\boldsymbol{s}$ is not constrained and can be applied along the straight interface Γ, even if $\sigma = 0$.

By calling N the total number of the interfaces Γ, and $v(0)$, $v(1)$, $w(0)$, $w(1)$ the normal and tangential components of the relative displacements at the ends 0, 1 of any segment Γ, restrictions $(128)^1$ and $(128)^2$ are equivalent to the $2N$ inequalities

$$v(0) \geq 0, v(1) \geq 0 \tag{3.133}$$

and the $2N$ equalities

$$w(0) = 0, w(1) = 0 \tag{3.134}$$

respectively. Equation (3.134) model the no-sliding condition. Even though shear cracks are important, they are not the most common in masonry structures (Como, 2013; Heyman, 1966). Particularly, when the structure is subjected to foundation displacements, the more frequent are tensile cracks causing the formation of a rigid macroblock partition of the structure and thus entailing relative movements between these parts. Furthermore, good masonry structures are constructed with blocks arranged to avoid shearing, as pointed out by many authors, such as, e.g. Giuffré (1991). Shear cracks of small size are frequent and larger sliding displacements can sometimes be seen locally either because of a bad masonry design or in panels where the crushing strength is locally exceeded.

Restrictions (3.133)–(3.134) can be easily expressed in terms of $\boldsymbol{U}$, in matrix form as:

$$\begin{aligned} \boldsymbol{A}_{ub}\boldsymbol{U} &\geq \boldsymbol{0} \\ \boldsymbol{A}_{eq}\boldsymbol{U} &= \boldsymbol{0}. \end{aligned} \tag{3.135}$$

where $\boldsymbol{A}_{ub}$ collects the unilateral restrictions and $\boldsymbol{A}_{eq}$ the no sliding equations. The boundary conditions can be expressed as:

$$\boldsymbol{A}_{s,ub}\boldsymbol{U} \geq \overline{\boldsymbol{U}}_n,\ \boldsymbol{A}_{s,eq}\boldsymbol{U} = \overline{\boldsymbol{U}}_t \tag{3.136}$$

where matrices $\boldsymbol{A}_{s,ub}$ and $\boldsymbol{A}_{s,eq}$ enforce the prescribed displacements on the boundary. Relations (3.135)–(3.136) define the set K^M_{PRD} of admissible displacement fields. The minimum problem (3.127), which approximates the minimum problem (3.112), is thus transformed into a linear programming problem:

$$E\left(\boldsymbol{U}^o_{PRD}\right) = \min_{U \in K^M_{PRD}} E(\boldsymbol{U}) \tag{3.137}$$

with:

$$K^M_{PRD} = \left\{\boldsymbol{U} \in R^{3M} / \boldsymbol{A}_{ub}\boldsymbol{U} \geq \boldsymbol{0},\ \boldsymbol{A}_{eq}\boldsymbol{U} = \boldsymbol{0},\ \boldsymbol{A}_{s,ub}\mathbf{U} \geq \overline{\boldsymbol{U}}_n,\ \mathbf{A}_{s,eq}\boldsymbol{U} = \overline{\boldsymbol{U}}_t\right\} \tag{3.138}$$

Once the minimizer $\boldsymbol{U}^o_{PRD}$ has been obtained, one can easily construct the deformed configuration of the structure on which the relative displacements among blocks represent fractures. Generally, the moving part is formed by macro-blocks mutually articulated among each other through ideal hinges. Their kinematics can be idealised as a kinematical chain controlled by the given settlements, the moving structure being statically determined.

The approximate minimization problem (3.137) transforms the original continuum minimum problem (3.112) into a minimum problem for a structure composed of rigid parts subjected to given loads, given settlements and unilateral contact conditions along the interfaces. Problem (3.137) is a standard linear finite-dimensional minimisation problem, since $E(\boldsymbol{U})$ is a linear function of the $3M$ dimensional vector $\boldsymbol{U}$ and the constraints (3.135)–(3.136) are linear relations.

Such a linear minimization problem is usually formalized as a problem of Linear Programming, as follows:

$$\begin{aligned} &\text{Minimise } -\boldsymbol{c}\cdot\boldsymbol{U} \\ &\text{s.t } A_{,-1}U > -\delta_n \\ &\boldsymbol{A}_{eq}\boldsymbol{U} = -\boldsymbol{\delta}_t \end{aligned} \tag{3.139}$$

The existence of the solution to this approximated problem is trivially guaranteed if the original problem is bounded from below. For a small number of variables, it can be solved precisely with the simplex method. For large problems, several well-known and efficient, approximate fast alternatives exist based, for example, on the interior-point algorithm Fig. 3.28.

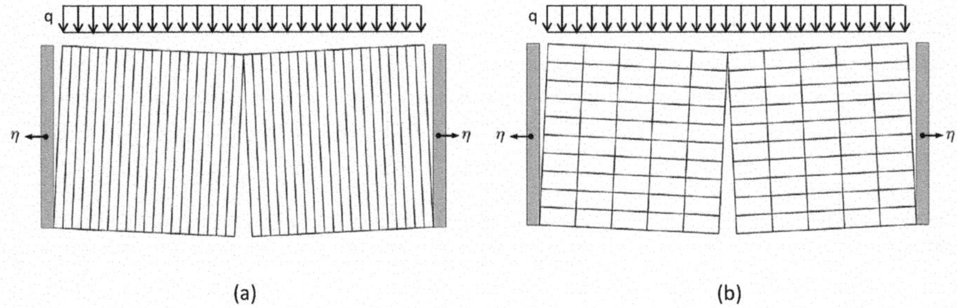

Fig. 3.28 PRD solutions adopting two different discretisation of the lintel. The lintel articulates into two rigid macroblocks

It is worth pointing out that K_{PRD} is a particular subset of the space SBV, that is of the Special Bounded Variation functions (Ambrosio et al., 2000). SBV functions can exhibit jumps on 1d sets, such as a finite collection of regular curves. The corresponding strain is singular and is concentrated on the jump set in the form of line Dirac deltas. In the PRD approximation, these Dirac deltas live over the skeleton of the mesh. Therefore, the crack pattern is identified by the set composed by the interfaces having non-zero singular strains, that is, by gap "openings" between adjacent rigid blocks in the deformed configuration. For more information and examples, the reader is referred to the following contributions: Iannuzzo (2017), Iannuzzo et al. (2018, 2020, 2021a).

PRD solution of the lintel problem. Here, the lintel problem is analysed with the PRD method adopting two different discretisations, as shown in Fig. 3.28. In both cases, the lintel structural domain articulates into two rigid macroblocks preserving the location of the hinges. As the only load considered is the line load q, the discretisation of Fig. 3.14b returns the same solution of Fig. 3.28a.

3.4.2 Stress-Based Approaches: Discretisation of the Total Complementary Energy

This Section illustrates two approaches capable of directly defining statically admissible stress states that may develop inside masonry structures due to imposed loads and given boundary displacements. These numerical methods stem from the variational formulation of minimum complementary energy. Similarly to Sect. 3.4.1, two numerical strategies are illustrated to discretise the minimum problem. Specifically, the first is a pure continuous approach, for which the stresses are continuously distributed inside the masonry domain; the second is a discrete approach, where the stresses are modelled as singular stress on the interfaces of the blocks in which the continuous domain is discretised.

The first method, namely the Continuous Airy-based Stress for Singularities (CASS) (Montanino et al., 2022), is summarised in Sect. 3.4.2.1: in this case, the stress state does not admit strong singularities, and possible singularities in the exact solution are approximated through narrow stress bands that develop inside the elements of the discretisation, in much the same way phase fields approximate cracks. It is to be pointed out that such narrow bands, which approximate the support of the singularities, are mesh independent.

The second approach is the dual-PRD method, which represents an application of the principle of duality of the primal PRD method, seen as a linear programming problem. In this case, the contact forces among elements can be directly obtained as the Lagrange multipliers of the Lagrangian formulation's minimum problem of Total Potential Energy when minimization is restricted to the set K^M_{PRD}.

3.4.2.1 Continuous Approach: CASS Method

The Continuous Airy-based Stress for Singularities (CASS) method (Montanino et al., 2022) looks for equilibrated solutions through the principle of minimum Total Complementary Energy. The problem is reformulated in terms of the Airy stress function F defined in (3.104), to enforce the internal equilibrium automatically.

With the CASS method, the domain Ω is discretised into a finite number of quadrangular plate-like C^1-finite elements (Bathe & Wilson, 1976), as depicted in Fig. 3.29.

To each node of the element, three different nodal variables are associated, namely F^i, F^i_x, and F^i_y, with $i = 1, \ldots, 4$, hence, a 12-term polynomial is needed to approximate the Airy stress function over each element, i.e.,

$$\begin{aligned} F =& a_0 + a_1 x + a_2 y + a_3 x^2 + a_4 xy + a_5 y^2 \\ &+ a_6 x^3 + a_7 x^2 y + a_8 xy^2 + a_9 y^3 + a_{10} x^3 y + a_{11} xy^3 \end{aligned} \tag{3.140}$$

The coefficients $a_0, \ldots, a_{11}$ are found by imposing the nodal conditions

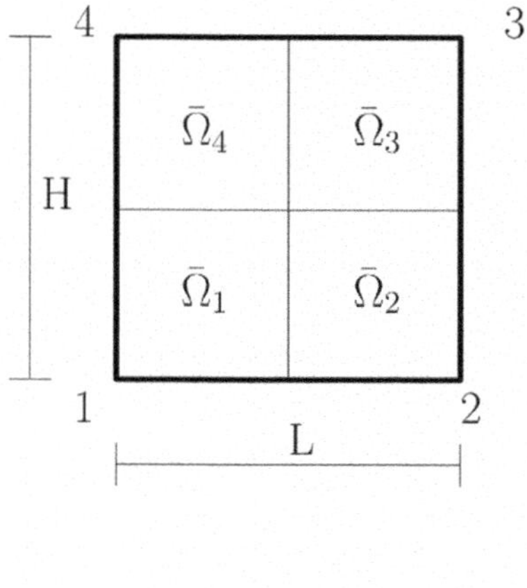

Fig. 3.29 The plate-like finite element adopted in the CASS method

$$F|_{x_i,y_i} = F^i,\ \frac{\partial F}{\partial x}|_{x_i,y_i} = F_x^i,\ \frac{\partial F}{\partial y}|_{x_i,y_i} = F_y^i,\ i = 1,..,4 \tag{3.141}$$

We remark that interpolation (3.141) is C^1-continuous, leading to two important consequences: the first is that possible stress singularities, expected on masonry structures undergoing settlements, cannot be reproduced sharply since the adopted continuity is not able to model folds in the Airy stress potential; secondly, the stress components, obtained as

$$\begin{aligned}
T_{xx}(x,y) &= \frac{\partial^2}{\partial y^2} F(x, y; F^1, F_x^1, F_y^1, \ldots, F^4, F_x^4, F_y^4),\\
T_{yy}(x,y) &= \frac{\partial^2}{\partial x^2} F(x, y; F^1, F_x^1, F_y^1, \ldots, F^4, F_x^4, F_y^4),\\
T_{xy}(x,y) &= -\frac{\partial^2}{\partial x \partial y} F(x, y; F^1, F_x^1, F_y^1, \ldots, F^4, F_x^4, F_y^4),
\end{aligned} \tag{3.142}$$

are not continuous across the element boundary. This implies that the conditions on the stress admissibility cannot be enforced on the element nodes but must be imposed at the element's interior. Thus, we divide each element into four subdomains $\overline{\Omega}_i$, as shown in Fig. 3.29, and compute the averaged stress as

$$\boldsymbol{T}^{\bar{i}} = \frac{4}{HL} \int_{\Omega_i} \boldsymbol{T} d\Omega \tag{3.143}$$

Equation (3.142) can be introduced in (3.143) and reported in matrix form as

$$\begin{aligned}
\overline{T}_{xx} &= K_{xx}\hat{F},\\
\overline{T}_{yy} &= K_{yy}\hat{F},\\
\overline{T}_{xy} &= K_{xy}\hat{F},
\end{aligned} \tag{3.144}$$

where $\boldsymbol{K}_{xx}$, $\boldsymbol{K}_{yy}$, and $\boldsymbol{K}_{xy}$ are matrices depending only on the finite element geometry, and $\widehat{\boldsymbol{F}}$ is the vector of the nodal degrees of freedom. The conditions (3.109) are therefore transformed in the discrete version

$$\begin{aligned}
&(\boldsymbol{K}_{xx} + \boldsymbol{K}_{yy})\widehat{\boldsymbol{F}} \le 0,\\
&\left(\boldsymbol{K}_{xx}\widehat{\boldsymbol{F}}\right) \odot \left(\boldsymbol{K}_{yy}\widehat{\boldsymbol{F}}\right) - \left(\boldsymbol{K}_{xy}\widehat{\boldsymbol{F}}\right) \odot \left(\boldsymbol{K}_{xy}\widehat{\boldsymbol{F}}\right) \ge \boldsymbol{0}
\end{aligned} \tag{3.145}$$

where $\odot$ denotes the Hadamard product, defined for two vectors as $(A \odot B)_i = A_i B_i$

The boundary conditions (3.105), restricting the values of the Airy stress function and its derivatives, can be imposed directly on the Neumann part of the boundary and translated in discrete form as

$$\boldsymbol{K}_{\partial\Omega}\widehat{\boldsymbol{F}} = \boldsymbol{b}_{\partial\Omega} \tag{3.146}$$

The last step is the discretisation of the Total Complementary Energy (3.96), which is made based on the stress approximation (3.144). Expanding (3.96), we get

$$\begin{aligned} E_c(T) = &-\int\limits_{\partial\Omega_D} \left[\delta_x n_x T_{xx} + \delta_y n_y T_{yy} + (\delta_y n_x + \right. \\ &\left. \delta_x n_y) T_{xy}\right] ds + \int\limits_{\Omega} (T_{xx} + T_{yy})\bar{\mathrm{E}} da, \end{aligned} \tag{3.147}$$

where δ_x, δ_y are the components of the given settlements $\overline{\mathbf{u}}$ on the constrained boundary $\partial\Omega_D$, n_x, n_y are the components of the outward normal $\boldsymbol{n}$, and $\bar{\mathrm{E}}$ is the magnitude of the eigenstrain $\overline{\mathbf{E}}$, on assuming $\overline{\mathbf{E}} = \bar{\mathrm{E}}\mathbf{I}$.

Introducing (3.144) in (3.147), we obtain

$$\begin{aligned} E_c\left(\widehat{\boldsymbol{F}}\right) = &-\sum_{i\in\partial\Omega_D}\sum_{j=1}^{N}\left[\delta_x^i n_x^i K_{xx}^{ij} + \delta_y^i n_y^i K_{yy}^{ij} + (\delta_y^i n_x^i + \delta_x^i n_y^i) K_{xy}^{ij}\right] l_i \hat{F}_j \\ &+ \sum_{i\in\Omega}\sum_{j=1}^{N}(K_{xx}^{ij} + K_{yy}^{ij})\bar{E}_i A_i \hat{F}_j \end{aligned} \tag{3.148}$$

where l_i is the length of the element side along the constrained boundary, A_i is the area of each subdomain $\overline{\Omega}_i$ in the domain Ω.

Expression (3.147) can be rewritten in linear form as

$$E_c\left(\widehat{\boldsymbol{F}}\right) = \boldsymbol{g}^T \widehat{\boldsymbol{F}} \tag{3.149}$$

being $\boldsymbol{g}$ the gradient of $E_c\left(\widehat{\boldsymbol{F}}\right)$, that can be expressed in components as

$$\begin{aligned} g_j = &-\sum_{i\in\partial\Omega_D}\left[\delta_x^i n_x^i K_{xx}^{ij} + \delta_y^i n_y^i K_{yy}^{ij} + (\delta_y^i n_x^i + \delta_x^i n_y^i) K_{xy}^{ij}\right]_{l_i} \\ &+ \sum_{i\in\Omega}(K_{xx}^{ij} + K_{yy}^{ij})\bar{E}_i A_i \end{aligned} \tag{3.150}$$

The discrete version of the minimization problem (3.111) is then

$$E_c\left(\widehat{\boldsymbol{F}}^{\circ}\right) = \min_{\widehat{\boldsymbol{F}} \in H_{\widehat{\boldsymbol{F}}}} \boldsymbol{g}^T \widehat{\boldsymbol{F}} \tag{3.151}$$

being

$$\begin{aligned} H_{\widehat{\mathbf{F}}} = \Big\{ & \widehat{\boldsymbol{F}} \in R^N \text{ s.t.} \left(\boldsymbol{K}_{xx} + \boldsymbol{K}_{yy}\right)\widehat{\mathbf{F}} \le \mathbf{0}, \\ & \left(\boldsymbol{K}_{xx}\widehat{\boldsymbol{F}}\right) \odot \left(\boldsymbol{K}_{yy}\widehat{\boldsymbol{F}}\right) - \left(\boldsymbol{K}_{xy}\widehat{\mathbf{F}}\right) \odot \left(\boldsymbol{K}_{xy}\widehat{\mathbf{F}}\right) \ge \mathbf{0} \\ & \boldsymbol{K}_{\partial\Omega}\widehat{\boldsymbol{F}} = \boldsymbol{b}_{\partial\Omega} \Big\}, \end{aligned} \tag{3.152}$$

the discrete version of the set of the admissible Airy stress potentials (3.110).

CASS solution of the lintel problem. Here we solve the lintel problem, presented in Fig. 3.22, with the CASS method. The boundary conditions, in terms of the contact moment $m(s)$ and of the axial contact force $n(s)$ (see Sect. 3.2.1), are reported in Fig. 3.30.

The domain is discretised into 120×240 square elements, achieving a total number of 87,483 degrees of freedom. The unilateral constraints on the stress (3.9) are represented by 115,200 cones type Eq. (3.145). While, the boundary conditions are imposed on 478 nodes, resulting in the linear system (3.146) composed of 1,434 equations.

Figure 3.31 shows the solution in terms of the Airy stress potential, while in Fig. 3.32, the corresponding principal stress field is represented. In particular, on the top edge of Fig. 3.32, a stress concentration that resembles the one depicted in Fig. 3.32c appear. Actually, the numerical solution obtained with the CASS method appears as a close approximation of a linear convex combination of the solutions represented graphically in Fig. 3.19a, c, closer to the one of Fig. 3.18c rather than to that of Fig. 3.32b. However, as expected, the energy level corresponding to the three solutions is the same.

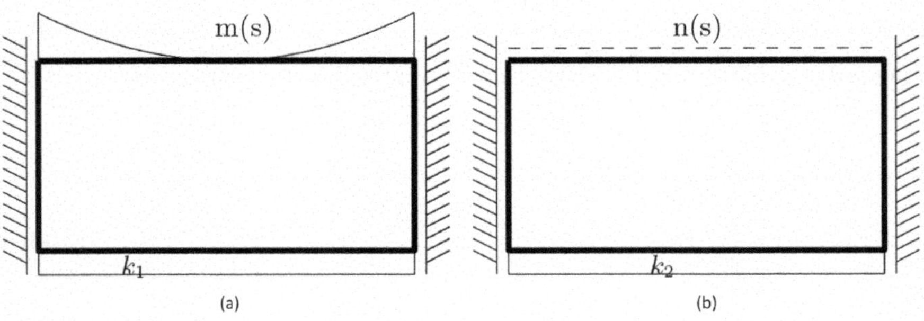

Fig. 3.30 Boundary conditions for solving the lintel problem through the CASS method. k_1 and k_2 are two unknown constants

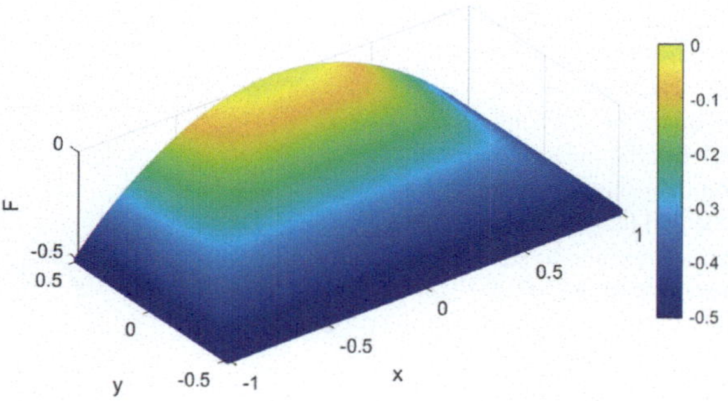

Fig. 3.31 Airy stress function resulting from the solution of minimisation problem (3.151)–(3.152) for the lintel of Fig. 3.22

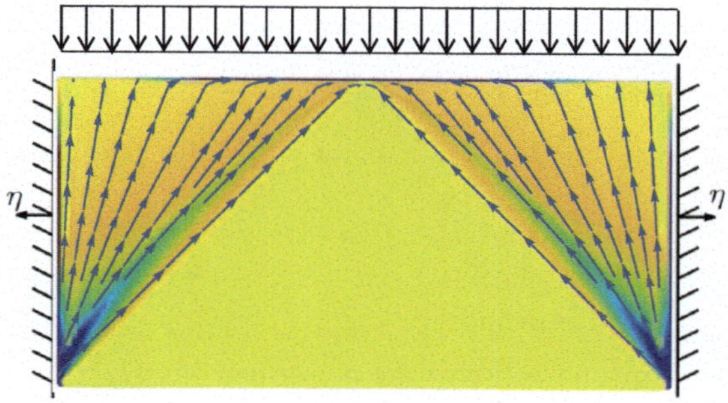

Fig. 3.32 Principal stress and isostatic lines for the problem of Fig. 3.8

3.4.2.2 Discontinuous Approach: Dual PRD Method, Rigid Blocks

This section illustrates the discontinuous approach based on the minimum of the total complementary energy. The DPRD approach is proposed referring to 3D structures in Iannuzzo and et al. (2020, 2021a), and used to benchmark the CASS solutions in Montanino et al. (2022). A 3D application on a spiral stair can be found in Olivieri et al. (2022).

It will be described by restricting the space of admissible stress fields to the convex Mohr–Coulomb cone $\mathbb{C} \in \mathrm{Sym}^-$ to consider an associated flow rule and a finite friction capacity. It is worth pointing out that in the case of infinite friction, the following discussion still holds, and it is the dual formulation of the PRD approach.

A 3D continuum masonry structure $\Omega \subset R^3$ is partitioned into a finite number of elements Fig. 3.33:

$$(\Omega_i)_{i\in\{1,\ldots,M\}} \tag{3.153}$$

The contact among the elements is unilateral and follows a Mohr–Coulomb friction-associated flow rule. The two open, disjoint parts$\partial\Omega_U$, $\partial\Omega_F$ represent the constrained and loaded part of the boundary ofΩ; while $\partial\Omega_I$ is the union of $\partial\Omega_{I_i}$ for$i \in \{1, \ldots, M\}$, i.e., the union of the interfaces shared by two adjacent elements of the partition (3.153). Prescribed displacements $\overline{\boldsymbol{u}}$ are applied on $\partial\Omega_U$:

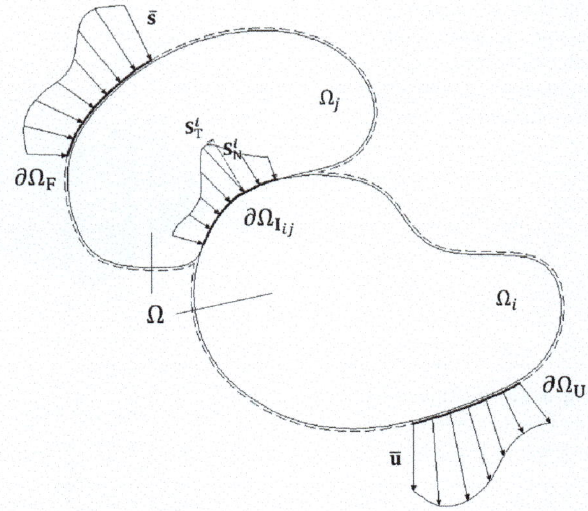

Fig. 3.33 An elastic body Ω subjected to body forces b, prescribed boundary displacements $\overline{u}$ on $\partial\Omega_U$, friction boundary condition on $\partial\Omega_I$ with given normal S_N and tangential S_T pressures, and to surface forces $\overline{s}$ on the $\partial\Omega_F$

$$\boldsymbol{u} = \overline{\boldsymbol{u}} \tag{3.154}$$

while given tractions $\overline{s}$ are prescribed on $\partial\Omega_F$:

$$\boldsymbol{s}_n(\boldsymbol{T}) = \overline{\boldsymbol{s}}, \tag{3.155}$$

where $\boldsymbol{s}_n(\boldsymbol{T})$ is the trace of the stress tensor $\boldsymbol{T}$ along the boundary, whose unit outward normal is $\boldsymbol{n}$. If $\boldsymbol{T}$ is regular, $\boldsymbol{s}_n(\boldsymbol{T}) = \boldsymbol{T}\boldsymbol{n}$. For more info on how to model the trace in the case of singular stresses, the reader is referred to (Angelillo et al., 2013, 2014; Fortunato Fabbrocino et al., 2018). Equilibrium also holds on internal interfaces. The space of admissible stresses and displacements must account for singular stresses and displacement jumps, both represented by Dirac delta distributions whose support is $\partial\Omega_I$. Indeed, to model the unilateral behaviour on the internal interfaces, it is required that the displacement field **u** can be discontinuous on each interface $\partial\Omega_i$, i.e., on $\partial\Omega_i$ **u** can have a finite jump Δu^i that can be decomposed in its normal $\Delta u_N^i = [v]$ and tangential $\Delta u_T^i = [w]$ components (see Sect. 3.4.1.2 and (Angelillo, 2014; Iannuzzo et al., 2018)). Therefore, it is assumed that an admissible displacement field belongs to

$$K = \{\boldsymbol{u} \in SBV(\Omega_i)\, \text{for}\, i = 1, .., n / \boldsymbol{u} = \overline{\boldsymbol{u}}\, \text{on}\, \partial\Omega_U\} \tag{3.156}$$

SBV being the set of special bounded variation functions (Ambrosio et al., 2004). Denoting with $|\Delta \boldsymbol{u}_T^i|$, $|\boldsymbol{S}_T^i|$ the $L^2(\partial\Omega_i)$-norm of the displacement jump and of the stress vector in the tangential direction **t**, and $|\boldsymbol{S}_N^i|$ the $L^2(\partial\Omega_i)$-norm of the stress vector in the normal direction **n**, the friction condition, on assuming an associated behaviour, reads:

$$\text{if}\left|\boldsymbol{S}_T^i\right| \leq \mu\left|\boldsymbol{S}_N^i\right|\text{then}\left|\Delta\boldsymbol{u}_T^i\right| = 0 \tag{3.157}$$

$$\begin{aligned} &\text{if } \left|\boldsymbol{S}_T^i\right| = \mu\left|\boldsymbol{S}_N^i\right|, \text{ there exists } \lambda \geq 0 \text{ s.t.} \Delta\boldsymbol{u}_T^i = -\lambda\boldsymbol{S}_T^i \text{ and} \\ &\Delta\boldsymbol{u}_N^i = -\lambda\mu\boldsymbol{S}_T^i, \end{aligned} \tag{3.158}$$

where μ is the friction coefficient, and $\boldsymbol{S}_N^i$ and $\boldsymbol{S}_T^i$ are the unknown, normal and tangential stress vector distributions on Γ_I. The stress tensor is symmetric and fulfils the following requirement:

$$\boldsymbol{T} \in \mathbb{C} \tag{3.159}$$

with $\mathbb{C}$ is the convex Mohr–Coulomb cone. To allow for singular stresses, it is assumed that $\boldsymbol{T}$ belongs to the special bounded measures (SBM) (Ambrosio et al., 2004; Angelillo, 2014; Lucchesi et al., 2006). Therefore, the set of admissible stress fields is:

$$H = \left\{\boldsymbol{T} \in SBM \text{s.t.} div\boldsymbol{T} + \boldsymbol{b} = 0, \boldsymbol{T} \in \mathbb{C}, \boldsymbol{s}(\boldsymbol{T}) = \overline{s} \text{ on } \partial\Omega_F\right\} \tag{3.160}$$

A solution of the boundary value problem (BVP) is thus represented by a stress $\boldsymbol{T} \in H$ and a displacement $\boldsymbol{u} \in K$ fulfilling Eqs. (3.157)–(3.158) also. The Gauss–Green theorem for an elastic material reads:

$$\begin{aligned} a\left(\boldsymbol{u}, \boldsymbol{u} - \boldsymbol{u}^*\right) = &\int_\Omega \boldsymbol{b} \cdot \left(\boldsymbol{u} - \boldsymbol{u}^*\right) dv + \int_{\partial\Omega_F} \overline{s} \cdot \left(\boldsymbol{u} - \boldsymbol{u}^*\right) da \\ &+ \sum_{i=1}^{M} \int_{\partial\Omega_I} \left[\boldsymbol{S}_N^i \cdot \left(\Delta\boldsymbol{u}_N^i - \Delta\boldsymbol{u}_N^{i,*}\right) + \boldsymbol{S}_T^i \cdot \left(\Delta\boldsymbol{u}_T^i - \Delta\boldsymbol{u}_T^{i,*}\right)\right] da \end{aligned} \tag{3.161}$$

with $a(\boldsymbol{u}^*, \boldsymbol{u})$ a quadratic functional doubling the elastic energy and $\boldsymbol{u}^*$ the displacement solving the BVP. It can be easily shown that for the problem at hand, the following variational criterion can be derived:

$$A\left(\boldsymbol{T}^*, \boldsymbol{T} - \boldsymbol{T}^*\right) \geq \int_{\partial\Omega_U} \overline{\boldsymbol{u}} \cdot \left(\boldsymbol{T}\boldsymbol{n} - \boldsymbol{T}^*\boldsymbol{n}\right) da, \forall \boldsymbol{T} \in H_1 \tag{3.162}$$

where A doubles the complementary elastic energy, and $\boldsymbol{T}^*$ is the stress tensor solving the BVP. Relation (3.162) is equivalent to the following minimisation problem:

$$T^* = \arg\min\{\Pi_c(T)_{T\in H_1}\} \tag{3.163}$$

With

$$\Pi_c(T) = \frac{1}{2}A(T, T) - \int_{\partial\Omega_U} \overline{u} \cdot Tnda \tag{3.164}$$

where $\Pi_c(T)$ the total complementary energy (TCE). In the case of an associated flow rule, the dual problem is represented by the total potential energy.

DPRD solution of the lintel problem. This section illustrates how the variational criterion based on the total complementary energy can select internal stress states compatible with given boundary displacements and external loads. In particular, it will be illustrated referring to four different discretisations grouped in two main cases. The first group looks at Heymanian solutions of the problems already solved with the PRD approach in Sect. 3.4.1.2. The second group collects examples where a finite friction capacity is considered. In all cases, displacements are retrieved from the dual solutions of the equilibrium problem. Interface forces are then mapped on the deformed configurations.

Figure 3.34 illustrates the TCE solutions of the two problems solved with the PRD method in Sect. 3.4.1.2. Solutions are obtained considering an infinite friction capacity which forces the sliding among blocks to be zero. Figure 3.34a, b depict the interface normal and tangential forces obtained from the nodal forces solving the variational problem. Figure 3.34a, b summarises the internal stress state through the interface resultants that combine the previous interface reactions. The dual solution provides displacements used to retrieve the deformed configurations, which results identical to the ones directly obtained with the TPE minimum.

In both cases, the lintel nucleates into two rigid bodies hinged in three points. The interface resultants of Fig. 3.34c show a typical trend of a thrust line, which can be achieved by further increasing the number of vertical cuts. Conversely, the discretisation of Fig. 3.34d entails a different internal flow of forces necessary to keep all the blocks together to form a three-hinge mechanism.

Figure 3.36 illustrates the solutions to the same mechanical problem using an interlocked discretisation and considering a finite friction capacity. Remarkably, the contact between the lintel and external supports is modelled in two ways, as illustrated in Fig. 3.35. The model of Fig. 3.35a assumes an infinite friction capacity over the interfaces among supports and lintel; conversely, the model of Fig. 3.35a considers two L-shape supports whose contact with the blocks is modelled with the same friction coefficient used over all the internal lintel's interfaces. Both models are analysed considering two different loading conditions: a uniform constant line load applied on the top edge, as shown Fig. 3.35a, b; and a uniform self-weight lumped and applied to each block's centroid. The value of the load q is such that to lump the lintel self-weight over the top edge. As the reader may note, in both cases with the linear load q, the lintel nucleates in a mechanism defined by three rigid macroblocks: two external triangles forming a three-hinge mechanism, while the middle pyramid

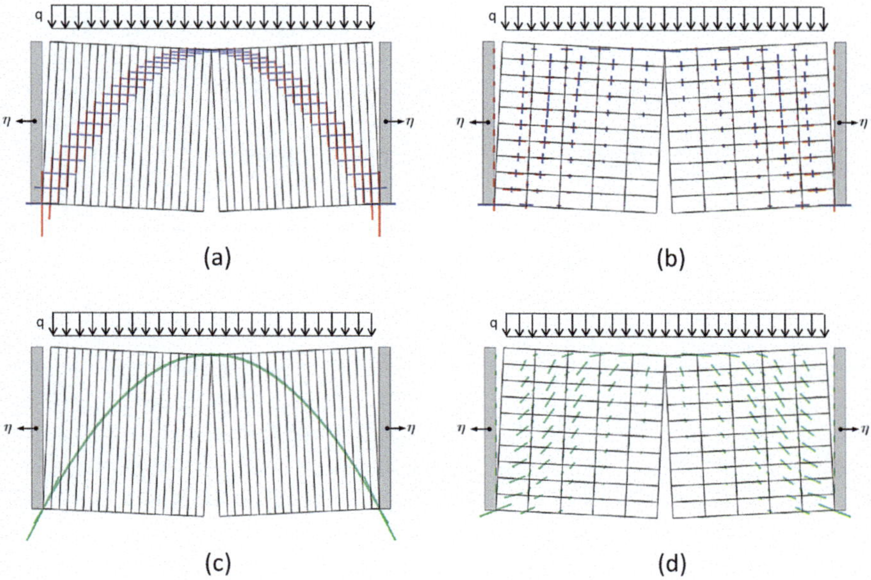

Fig. 3.34 Solution of the minimum of the complementary energy applied to the discretisations solved with the PRD method in Sect. 3.4.1.2. The first row depicts the interface normal and tangential forces obtained from the nodal forces solving the TCE minimum. The second row shows the interface resultants obtained by combining the previous interface reactions. Additionally, forces are displayed on the deformed configurations, which are retrieved directly from the values of the dual problem. Displacements are identical to the PRD ones, as TPE and TCE are dually connected

triangle displaces down. In the case of self-weight, the mechanism is more spread all over the lintel, but it still shows the three-hinge mechanism.

Notably, in the case of Fig. 3.36d, dilatancy effects can be seen and are mainly due to the supports shape and the diffuse sliding among blocks. Referring to the first column analyses of Fig. 3.36, it is worth noting that a sufficiently high friction value (say greater than 50°) would be sufficient to prevent the uniform sliding down of

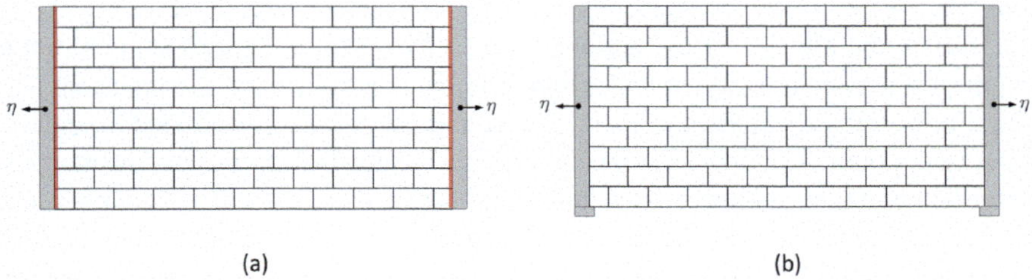

Fig. 3.35 The interlocked discretisation of the lintel with two strategies to avoid sliding among lintel and supports: an infinite friction capacity along the red lines (**a**) or two L-shape supports (**b**)

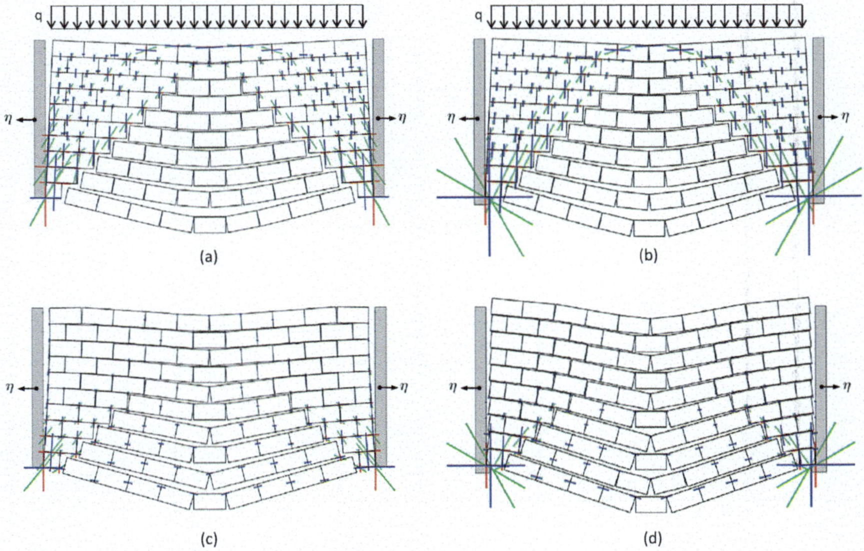

Fig. 3.36 Solution of the TCE minimum considering a finite friction capacity among the lintel's blocks. Notably, in the first column, infinite friction between blocks and the external supports is considered to prevent a uniform sliding down of the lintel; in the second row, the supports have an L-shape. Lastly, the first row shows the solutions when only a uniform load is applied, while in the second one, only a self-weight of the block is considered

the lintel, as the reader can recognise looking at the interface resultant inclinations (green lines).

3.5 Application to a Practical Case Study

The numerical methods shown in the previous section are here applied to the case study of an unsymmetric masonry façade under different static and kinematic conditions. The geometry of the case study, with its dimensions, is provided in Fig. 3.37. The load on the top side of the masonry is $q = 25$ kN/m, the body load is $b = 18$ kN/m^2. The façade is analysed under two different conditions: a settlement analysis with a uniform vertical displacement η affecting part of the constrained boundary in between points A and B, as shown in Fig. 3.37; a load-bearing capacity analysis performed applying separately two horizontal forces F = 200 kN to the upper left and right corners, which are then increased through two scalar load multipliers λ_1 and λ_2 till the structural collapse. Deformed configuration, internal stress state, fracture and strain obtained through the proposed energy methods are shown and discussed.

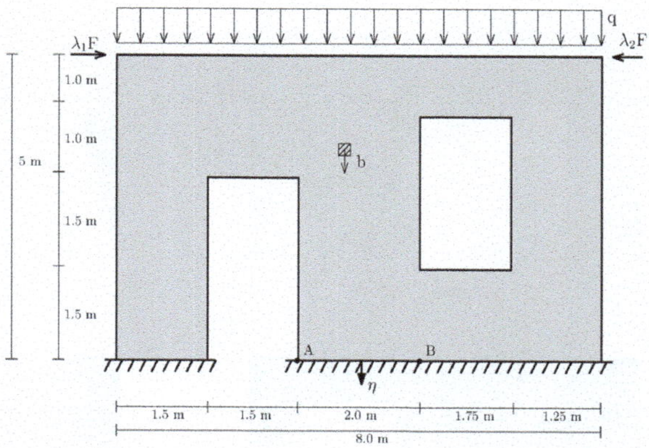

Fig. 3.37 Irregular 2D masonry façade: geometry, loads and boundary conditions adopted for the settlement and load-bearing capacity analyses

3.5.1 *Effects of the Settlement*

In this section, the effects of a settlement η affecting the central baseline part of the facade are studied using displacement- and force-based approaches. Figure 3.38 shows the deformed configurations obtained with the CDF method and with an interlocked discretisation using the DPRD analysis. The same mechanism involving the rotation of a triangular part of the masonry façade can be seen. The interface forces' flow also shows how the internal stress state changes and adapts to the new boundary conditions.

Figure 3.39 shows the results obtained with the PRD, CASS and CDF methods. In particular, a comparison between Fig. 3.39a, b shows the congruency of the force paths obtained with the DPRD and CASS methods, respectively. It can be seen that some stress bands perfectly overlap the interface resultants. Figure 3.39c, d shows the norm of the symmetric and non-symmetric part of the displacement gradient

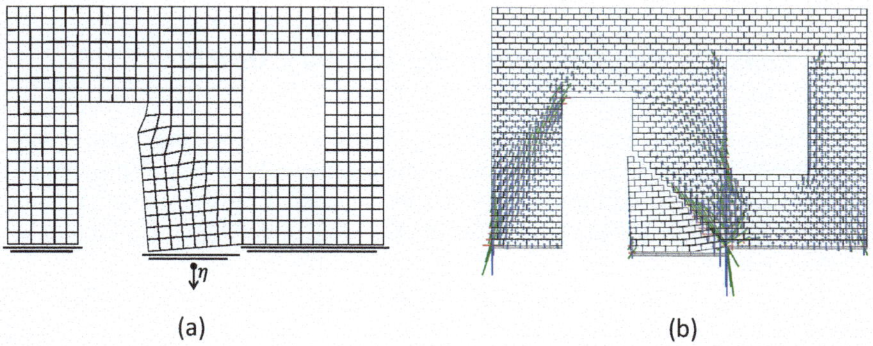

Fig. 3.38 Settlement analysis: **a** CDF deformed configuration; **b** deformed configuration and interface contact forces obtained with a DPRD analysis on an interlocked discretisation; blue, red and green lines depict normal, tangential and resultant interface forces, respectively

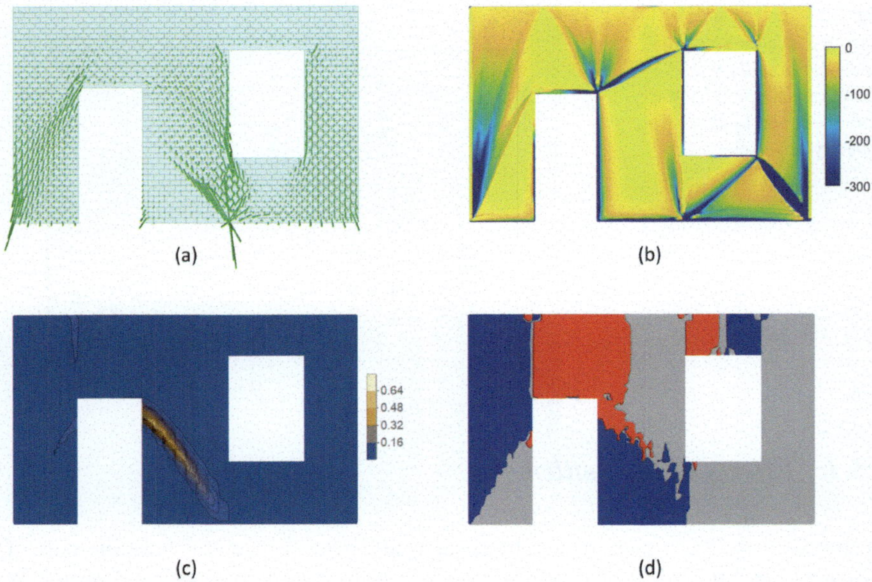

Fig. 3.39 Settlement analysis obtained with DPRD, CASS and CDF methods. **a** DPRD interface resultants displayed on the undeformed configuration; **b** CASS minimum principal stress distribution; **c** norm of the diffuse deformation obtained with the CDF method; **d** rigid macroblock identification obtained depicting the CDF rotation field (blue and red denote counter clockwise and clockwise rotations, respectively)

obtained with the CDF method. The analysis of Fig. 3.39c, d reveals that the preferential deformation pattern of this structure is the development of concentrated deformation, referred to as fractures, and the creation of rigid macroblocks, characterised by uniform in-sign rotation values, as depicted in Fig. 3.39d.

3.5.2 Load Bearing Capacity of the Masonry Structure

Here, all different approaches are applied to perform two load-bearing capacity analyses of the façade. The façade is subjected to homogeneous boundary displacements, while self-weight and uniform loads are still considered. Two loading cases are considered, where an additional horizontal force is separately applied to the upper left (Figs. 3.40 and 3.41) and right (Figs. 3.42 and 3.43) corners.

The force values are increased till structural collapse is achieved. The solution in terms of displacements, deformation, internal stress states are then shown, referring to the onset of motion. In DPRD analyses, two additional (grey) blocks to engage the lintel response are considered to apply the horizontal loads to the two upper corners.

In both cases, static and kinematic approaches are used to find the maximum load multipliers. The CASS and DPRD approaches are implemented in mathematical

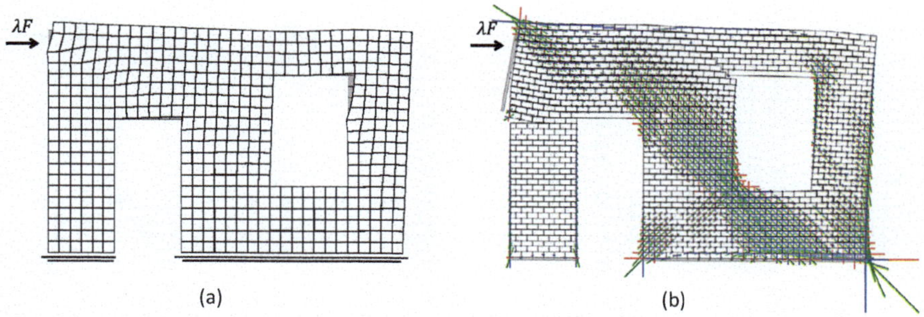

Fig. 3.40 Load bearing capacity of a masonry façade under an increasing horizontal load applied to the left upper corner till the structural collapse: **a** CDF deformed configuration corresponding to $\lambda_1^{CDF} = 205$; **b** DPRD deformed configuration and interface forces path corresponding to $\lambda_1^{DPRD} = 188$ that actives the mechanism

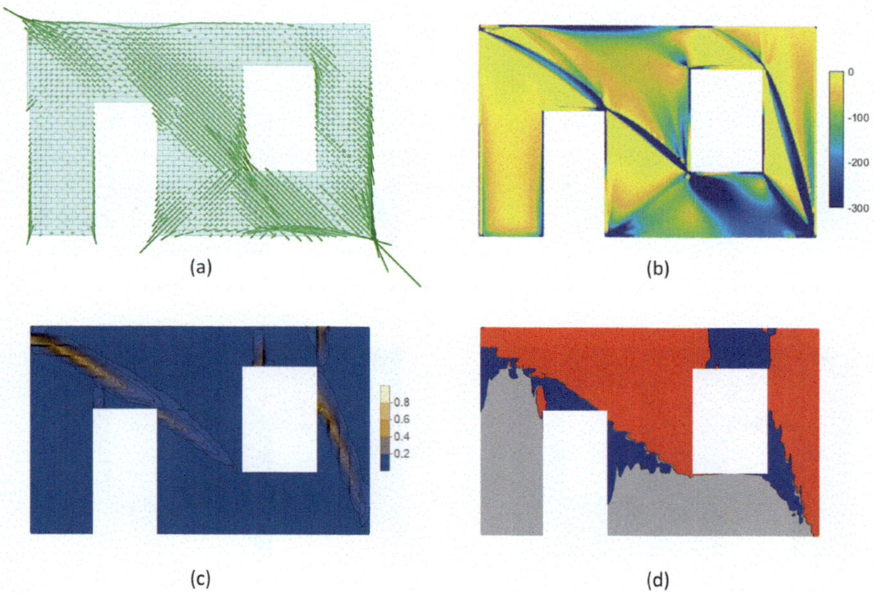

Fig. 3.41 Load bearing capacity of a masonry façade under an increasing horizontal load applied to the left upper corner and driving the structure to the collapse: **a** DPRD interface resultants on the undeformed configuration; **b** CASS minimum principal stress distribution; **c** CDF norm of the deformation and **d** CDF macroblock identification

programming problems that find the ultimate statically admissible solution for which equilibrium is still achieved. With the CDF energy method, the load is increased until the potential energy is no longer bounded from (Fig. 3.41).

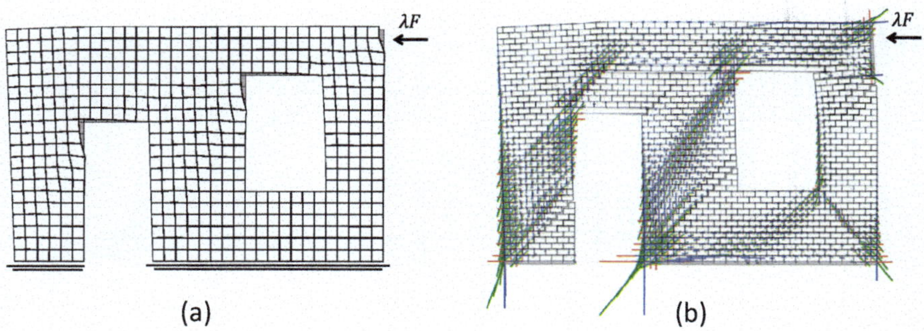

Fig. 3.42 Load bearing capacity of a masonry façade under an increasing horizontal load applied to the right upper corner till the structural collapse: **a** CDF deformed configuration corresponding to $\lambda_1^{CDF} = 175$; **b** DPRD deformed configuration and interface forces path corresponding to $\lambda_1^{DPRD} = 158$ that actives the mechanism

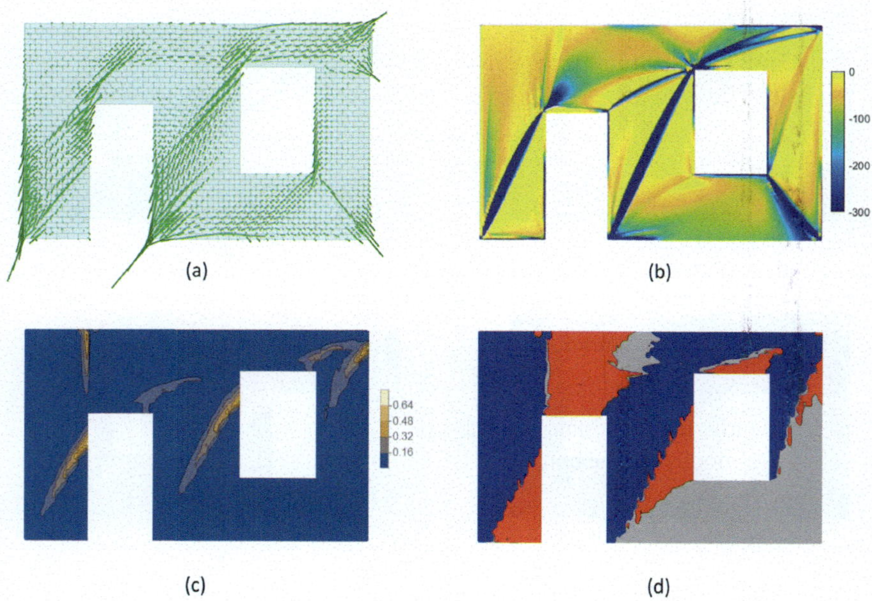

Fig. 3.43 Load bearing capacity of a masonry façade under an increasing horizontal load applied to the right upper corner and driving the structure to the collapse: **a** DPRD interface resultants on the undeformed configuration; **b** CASS minimum principal stress distribution; **c** CDF norm of the deformation and **d** CDF macroblock identification

In both cases, the horizontal loads are set to $F = 1$kN; the results in terms of the limit load multipliers are from the left side:

- $\lambda_1^{DPRD} = 188$;
- $\lambda_1^{CDF} = 205$;

- $\lambda_1^{CASS} = 191$;

while from the right side:

- $\lambda_2^{DPRD} = 158$;
- $\lambda_2^{CDF} = 175$;
- $\lambda_2^{CASS} = 140$;

Figures 3.40a and 3.42a show the CDF mechanisms corresponding to the limit exceeding the limit loads applied to the left and right corners, respectively. Similarly, Figs. 3.40b and 3.42b show the deformed configurations and interface force paths obtained through the dual PRD analyses corresponding to the limit load conditions.

Figures 3.41 and 3.43 show different kinematical and statical results corresponding to the collapse loads. DPRD interface resultants on the undeformed configuration and adopting an interlocked discretisation with a friction angle of 40° are shown in Figs. 3.41a and 3.43a. The force paths are similar to the CASS principal stress distribution shown in Figs. 3.41b and 3.43b. CASS stress patterns are coherent with the CDF deformation fields shown in Figs. 3.41c and 3.43c. Indeed, comparing stress and strain, the normality rule is fulfilled. It is worth noting that even using continuous approaches (i.e., CASS and CDF methods), the tendency of masonry structures to develop concentrated stress and deformations is reached. This mechanical behaviour also shows how masonry structures tend to develop rigid macroblocks in limit conditions, as also highlighted by the rotation fields depicted in Figs. 3.41d and 3.43d.

Figures 3.40, 3.41, 3.42 and 3.43 show the extreme ability of the proposed methods to deal with singularities that develop inside masonry structures as effects of soil settlements and a consequence of the application of limit loads.

References

Ambrosio, L., Miranda, M., & Pallara, D. (2004). Special functions of bounded variation in doubling metric measure spaces. *Calculus of Variations: Topics from the Mathematical Heritage of E. De Giorgi, 14*, 1–45.

Ambrosio, L., Fusco, N., & Pallara, D. (2000). Functions of bounded variation and free discontinuity problems. Courier Corporation.

Angelillo, M., Fortunato, A., Montanino, A., & Lippiello, M. (2014). Singular stress fields in masonry structures: Derand was right. *Meccanica, 49*(5), 1243–1262.

Angelillo, M., Babilio, E., & Fortunato, A. (2013). Singular stress fields for masonry-like vaults. *Continuum Mechanics and Thermodynamics, 25*(2–4), 423–441.

Angelillo, M. (1993). Constitutive relations for no-tension materials. *Meccanica, 28*(3), 195–202.

Angelillo, M., & Fortunato, A. (2001). Global constraints for stress constrained materials: The problem of Saint Venant. *Meccanica, 36*(5), 497–524.

Angelillo, M. (2014). Practical applications of unilateral models to Masonry equilibrium. In *Mechanics of masonry structures* (pp. 109–210). Springer.

Angelillo, M., Cardamone, L., & Fortunato, A. (2010). A numerical model for masonry-like structures. *Journal of Mechanics of Materials and Structures, 5*(4), 583–615.

Baggio, C., & Trovalusci, P. (1993). Discrete models for jointed block masonry walls. *The Sixth North American Masonry Conference, 2*, 939–949.

Bathe, K. J., & Wilson, E. L. (1976). *Numerical methods in finite element analysis*. Prentice Hall.

Batten, M. I. (1936). The architecture of Dr. Robert Hooke, FRS. *The Volume of the Walpole Society, 25*, 83–113.

Como, M. (1992). Equilibrium and collapse analysis of masonry bodies. *Masonry Construction, 27*, 185–194.

Como, M. (2013). *Statics of historic masonry constructions*. Springer.

Couplet, P. (1729, 1730). De la poussée des voûtes. Histoire de l'Académie Royale des Sciences 79–117, 117–141.

Cusano, C., Montanino, A., Cennamo, C., Zuccaro, G., & Angelillo, M. (2021a). Geometry and stability of a double-shell dome in four building phases: The case study of Santa Maria Alla Sanità in Naples. *International Journal of Architectural Heritage,* 1–27.

Cusano, C., Montanino, A., Olivieri, C., Paris, V., & Cennamo, C. (2021b). Graphical and analytical quantitative comparison in the domes assessment: The case of San Francesco di Paola. *Applied Sciences, 11*(8), 3622.

Danyzy, A. A. H. (1732). Méthode générale pour déterminer la résistance qu'il faut opposer à la poussée des voûtes. *Histoire de la Société Royale des Sciences établie à Montpellier, 2*(1718–1745), 40–56.

De Chiara, E., Cennamo, C., Gesualdo, A., Montanino, A., Olivieri, C., & Fortunato, A. (2019). Automatic generation of statically admissible stress fields in masonry vaults. *Journal of Mechanics of Materials and Structures, 14*(5), 719–737.

Di Pasquale, S. (1984). Statica dei solidi murari teorie ed esperienze. Dipartimento di Costruzioni. Universita di Firenze.

Fichera, G. (1973). Boundary value problems of elasticity with unilateral constraints. In *Linear theories of elasticity and thermoelasticity* (pp. 291–424). Springer.

Fortunato, A. (2010). Elastic solutions for masonry-like panels. *Journal of Elasticity, 98*(1), 87–110.

Fortunato, A., Fabbrocino, F., Angelillo, M., & Fraternali, F. (2018). Limit analysis of masonry structures with free discontinuities. *Meccanica, 53*(7), 1793–802.

Fraternali, F., Angelillo, M., & Fortunato, A. (2002). A lumped stress method for plane elastic problems and the discrete-continuum approximation. *International Journal of Solids and Structures, 39*(25), 6211–6240.

Frézier, A. F. (1737). La theorie et la pratique de la coupe des pierres et des bois, pour la construction des voutes et autres parties des bâtimens civils & militaires. Traité de stereotomie a l'usage de l'architecture (Vol. 1). Doulsseker.

Gagliardo, R., Portioli, F. P., Cascini, L., Landolfo, R., & Lourenço, P. B. (2021). A rigid block model with no-tension elastic contacts for displacement-based assessment of historic masonry structures subjected to settlements. *Engineering Structures, 111609*, 229.

Gesualdo, A., Iannuzzo, A., Monaco, M., & Penta, F. (2018). Rocking of a rigid block freestanding on a flat pedestal. *Journal of Zhejiang University-SCIENCE A, 19*(5), 331–345.

Giaquinta, M., & Giusti, E. (1985). Researches on the equilibrium of masonry structures. *Archive for Rational Mechanics and Analysis, 88*(4), 359–392.

Gilbert, M., & Melbourne, C. (1994). Rigid-block analysis of masonry structures. *Structural Engineer, 72*(21).

Giuffrè, A. (1991). *Letture sulla Meccanica delle Murature Storiche*. Kappa.

Gurtin, M. E. (1972). Handbuch der physik. In *The linear theory of elasticity*. Springer.

Heyman, J. (1966). The stone skeleton. *International Journal of Solids and Structures, 2*(2), 249–279.

Heyman, J. (1997). *The stone skeleton: Structural engineering of masonry architecture*. Cambridge University Press.

Iannuzzo, A. (2017). *A new rigid block model for masonry structures. Ph.D. Dissertation*. Università degli Studi di Napoli Federico II.

Iannuzzo, A., Angelillo, M., De Chiara, E., De Guglielmo, F., De Serio, F., Ribera, F., & Gesualdo, A. (2018). Modelling the cracks produced by settlements in masonry structures. *Meccanica, 53*(7), 1857–1873.

Iannuzzo, A. (2019). Energy based fracture identification in masonry structures: The case study of the church of "Pieta dei Turchini." *Journal of Mechanics of Materials and Structures, 14*, 683–702.

Iannuzzo, A., Olivieri, C., & Fortunato, A. (2019). Displacement capacity of masonry structures under horizontal actions via PRD method. *Journal of Mechanics of Materials and Structures, 14*, 703–718.

Iannuzzo, A., Van Mele, T., & Block, P. (2020). Piecewise rigid displacement (PRD) method: A limit analysis-based approach to detect mechanisms and internal forces through two dual energy criteria. *Mechanics Research Communications,* 107.

Iannuzzo, A., Dell'Endice, A., Van Mele, T., & Block, P. (2021a). Numerical limit analysis-based modelling of masonry structures subjected to large displacements. *Computers & Structures, 242*(2).

Iannuzzo, A., Block, P., Angelillo, M., & Gesualdo, A. (2021b). A continuous energy-based numerical approach to predict fracture mechanisms in masonry structures: CDF method. *Computers & Structures, 257*, 106645.

Jenkins, C., Haugen, F., & Spicher, W. H. (1998). Experimental measurement of wrinkling in membranes undergoing planar deformation. *Experimental Mechanics, 38*, 147–152.

Lions, J. L., & Stampacchia, G. (1967). Variational inequalities. *Communications on Pure and Applied Mathematics, 20*(3), 493–519.

Livesley, R. K. (1978). Limit analysis of structures formed from rigid blocks. *International Journal for Numerical Methods in Engineering, 12*(12), 1853–1871.

Lucchesi, M., Šilhavý, M., & Zani, N. (2006). A new class of equilibrated stress fields for no-tension bodies. *Journal of Mechanics of Materials and Structures, 3*(503–539), 1.

Malena, M., Angelillo, M., Fortunato, A., De Felice, G., & Mascolo, I. (2021). Arch bridges subject to pier settlements: Continuous versus piecewise rigid displacement methods. *Meccanica,* 56(10), 2487–2505.

Mansfield, E. H. (1969). Tension field theory, a new approach which shows its duality with inextensional theory. In *Applied Mechanics* (pp. 305–320). Springer.

Montanino, A., Olivieri, C., Zuccaro, G., & Angelillo, M. (2021). From stress to shape: Equilibrium of cloister and cross vaults. *Applied Sciences, 11*(9), 3846.

Montanino, A., De Gregorio, D., Olivieri, C., & Iannuzzo, A. (2022). The continuous airy-based for stress-singularities (CASS) method: An energy-based numerical formulation for unilateral materials. *International Journal of Solids and Structures, 256*.

Moreau, J. J. (1968). La notion de sur-potentiel et les liaisons unilatérales en élastostatique. *Comptes Rendus Hebdomadaires Des Séances De L'académie Des Sciences, 267*, 954–957.

Orduña, A., & Lourenço, P. B. (2005). Three-dimensional limit analysis of rigid blocks assemblages. Part I: Torsion failure on frictional interfaces and limit analysis formulation. *International Journal of Solids and Structures, 42*(18–19), 5140–5160.

Olivieri, C., Iannuzzo, A., Fortunato, A., & DeJong, M. J. (2022). The effect of concentrated loads on open-well masonry spiral stairs. *Engineering Structures, 272*, 114952.

Panagiotopoulos, P. D. (1985). *Inequality problems in mechanics and applications: Convex and nonconvex energy functions*. Springer Science & Business Media.

Portioli, F., & Cascini, L. (2017). Large displacement analysis of dry-jointed masonry structures subjected to settlements using rigid block modelling. *Engineering Structures, 148*, 485–496.

Temam, R., & Strang, G. (1980). Functions of bounded deformation. *Archive for Rational Mechanics and Analysis, 75*(1), 7–21.

Trusdell, C., & Noll, W. (2004). The non-linear field theories of mechanics. In *The non-linear field theories of mechanics* (pp. 1–579). Springer.

von Neumann, J., & Richtmyer, R. D. (1950). A method for the numerical calculation of hydrodynamic shocks. *Journal of Applied Physics, 21*(3), 232–237.

Chapter 4
Thrust Network Analysis for Masonry Assessment

Ricardo Maia Avelino, Tom Van Mele, and Philippe Block

Abstract This chapter presents Thrust Network Analysis (TNA), a lower-bound limit-analysis-based method to assess vaulted masonry structures. With TNA, admissible equilibrium states corresponding to compressive force networks within the structural geometry can be studied. A novel multi-objective optimisation framework is described, which enables finding particular admissible equilibrium solutions in 3D structures subjected to general loading and support displacements. The equilibrium solutions searched include the minimum and maximum horizontal thrust states, the minimum vault thickness, maximum vertical and horizontal collapse loads, and compatible internal stress states following support movements. A complete picture of the structure's stability level is obtained by combining these states, and the location where cracks arise at limit states is also highlighted. The method can be applied based on the surveyed geometric data of the structure's envelope, being well-suitable for practical engineering problems. Analysis of shallow cross vaults and a surveyed cathedral demonstrate the method's applicability and its relevance to assessing vaulted masonry structures.

4.1 Introduction

Masonry structures form a large part of the world's built heritage and serve as housing for millions worldwide. Moved by the pressing need for a more sustainable built environment and heritage protection, research on the analysis and preservation of existing masonry buildings has grown recently (Angelillo et al., 2021; Funari et al.,

R. Maia Avelino · T. Van Mele · P. Block (✉)
Block Research Group, Institute of Technology in Architecture, Department of Architecture, ETH Zurich, Zurich, Switzerland
e-mail: block@arch.ethz.ch

R. Maia Avelino
e-mail: maia@arch.ethz.ch

T. Van Mele
e-mail: van.mele@arch.ethz.ch

K. Bagi and M. Angelillo (eds.), *Discrete Computational Mechanics of Masonry Structures*, CISM International Centre for Mechanical Sciences 609,
https://doi.org/10.1007/978-3-031-32476-5_4

2021; Mehrotra et al., 2023). Novel analysis tools and methods have been conceived to determine their level of stability and resistance against the actions of exceptional actions such as seismic and wind loads (D'Altri et al., 2019; Tralli et al., 2014).

From a mechanics point of view, the highly nonlinear, discrete and unilateral behaviour of masonry structures precludes the use of general analysis tools without proper consideration of either material properties or specific modelling of the internal joints (Aita et al., 2015; Huerta, 2008; Shin et al., 2016). Consequently, different approaches have been adopted for the structural analysis of masonry, such as the Discrete Element Method (DEM) (Cundall, 1971; Lemos, 2007, 2019), nonlinear Finite Element Method (FEM) (Lourenço and Rots, 1997; Milani et al., 2008; Parisi et al., 2019; Smoljanović et al., 2013) and Limit Analysis (LA) based approaches (Heyman, 1966, 1995). This contribution focuses on the latter, which offers a framework for computing collapse loads or collapse mechanisms in structures supported by geometry-based information, i.e., without relying on mechanical properties, such as compressive strength and stiffness, which are often unknown or unknowable.

Limit Analysis is a branch of plastic analysis which searches for limit states in the structure instead of computing the elastic deformations throughout the material. It was initially applied to the design of steel frames (Baker et al., 1956; Prager, 1959) and extended to masonry by Heyman (1966) by assuming infinite compressive strength, null tensile capacity and no-sliding failure. Two approaches can be adopted to apply LA: lower-bound approaches, based on the Safe Theorem (Block and Ochsendorf, 2002; Fraddosio et al., 2019; Heyman, 1969), and upper-bound approaches (Chiozzi et al., 2017; Grillanda et al., 2019).

By applying the Safe Theorem to masonry, admissible stress states are characterised as compressive force paths within the structural geometry and in equilibrium with externally applied loads. For two-dimensional structures, these compressive force path result in the so-called thrust lines, applied to arches (Moseley, 1843), sliced domes (Poleni, 1748) and vaults (Ungewitter, 1890), and commonly applied in combination with graphic statics (e.g., Wolfe, 1921). Currently, this technique is known as Thrust Line Analysis (TLA), and it has been extended to a few three-dimensional problems, such as the dome under symmetric loading cases (Aita et al., 2019; Zessin et al., 2010), vaults (Smars, 2000) and spiral staircases (Angelillo et al., 2021). TLA is also the core of 2D masonry analysis tools such as LimitState (LimitState Ltd, 2020) and Archie-M (Obvis, 2016).

Nevertheless, extending lower-bound methods to fully three-dimensional structures is challenging due to the increased degree of indeterminacy and the complex geometry of masonry vaults (Huerta, 2008). Therefore, various lower-bound equilibrium formulations have recently been developed to assess three-dimensional masonry structures. These strategies can be divided into continuous and discrete approaches.

Continuous approaches model the internal forces of the structure as a continuous membrane with the equilibrium equations solved by assuming a Pucher formulation and considering the potential stress (or Airy) functions to describe the internal distribution of the stresses (Angelillo et al., 2013; Fraddosio et al., 2020; Fraternali, 2010).

Discrete approaches consider a singular internal stress field discretised by a network (O'Dwyer, 1999; Block and Ochesendorf, 2007). The compressive forces are modelled as axial forces along the network's edges, and loads are applied to its vertices. The equilibrium equations are linearised with discrete approaches, and geometric irregularities, concentrated loads, and openings can be directly investigated.

A discrete network approach was proposed initially in O'Dwyer (1999). In Block and Ochesendorf (2007), this approach was combined with graphic statics (Culmann, 1875), with the introduction of Thrust Network Analysis (TNA). With TNA, the spatial equilibrium of the networks is explored by the reciprocal relations of form and force diagrams analogous to the force polygons in graphic statics (Block, 2009). The problem can also be formulated based on the Force Density Method (Schek, 1974) in which the force densities at the edges of the network are variables of the equilibrium equations. Using this formulation, TNA has been coupled with optimisation procedures and applied to masonry assessment problems (Block and Lachauer, 2014; Bruggi, 2020; Maia Avelino et al., 2021a, 2022a). In Block and Lachauer (2014), networks with fixed horizontal projections are considered for which the degrees of freedom of the network can be explicitly defined with the independent edges (Van Mele and Block, 2014). This formulation has been applied to find the minimum and maximum thrust states (Bruggi, 2020), compute the structure's Geometric Safety Factor (GSF) (Maia Avelino et al., 2021a), and find specific internal stress states arising after differential foundation displacements (Maia Avelino et al., 2022a). Other researchers have also considered problems with disconnected networks (Fantin and Ciblac, 2016; Nodargi and Bisegna, 2022), which enlarge the solution space but require additional friction checks on the interface of the nodes.

This book chapter revisits the TNA formulation used to find admissible equilibrium states in masonry structures. It presents a modular multi-objective optimisation formulation (Maia Avelino, 2023), which enables finding multiple admissible equilibrium states with a unique approach. This modular framework can be applied to determine the structural safety levels and the structural response to external actions such as support displacements and vertical and horizontal live loads.

This paper is organised as follows: Sect. 4.2 presents TNA and the numerical formulation applied to explore the geometry of thrust networks, Sect. 4.3 presents the constrained optimisation framework to find particular equilibrium solutions, Sect. 4.4 shows applications of the method to vaulted masonry structures, and Sect. 4.5 concludes this paper by summarising the results presented and pointing to further research.

4.2 Thrust Network Analysis

This section describes a numerical formulation to find thrust networks. The main elements of TNA are defined in Sect. 4.2.1. The equilibrium equations for finding the geometry of the network based on its internal force densities are described in Sect. 4.2.2. The particular case of networks fixed in the plan is discussed in Sect. 4.2.3,

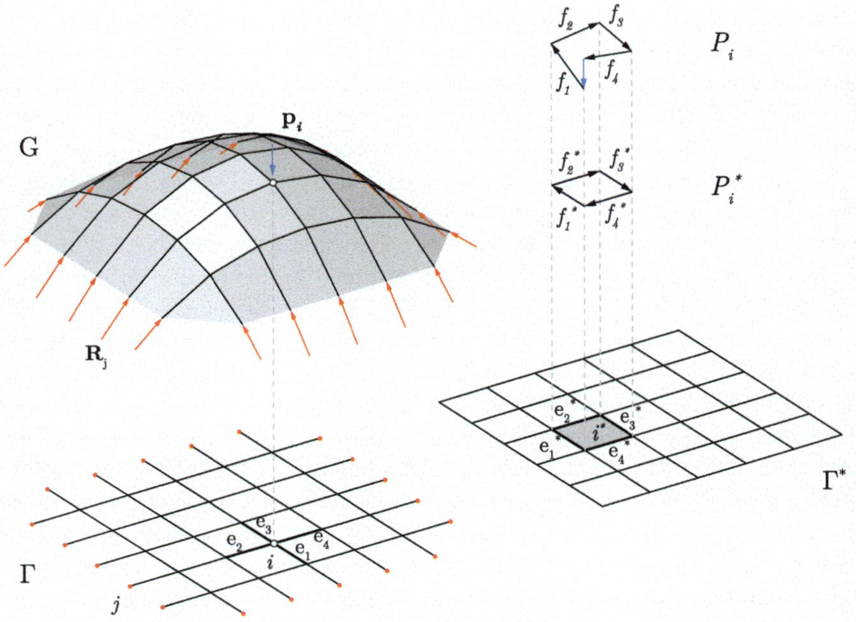

Fig. 4.1 Thrust Network (G) with its horizontal projection, the form diagram (Γ). Equilibrium P_i of node i subjected to the vertical applied load $\mathbf{p}_i$ is highlighted. The horizontal projection of P_i is the closed polygon P_i^*. Summing the polygons, the network's force diagram (Γ*) is obtained

and strategies to handle the degrees of freedom in these cases are presented in Sect. 4.2.4.

4.2.1 Main Elements

This section defines the main elements of Thrust Network Analysis (TNA), after (Block, 2009), which are illustrated in Fig. 4.1.

A thrust network (G) is a *directed* and *connected* spatial graph in which the edges of G represent the spatial compressive resultants (or thrusts) within the structure.

The form diagram (Γ) is a planar graph constructed from the projection of G. The form diagram stores the network's topology, connectivity, and planar coordinates. Each vertex i in the form diagram with coordinates (x_i, y_i) maps to a vertex in the thrust network with elevation z_i. Similarly, each edge e_i in the form diagram maps to an edge in the thrust network carrying axial force f_i.

Boundary conditions and loads are applied to the nodes of the network. For example, Fig. 4.1 highlights the support points in the form diagram with a red dot, where reaction forces $\mathbf{R}_j$ arise. An external vertical load $\mathbf{p}_i$ applied to node i is also highlighted. The equilibrium of the node can be verified through the construction of closed polygons of force vectors P_i indicated in Fig. 4.1.

The force diagram (Γ^*) is a graphical representation of the horizontal equilibrium of G. When all loads are parallel, the spatial equilibrium of a node i can be projected onto a plane perpendicular to the loads resulting in a two-dimensional graphic statics problem. In Fig. 4.1, the projected equilibrium of node i is the closed polygon P_i^* where the vertical applied load $\mathbf{p}_i$ vanishes. The horizontal equilibrium is then resolved with the horizontal components of the thrusts (f_i^*). The form (Γ) and force (Γ^*) diagrams are reciprocal, such that their corresponding edges e_i, e_i^* are parallel, and the length of e_i^* is proportional to the horizontal force f_i^*.

The geometry of thrust networks can be explored by searching and modifying the reciprocal form and force diagrams, as proposed in Block et al. (2019), Rippmann et al. (2012) using a parallelisation algorithm. Recently, new computational tools have been developed to provide interactive access to this process, such as RhinoVault (Block Research Group, 2021; Rippmann, 2016). This pipeline has been adopted to the conceptual design stage of several shell prototypes and projects, e.g., the Striatus Bridge (Bhooshan et al., 2022) and Armadillo Vault (Block et al., 2018).

Nevertheless, fitting this network within a (usually tight) structural envelope is required for masonry assessment, which requires robust optimisation strategies and suitable problem parametrisation. Furthermore, the graphic-statics-based TNA framework is limited to vertical loading cases, so horizontal forces can not be applied. The following section presents a numerical description of the network's equilibrium based on force densities which enables general loading cases and is suitable for numerical optimisation.

4.2.2 Equilibrium Equations

This section presents the equilibrium formulation in general force networks based on force densities (Schek, 1974).

A network composed of m edges and n vertices is considered, with n_b supported vertices and n_i free vertices, such that $n = n_\mathrm{i} + n_\mathrm{b}$. The nodal positions of the network are cast in the vectors $\mathbf{x}, \mathbf{y}, \mathbf{z}$ $[n \times 1]$, and the applied forces in each direction are collected in $\mathbf{p}_\mathrm{x}, \mathbf{p}_\mathrm{y}, \mathbf{p}_\mathrm{z}$ $[n \times 1]$. The equilibrium variables are the force densities $\mathbf{q}$ $[m \times 1]$ defined for each edge as the ratio between its axial force f_i and its length l_i.

The equilibrium equations can be written by introducing the connectivity matrix $\mathbf{C}$ $[m \times n]$ (see Schek, 1974) and defining the coordinate difference matrices $\mathbf{U} = \mathrm{diag}(\mathbf{Cx})$, $\mathbf{V} = \mathrm{diag}(\mathbf{Cy})$, $\mathbf{W} = \mathrm{diag}(\mathbf{Cz})$. The $3n_\mathrm{i}$ linear internal equilibrium equations become

$$\mathbf{C}_{\rm i}^{\rm T}\mathbf{U}\mathbf{q} = \mathbf{p}_{\rm x,i}, \tag{4.1a}$$

$$\mathbf{C}_{\rm i}^{\rm T}\mathbf{V}\mathbf{q} = \mathbf{p}_{\rm y,i}, \tag{4.1b}$$

$$\mathbf{C}_{\rm i}^{\rm T}\mathbf{W}\mathbf{q} = \mathbf{p}_{\rm z,i}, \tag{4.1c}$$

where, $\mathbf{C}_{\rm i}$ $[m \times n_{\rm i}]$ and $\mathbf{p}_{\rm x,i}$, $\mathbf{p}_{\rm y,i}$, $\mathbf{p}_{\rm z,i}$, $[n_{\rm i} \times 1]$ are slices of the connectivity matrix and applied loads in the $n_{\rm i}$ free nodes.

At the supported boundaries, the emerging reaction forces $\mathbf{R}_i = [R_{{\rm x},i}; R_{{\rm y},i}; R_{{\rm z},i}]$ can be retrieved from the reaction components $\mathbf{R}_{\rm x}$, $\mathbf{R}_{\rm y}$, $\mathbf{R}_{\rm z}$ $[n_{\rm b} \times 1]$ calculated as

$$\mathbf{R}_{\rm x} = \mathbf{C}_{\rm b}^{\rm T}\mathbf{U}\mathbf{q} - \mathbf{p}_{\rm x,b}, \tag{4.2a}$$

$$\mathbf{R}_{\rm y} = \mathbf{C}_{\rm b}^{\rm T}\mathbf{V}\mathbf{q} - \mathbf{p}_{\rm y,b}, \tag{4.2b}$$

$$\mathbf{R}_{\rm z} = \mathbf{C}_{\rm b}^{\rm T}\mathbf{W}\mathbf{q} - \mathbf{p}_{\rm z,b}. \tag{4.2c}$$

where $\mathbf{C}_{\rm b}$ $[m \times n_{\rm b}]$, and $\mathbf{p}_{\rm x,b}$, $\mathbf{p}_{\rm y,b}$, $\mathbf{p}_{\rm z,b}$ $[n_{\rm b} \times 1]$ are slices of the connectivity matrix and applied loads in the $n_{\rm b}$ supported nodes.

With this formulation, the infinite space of equilibrium networks for a given topology (or connectivity) can be explored in terms of the position of the supports $\mathbf{x}_{\rm b}$, $\mathbf{y}_{\rm b}$, $\mathbf{z}_{\rm b}$ and values of the force densities in the edges of the network $\mathbf{q}$, resulting in $3n_{\rm b} + m$ parameters.

A subspace of this equilibrium is explored with TNA, considering the cases where the horizontal projection of the network, i.e., the form diagram, is fixed, as introduced in Block and Lachauer (2014). The following section presents the consequence of this assumptions.

4.2.3 The Case of a Fixed Form Diagram

By considering the form diagram fixed in the analysis, the planar coordinates of the network $\mathbf{x}$, $\mathbf{y}$ are known, and the horizontal equilibrium equations (4.1a) and (4.1b) can be rearranged, introducing the horizontal equilibrium matrix $\mathbf{E}$ $[2n_{\rm i} \times m]$ and the vector of applied horizontal forces in the internal nodes $\mathbf{p}_{\rm h,i}$ $[2n_{\rm i} \times 1]$,

$$\mathbf{E}\mathbf{q} = \mathbf{p}_{\rm h,i}, \quad \text{with: } \mathbf{E} = \begin{bmatrix} \mathbf{C}_{\rm i}^{\rm T}\mathbf{U} \\ \mathbf{C}_{\rm i}^{\rm T}\mathbf{V} \end{bmatrix}, \mathbf{p}_{\rm h,i} = \begin{bmatrix} \mathbf{p}_{\rm x,i} \\ \mathbf{p}_{\rm y,i} \end{bmatrix}. \tag{4.3}$$

Assuming that the form diagram is fixed imposes additional constraints to the vector of force densities, such that its components can not be chosen freely. Indeed, the number of force densities that can be chosen freely in Eq. 4.3 corresponds to the number k of degrees of freedom (DOF), or the degree of statical indeterminacy, of the network.

As shown in Van Mele and Block (2014), the number of DOF is equal to the rank deficiency of the matrix **E**. The free parameters are denoted *independent force densities*, and they relate to specific *independent edges* in the network. The independent force densities $\mathbf{q}_{\mathrm{id}}$ $[k \times 1]$ are then used to describe the force densities $\mathbf{q}$ in the network through the linear transformation

$$\mathbf{q} = \mathbf{B}\mathbf{q}_{\mathrm{id}} + \mathbf{d}, \ \text{with: } \mathbf{B} = \begin{bmatrix} -\mathbf{E}_{\mathrm{d}}^{\dagger}\mathbf{E}_{\mathrm{id}} \\ \mathbf{I}_{\mathrm{k}} \end{bmatrix}, \mathbf{d} = \begin{bmatrix} \mathbf{E}_{\mathrm{d}}^{\dagger}\mathbf{p}_{\mathrm{h,i}} \\ \mathbf{0} \end{bmatrix}, \tag{4.4}$$

where $\mathbf{E}_{\mathrm{d}}$ $[2n_{\mathrm{i}} \times (m-k)]$ and $\mathbf{E}_{\mathrm{id}}$ $[2n_{\mathrm{i}} \times k]$ are slices of **E** related to the dependent and independent edges, respectively, $\mathbf{E}_{\mathrm{d}}^{\dagger}$ is the generalised inverse or Moore-Penrose pseudo-inverse of $\mathbf{E}_{\mathrm{d}}$, and $\mathbf{I}_{\mathrm{k}}$ is the identity matrix of size k.

After such variable reduction, the elevation of the free nodes in the network $\mathbf{z}_{\mathrm{i}}$ is a function of $\mathbf{q}_{\mathrm{id}}$ and $\mathbf{z}_{\mathrm{b}}$, resulting in

$$\mathbf{z}_{\mathrm{i}}\left(\mathbf{q}_{\mathrm{id}}, \mathbf{z}_{\mathrm{b}}\right) = \mathbf{D}_{\mathrm{i}}^{-1}\left(\mathbf{p}_{\mathrm{z,i}} - \mathbf{D}_{\mathrm{b}}\mathbf{z}_{\mathrm{b}}\right), \tag{4.5}$$

where $\mathbf{D}_{\mathrm{i}} = \mathbf{C}_{\mathrm{i}}^{\mathrm{T}}\mathbf{Q}\mathbf{C}_{\mathrm{i}}$ $[n_{\mathrm{i}} \times n_{\mathrm{i}}]$, $\mathbf{D}_{\mathrm{b}} = \mathbf{C}_{\mathrm{i}}^{\mathrm{T}}\mathbf{Q}\mathbf{C}_{\mathrm{b}}$ $[n_{\mathrm{i}} \times n_{\mathrm{b}}]$ and $\mathbf{Q} = \mathrm{diag}(\mathbf{q})$ $[m \times m]$.

With this approach, the number of variables of the problem reduces to $n_{\mathrm{b}} + k$. Equation 4.5 will be used to compute the free nodal elevations enabling the constraint optimisation framework described in Sect. 4.3.

4.2.4 Independent Edges in Form Diagrams

As presented in Sect. 4.2.3, the number of independent force densities k that can be chosen freely in Eq. 4.3 correspond to the dimension of the nullspace of **E** computed as

$$k = m - \mathrm{rank}(\mathbf{E}). \tag{4.6}$$

Therefore, finding a set of independent edges in the network is equivalent to finding one base of the nullspace of **E**. A sequential Singular Value Decomposition (SVD) approach is applied, as described in Maia Avelino et al. (2021a), to find a base of k linearly independent columns of the nullspace **E** that relate to k independent edges in the network. As discussed in Liew et al. (2019), these groups are not unique and can not be selected randomly for non-triangulated topologies.

The result of this process is illustrated in four networks in Fig. 4.2, where one of the infinite possible groups of independent edges for each topology is depicted with $k < m$.

In Fig. 4.2, (a) the orthogonal form diagram with $n_{\mathrm{d}} = 6$ divisions has $m = 60$ edges and $k = 10$ independent edges, (b) the radial diagram with three hoops and 12 meridians has $m = 84$ and $k = 13$, (c) the cross diagram with $n_{\mathrm{d}} = 6$ divisions has $m = 96$ and $k = 8$, and (d) the three-sided diagram presents $m = 42$ and $k = 4$.

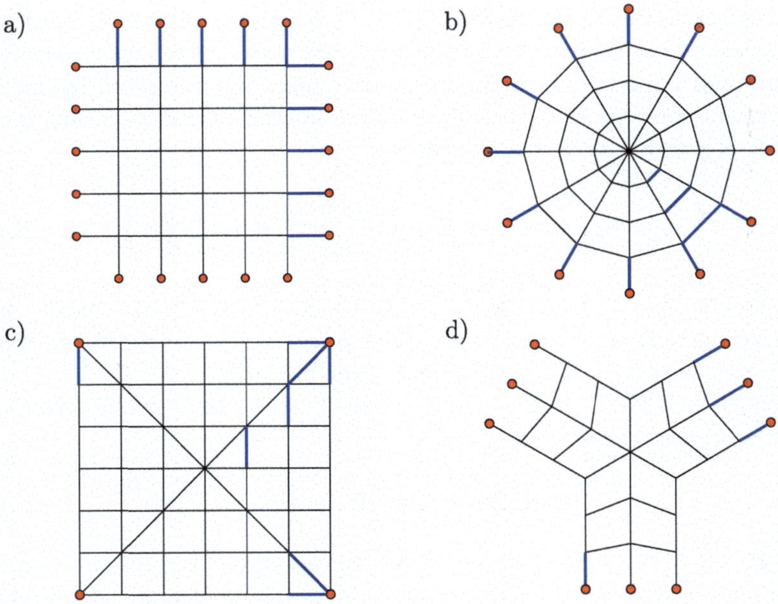

Fig. 4.2 Independent edges highlighted in blue for different topologies: **a** an orthogonal grid, **b** a radial arrangement, **c** a cross diagram, and **d** a three-sided diagram

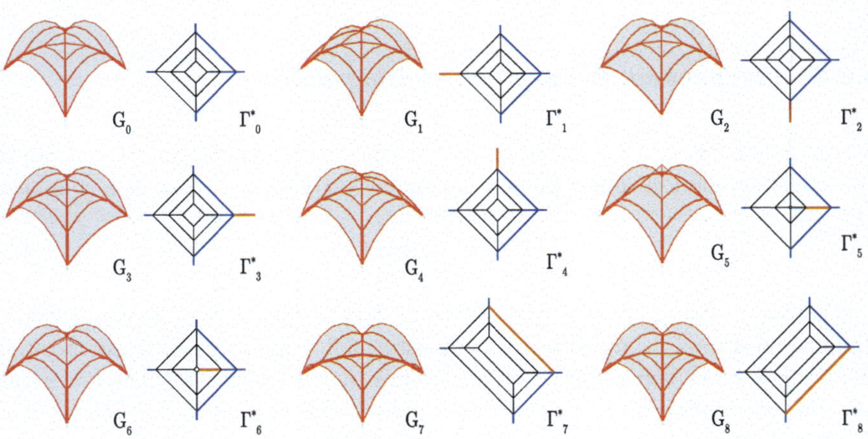

Fig. 4.3 Indivudual effect of increasing the force magnitude of the independent edges on the cross form diagram. Modified thrust network (G_i) and force diagrams (Γ_i^*) are shown, highlighting the ith independent modified

The infinite space of equilibrium networks having a fixed horizontal projection can be explored using this formulation by combining the effect of each independent edge in the final geometry. For example, Fig. 4.3 shows the effect of increasing the force in the ith independent edge in the thrust network geometry (G_i) for the cross form diagram (Fig. 4.2c). The corresponding force diagram (Γ_i^*) is depicted for each (G_i). In Γ_i^*, the force increase is evident by the stretch of the ith reciprocal independent edge, highlighted in orange (see Fig. 4.3).

The following section introduces the search of admissible thrust networks as a constrained nonlinear optimisation problem assuming the parametrisation in independent edges.

4.3 Constrained Equilibrium

This section computes admissible thrust networks through a constrained nonlinear optimisation framework that takes the general form below:

$$\underset{\tilde{\mathbf{x}}}{\text{minimise}} \quad f_{\text{obj}}(\tilde{\mathbf{x}}) \tag{4.7a}$$

$$\text{subject to} \quad g_i(\tilde{\mathbf{x}}) \geq 0, \qquad \text{for} \quad i = [1, \ldots n_c], \tag{4.7b}$$

where the n_c inequality constraints translate the limit analysis assumptions for admissible networks, as listed in Sect. 4.3.1. Different objective functions f_{obj} are implemented, as defined in Sect. 4.3.2. The optimisation variables $\tilde{\mathbf{x}}$ correspond to the independent force densities $\mathbf{q}_{\text{id}}$ and the support elevations $\mathbf{z}_{\text{b}}$ (see Sect. 4.2.3) followed by occasional auxiliary variables depending on the selected objective (see Sect. 4.3.2).

4.3.1 Constraints from Limit Analysis

As described in Heyman (1966), limit analysis can be applied to masonry structures given that infinite compressive and null tensile strength are assumed, and no sliding failure can occur. In continuum mechanics, these assumptions are generalised in applying the normal, rigid, no-tension material (NRNT) described in Angelillo et al. (2018). For the present formulation, these assumptions translate into no-tensile axial forces at the network edges and in constraining the network to remain within the structural section. The former enforces negative axial force densities q_i in the edges (Eq. 4.8a). The latter is implemented by constraining the nodal elevations z_i to lay between the elevations of intrados z_i^{LB} and extrados z_i^{UB} (Eq. 4.8b).

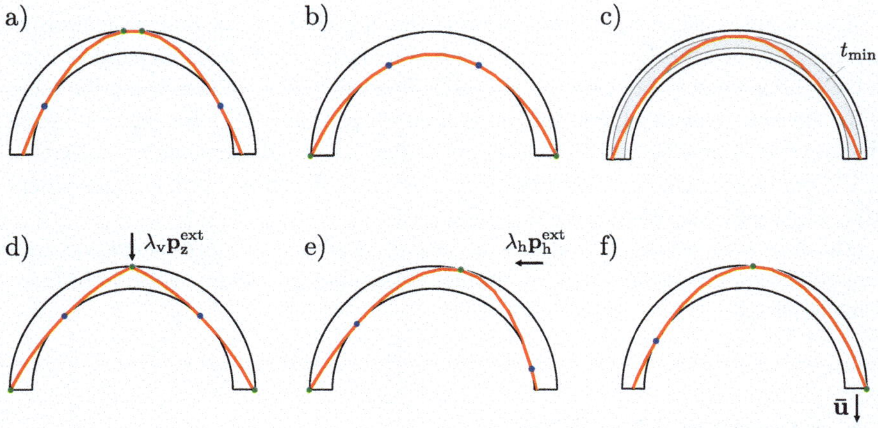

Fig. 4.4 Examples of different objective functions on a semicircular arch: **a** minimum thrust, **b** maximum thrust, **c** minimum thickness $t_{\min}$, **d** maximum vertical load multiplier λ_v associated to an external vertical load $\mathbf{p}_z^{ext}$, **e** maximum horizontal load multiplier λ_h associated to an external horizontal load $\mathbf{p}_h^{ext}$ and (f) minimum complementary energy for boundary displacement $\bar{\mathbf{u}}$

$$q_i \leq 0 \qquad \text{for} \qquad i = [1, \ldots, m], \tag{4.8a}$$

$$z_i^{LB} \leq z_i \leq z_i^{UB} \qquad \text{for} \qquad i = [1, \ldots, n]. \tag{4.8b}$$

Additional constraints can be applied, i.e. by limiting the extension of the reaction forces to cross the extrados, as detailed in Maia Avelino et al. (2021a), which is required to model problems such as domes and buttressed vaults.

4.3.2 Objective Functions

Different objective functions (f_{obj}) are coupled to the optimisation problem in Eqs. 4.7. Figure 4.4 illustrates the different objective functions on a semi-circular arch.

In Fig. 4.4, the points in which the thrust line touches the extrados (resp. intrados) are marked in green (resp. blue). They indicate the location where cracks are expected in the solution. When the thrust touches the intrados (resp. extrados), a crack will form in the extrados (resp. intrados).

The objective functions implemented and illustrated in Fig. 4.4 are defined mathematically in the following subsections below.

(a) Minimum horizontal thrust: Minimises the horizontal component of the emerging reactions $\mathbf{R}_x$, $\mathbf{R}_y$, such that the objective function becomes

$$f_{\min} = \sum_{i=1}^{n_b} \sqrt{R_{x,i}^2 + R_{y,i}^2}. \tag{4.9}$$

(b) Maximum horizontal thrust: Maximises the horizontal component of the emerging reactions $\mathbf{R}_x$, $\mathbf{R}_y$, which is equivalent to minimising the opposite of $f_{\min}$, resulting in

$$f_{\max} = -\sum_{i=1}^{n_b} \sqrt{R_{x,i}^2 + R_{y,i}^2}. \tag{4.10}$$

(c) Minimum structural thickness: Minimises the additional variable $t > 0$ representing the structural thickness in the vault computed orthogonally to the masonry's middle surface as

$$f_{\text{thk}} = t. \tag{4.11}$$

Consequently, the additional variable will influence the bounds of the intrados and extrados, modifying the constraint Eq. 4.8b, which will take the form of

$$z_i^{\text{LB}}(t) \leq z_i \leq z_i^{\text{UB}}(t) \quad \text{for} \quad i = [1, \ldots, n]. \tag{4.8b}$$

(d) Maximum vertical load multiplier: Maximises the vertical load multiplier ($\lambda_v \geq 0$) for a given vertical applied load case $\mathbf{p}_z^{\text{ext}}$. The objective function becomes

$$f_v = -\lambda_v. \tag{4.12}$$

The external load and multiplier are then added to the computation of the nodal network elevations in Eq. 4.5, resulting in

$$\mathbf{z}_i(\mathbf{q}_{\text{id}}, \mathbf{z}_b, \lambda_v) = \mathbf{D}_i^{-1}\left((\mathbf{p}_{z,i} + \lambda_v \mathbf{p}_{z,i}^{\text{ext}}) - \mathbf{D}_b \mathbf{z}_b\right). \tag{4.5}$$

(e) Maximum horizontal load multiplier: Maximises the horizontal load multiplier ($\lambda_h \geq 0$) for a given horizontal external load case $\mathbf{p}_h^{\text{ext}}$, such that the objective function becomes simply

$$f_h = -\lambda_h. \tag{4.13}$$

As a collateral effect, adding this variable modifies the horizontal equilibrium of the structure (Eq. 4.3), now computed as

$$\mathbf{q}(\mathbf{q}_{\text{id}}, \lambda_h) = \mathbf{B}\mathbf{q}_{\text{id}} + \lambda_h \mathbf{E}_d^{\dagger} \begin{bmatrix} \mathbf{p}_{h,i}^{\text{ext}} \\ \mathbf{0} \end{bmatrix}. \tag{4.3}$$

Furthermore, as discussed in Bruggi (2020), the application of horizontal loads at fixed form diagrams requires that the loads can be transferred appropriately to the supports, which can be verified mathematically by ensuring that the rank of $\mathbf{E}$ is not modified by stacking the horizontal load case $\mathbf{p}_{h,i}^{\text{ext}}$.

(f) Minimum complementary energy: Minimises the complementary energy W_{c} for a general foundation settlement $\bar{\mathbf{u}}$ $[n_{\mathrm{b}} \times 3]$ which is equal to minimising the opposite work of the reaction forces (Iannuzzo et al., 2020). The objective function becomes

$$f_{\mathrm{c}} = W_{\mathrm{c}} = -\sum_{i}^{n_{\mathrm{b}}} \mathbf{R}_i \cdot \bar{\mathbf{u}}_i. \tag{4.14}$$

These different objective functions can be added to the optimisation framework in Eq. 4.7. One application for each objective is illustrated in Sect. 4.4.

4.3.3 Solving Strategies and Starting Points

The optimisation problem is solved using interior point methods, using the open-source solver IPOPT (Wächter and Biegler, 2006). All analyses presented in this chapter have been conducted with the Python-package *compas_tno* (Maia Avelino, 2022). *compas_tno* formulates the constrained optimisation problem (Eq. 4.7), calls the solver, and post-process the final results back into the new geometry of the network.

The starting point of the analysis is the compression-only load-path optimisation described in Liew et al. (2018). This problem corresponds to a convex problem described in Maia Avelino (2023). The result of this pre-conditioning optimisation is a compression-only network that does not yet fit the bounds of the structure. By applying the constraints defined in Sect. 4.3.1, an admissible network is retrieved if the problem is feasible.

The Jacobian matrix and gradients of the constraints and objective functions are defined analytically and passed to IPOPT during the solving process. Jacobian construction and different gradient functions are defined in Maia Avelino (2023).

4.4 Applications

This Section shows applications of the modular multi-objective optimisation framework described in vaulted masonry structures. Section 4.4.1 defines the geometry of a shallow cross vault that will be used throughout Sects. 4.4.2–4.4.6, where different assessment problems are tackled by setting different objective functions in the framework. Finally, Sect. 4.4.7 discussed the application of the method from a geometry extracted from point clouds.

4.4.1 Vault Geometry

A square shallow cross vault is considered based on the parametric study from Maia Avelino et al. (2021b). The geometry is defined by the vault's base length (l_0), (central) radius (R), springing angle (β) and thickness (t). The ratio r/l_0 generates pointed ($r/l_0 > 0.5$) or rounded cross vaults ($r/l_0 = 0.5$), and based on the springing angle β, shallower cross vaults can be considered. Whenever $\beta > 0$, the effective span (s) marks the distance among the springs. The thickness of the structure is referred to by the adimensional t/s. The parameters of the model are highlighted in Fig. 4.5a. Figure 4.5b shows the cross vault selected for this study with $r/l_0 = 0.5$, $\beta = 30°$, $t = 0.5$ m and $s = 10.0$ m.

A density of $\rho = 20$ kN/m^3 is assumed for the masonry, resulting in a total self-weight of the structure equals $W = 1088$ kN. The form diagram selected in the following analysis is the cross orthogonal topology presented in Fig. 4.2c with a level of discretisation $n_d = 16$, representing 16 divisions of the diagonal.

4.4.2 Computing the Extremes of Thrust

This section analyses the vaulted geometry defined in Sect. 4.4.1, finding its minimum and maximum horizontal thrust states. The horizontal thrust is transferred only through the corners of the vault, such that no thrust is transmitted along its open edges.

To obtain the minimum thrust solution, the optimisation in Eq. 4.7 is solved by selecting f_{min} as the objective function (Eq. 4.9). The solution is depicted in Fig. 4.6a. In the plots, the thickness of the edges in the thrust network is proportional to the force carried such that edges having zero force are not shown. Furthermore, the points where the network touches the extrados of the vault are highlighted in green, and the points where it touches the intrados are highlighted in blue. At the minimum thrust state, the deepest network is obtained. The network touches the extrados along the

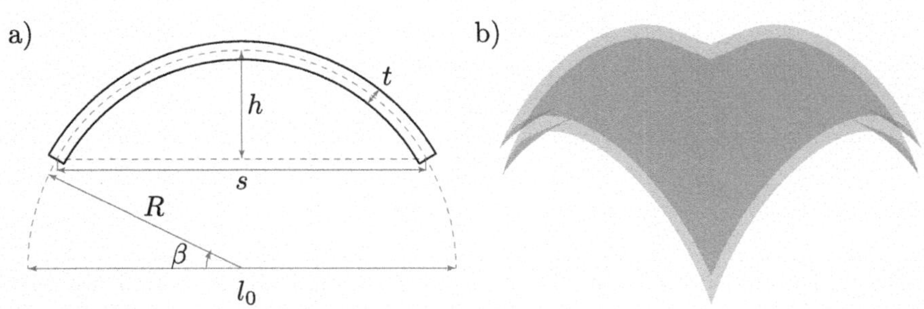

Fig. 4.5 **a** Parameters used to define the cross vault. **b** Cross vault obtained with $t/s = 0.05$, $r/l_0 = 0.5$ and $\beta = 30°$

midspan of the vault, indicating cracks in the intrados of the structure. The value of overall thrust-over-weight obtained is $T_{\min}/W = 0.97$, resulting in a horizontal thrust exerted of $T_{\min} = 1055$ kN for this problem.

Conversely, the maximum thrust state is found by taking as the objective function $f_{\max}$ (Eq. 4.10). The solution is depicted in Fig. 4.6b. The network assumes its shallowest geometry touching the intrados along the midspan, indicating a crack appearing in the extrados of the cross vault. As a result, the value of the thrust-over-weight is $T_{\max}/W = 1.57$, resulting in a horizontal thrust exerted of $T_{\max} = 1708$ kN.

4.4.3 Stability Domain

This section computed the Geometric Safety Factor (GSF) and the stability domain of the shallow cross vault described in Sect. 4.4.1, measuring the level of stability in the structure. After Heyman (1969), the GSF is computed as the ratio between the actual structural thickness (t_0) and the minimum thickness of the structure ($t_{\min}$).

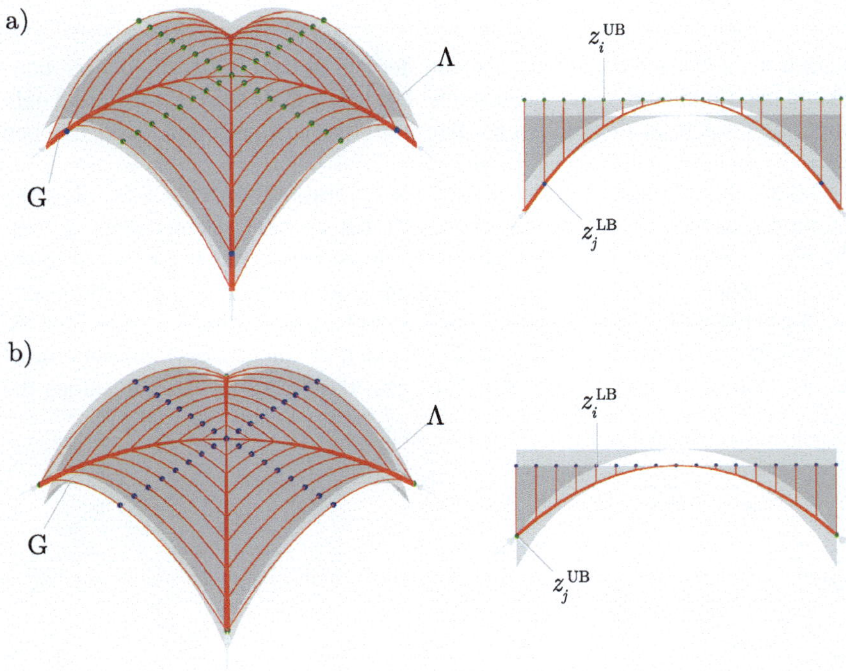

Fig. 4.6 **a** Minimum thrust solution for the shallow vault with $T/W = 0.97$; **b** Maximum thrust solution for the shallow vault with $T/W = 1.57$. Points where the network touches the intrados (z_i^{LB}) and extrados (z_j^{UB}) are highlighted in blue and green, respectively

To obtain the GSF, the constrained optimisation problem in Eq. 4.7 is solved with the minimum thickness f_{thk} objective (Eq. 4.11). The minimum thickness value of $t_{\min} = 0.151$ m is obtained, resulting in the network and tightened structural envelope depicted in Fig. 4.7. The GSF is computed, resulting in GSF $= t_0/t_{\min} = 3.3$. From Heyman (1969), masonry is considered to present a sufficient safety level if the thrust line (or network) is found within the middle third of the structural section (i.e., GSF ≥ 3.0). From this benchmark, we can assume that this shallow vault with $t_0/s = 0.05$ is safe.

Beyond computing the GSF, an additional measure of stability can be obtained by looking at the structure's stability domain (Maia Avelino et al., 2021a, 2021b). This domain serves as a measure of the infinite space of admissible stress states arising at the structure. It is obtained by computing the minimum and maximum horizontal thrusts (as in Sect. 4.4.2) for decreasing thicknesses of the structure until the minimal thickness $t_{\min}$. The stability domain for the cross vault analysed is depicted in Fig. 4.8.

Indeed, the area among the maximum and minimum thrust lines stores all possible internal states for the structure that can be evaluated with a given form diagram. As

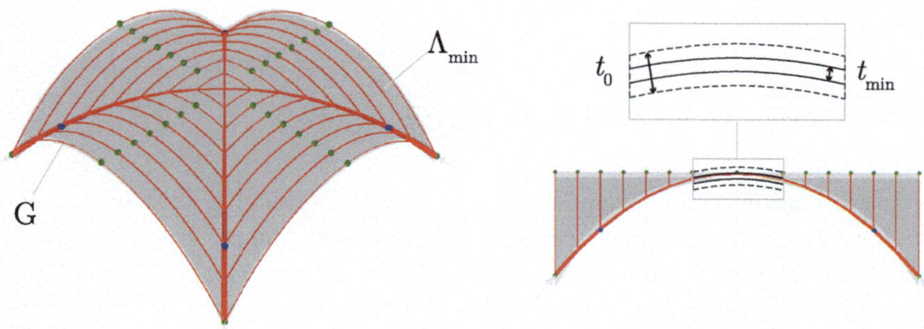

Fig. 4.7 Minimum thickness solution for the shallow vault with thickness $t_{\min} = 0.151$ m resulting in a GSF of 3.3

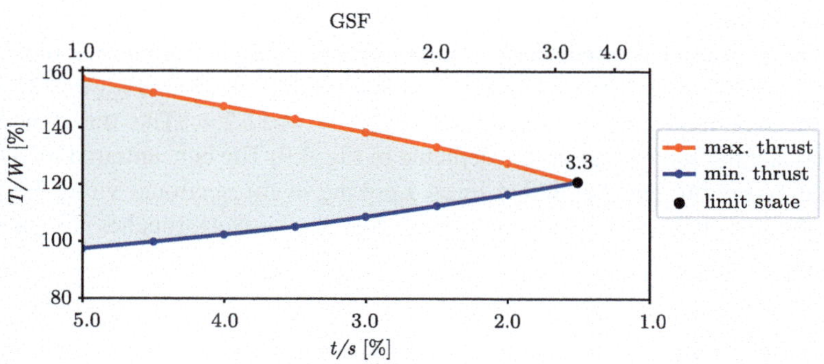

Fig. 4.8 Stability domain obtained for the shallow cross vault plotting the normalised thrust-over-weights (T/W) for decreasing normalised structural thicknesses (t/s)

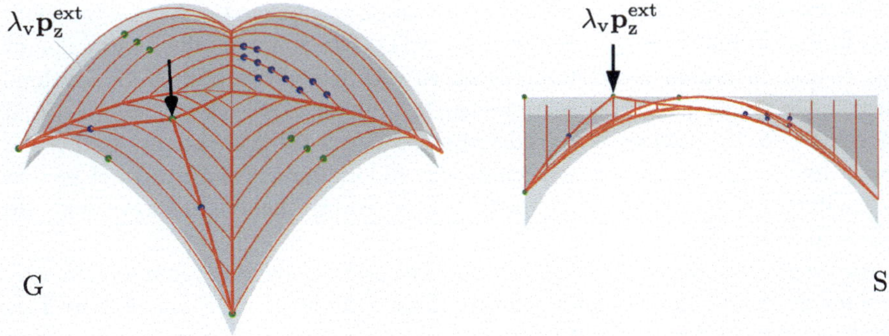

Fig. 4.9 Results for optimising a concentrated vertical load applied to the middle of the web for an orthogonal form diagram added with a direct force path to the supports. The maximum vertical load multiplier is $\lambda_v^{max} = 121.8$ resulting in a load-over-total-weight of P_{max}/W=11.2%

discussed in Maia Avelino et al. (2021b), Nodargi and Bisegna (2022), different form diagrams can be compared to enlarge this stability domain and obtain a larger GSF.

4.4.4 Maximum Vertical Load Multiplier

The maximisation of a concentrated vertical load is studied in this Section. The cross vault is subjected to an off-centred unitary pointed load $\mathbf{p}_z^{ext}$ $[n \times 1]$ applied in the middle of the structural web $(x_p, y_p) = (5.0, 2.5)$. The auxiliary variable $\lambda_v \geq 0$ representing the vertical load multiplier is considered to model this problem. For the present problem, $\mathbf{p}_z^{ext}$ has zero entries to all but the position related to the vertex (x_p, y_p), which has entry -1.0 representing an additional load pointing downwards.

To account for the additional load, the form diagram used in Sects. 4.4.2–4.4.3 is modified by adding a direct path from the point of load application (x_p, y_p) to the adjacent supports.

The optimisation is performed to minimise Eq. 4.12. It results in a maximum applied load multiplier $\lambda_v^{max} = 121.8$, i.e., a maximum applied load $P_{max} = 121.8$ kN, resulting in a load-over-total-weight of $P_{max}/W = 11.2\%$. The thrust network obtained and the force diagram are depicted in Fig. 4.9. The concentrated load flows to the supports through the added lines. Looking at the sectional view, the point where the load is applied is discontinuous, and the network touches the extrados (green dot).

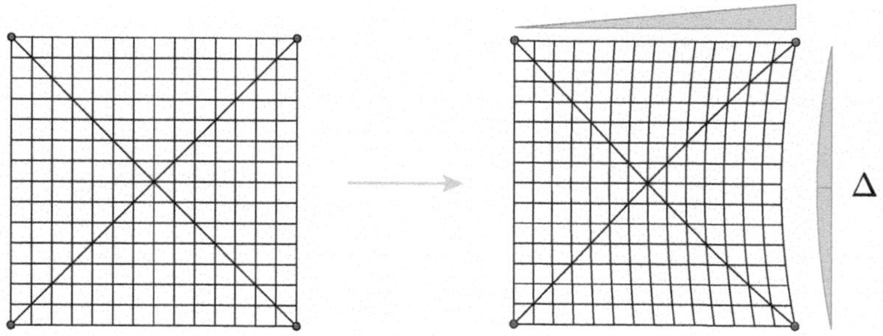

Fig. 4.10 Diagram modification by applying a tapered curved profile to the nodes of the form diagram to enable the application of horizontal loads with $\Delta = 5\%$ of the span

4.4.5 *Maximum Horizontal Load Multiplier*

This section presents the problem of computing the maximum horizontal load multiplier at the shallow cross vault. The vector defining the loading case $\mathbf{p}_{\mathrm{h}}^{\mathrm{ext}}$ assigns a force in the x-direction at every node with the same magnitude as the lumped self-weight ($p_{z,i}$) applied to that node. An additional variable $\lambda_{\mathrm{h}} \geq 0$ is considered to represent the horizontal load multiplier, i.e. a fraction of the vertical load applied horizontally.

The form diagram used for the analysis is modified with a curved profile and maximum magnitude equal to $\Delta = 5\%$ of the span. This sliding transformation can be seen as the equivalent of tilting the planar diagram to respond to the horizontal loads. This transformation is depicted in Fig. 4.10.

The optimisation problem is solved again using the objective function in Eq. 4.13. The maximum value obtained for this problem is $\lambda_{\mathrm{h}}^{\mathrm{max}} = 0.179$, meaning that the maximum horizontal-load-over-weight applied to the cross vault is equal to $P_{\mathrm{max}}/W = 17.9\%$. The solution is depicted in Fig. 4.11.

This problem is relevant to modelling, e.g., the static equivalent of seismic forces in masonry vaults (Dejong, 2009). It also corresponds to tilting the vault by a giving rotation angle such that a fraction of the weight is applied horizontally (Zessin, 2012). The maximum horizontal load multiplier obtained depends on the form diagram chosen and the sliding magnitude Δ applied. Hence, further modifications and new topologies should be coupled to this problem, considering alternative internal force distributions.

4.4.6 *Internal State Under Support Displacements*

This section discusses the problem of subjecting the vault to a support settlement. By minimising the complementary energy (Eq. 4.14) of the structure subjected to a given support displacement $\bar{\mathbf{u}}$, a stress state can be found compatible with $\bar{\mathbf{u}}$ (Angelillo,

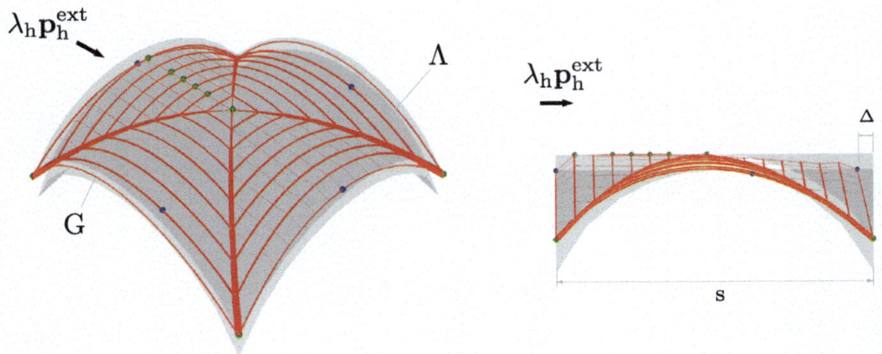

Fig. 4.11 Results for maximising the horizontal load multiplier $\lambda_h^{max} = 0.179$, resulting in a normalised horizontal force $P_{max}/W = 17.9\%$ applied to the cross vault with a horizontal slide of the nodes of the pattern equal to $\Delta/s = 5\%$

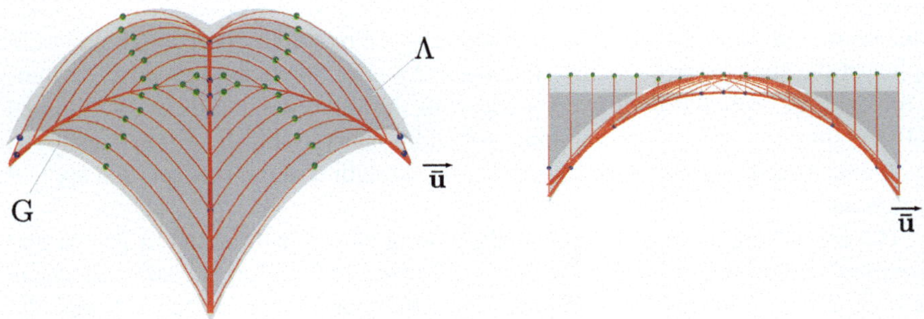

Fig. 4.12 Thrust network (G) obtained for the optimisation minimising the complementary energy subjected to a unit diagonal settlement in the highlighted support $\bar{\mathbf{u}}$. The objective function value is $W_c = 266.9$ kNm

2014; Iannuzzo et al., 2018, 2020). The shallow cross vault is subjected to a unitary outward diagonal displacement applied to one of the vault's support $\bar{\mathbf{u}}$ (see Fig. 4.12). The thrust network obtained and extreme points touching intrados and extrados are depicted in Fig. 4.12.

The optimal complementary energy value is $W_c = 266.9$ kNm i.e., $W_c/W = 20.4\%$. The outward diagonal displacement reflects a spreading of the webs crossing the pulled diagonal. This spreading reflects into two crack lines crossing that diagonal, obtained by connecting adjacent vertices touching the extrados (green). Qualitatively, the crack lines in the webs described follow the crack pattern obtained in cross vaults with Discrete Element Modelling (DEM) in McInerney and DeJong (2015).

Minimising the complementary energy shows how cracks are induced at the onset of support displacements. Future work can rely on TNO analysis to study the inverse problem of determining the foundation displacements that resulted in the cracked configuration observed in general vaulted masonry structures. The crack location can then be used to inform reinforcing strategies if necessary (De Santis et al., 2019; Fugger et al., 2022).

4.4.7 Towards the Application of TNA in Surveyed Buildings

This final section highlights the potential of applying the TNA framework directly to the scanned geometries of surveyed buildings. The case of St. Angelo Church in Anagni, Italy (Maia Avelino et al., 2022b) is revisited and summarised in Fig. 4.13. Given TNA's simple input requirement, the analysis can be performed based on the geometry acquired, and the pipeline for the analysis is described herein.

The geometry is acquired through photogrammetry or laser scanning surveys, and the vault's intrados and extrados surfaces are extracted to generate the optimisation constraints (see Eqs. 4.8). Finding a suitable form diagram for an existing, distorted, highly constrained structural envelope is challenging. Currently, the patterning of such diagrams follows the designer's experience following the structure's geometric features, curvature, openings and cracks. In Fig. 4.13b, five different diagram

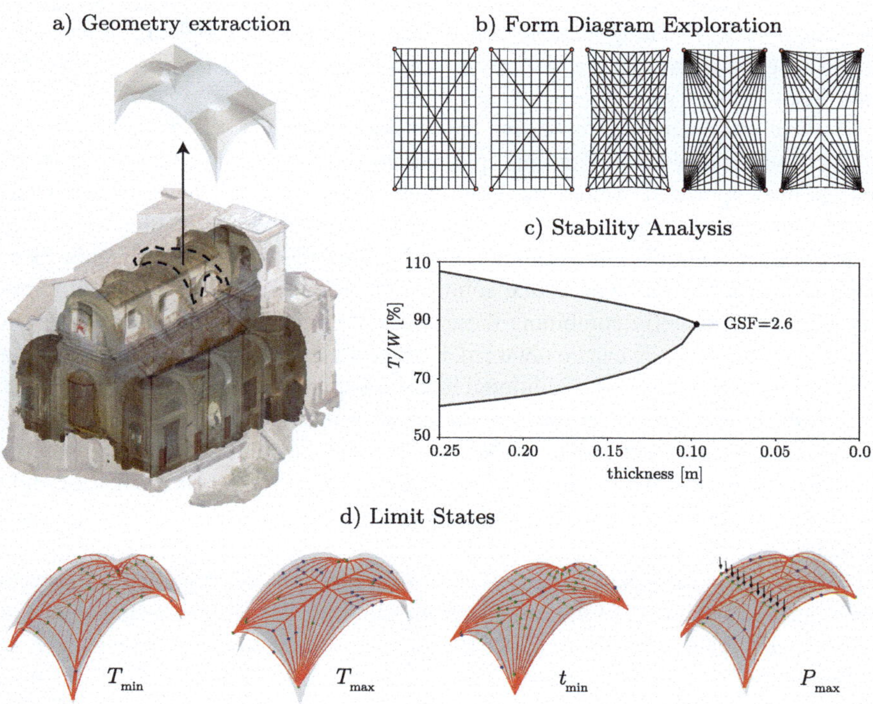

Fig. 4.13 Pipeline for the TNA stability analysis from geometry obtained from scanned data. Case study at the St. Angelo Church, in Anagni, Italy

topologies are used to investigate the structure's stability domain. The domain is computed considering the envelope of these diagrams as shown in Fig. 4.13c. The stability domain obtained is a lower bound of the structure's actual space of allowable stresses. Further form diagrams can be added to the analysis to expand such domains. A series of limit states can also be computed, e.g., the minimum ($T_{\min}$) and maximum ($T_{\max}$) thrust states, the minimum thickness ($t_{\min}$) and collapse load cases ($P_{\max}$).

The modular multi-objective framework presented in this book chapter enables these analyses that could be readily applied to isolated vaults of scanned heritage buildings. Future research will focus on improving or automating the definition of form diagrams to the analyses and considering the global structural stability of the building, taking into account its laterally supporting structure and the interaction between adjacent vaulted elements in the structure.

4.5 Conclusions

This chapter reviews Thrust Network Analysis (TNA) approaches for masonry structures based on limit analysis. TNA enables finding admissible stress states in masonry structures by formulating suitable constrained optimisation problems. Admissible stress states in masonry structures correspond to compressive networks within the structure's geometry.

A novel modular multi-objective optimisation framework is presented in this chapter, which enables finding multiple particular solutions in masonry structures with a single approach. Among these solutions are the minimum and maximum horizontal thrusts, the minimum structural thickness (or maximum Geometric Safety Factor), the maximum vertical and horizontal collapse loads, and the effect of foundation settlements. These particular solutions are obtained by considering different objective functions. By combining these states, a complete picture of the level of stability of the structure can be obtained. Furthermore, it can give information about crack patterns after applying additional loads or support movements.

Finally, recent activity on lower-bound equilibrium methods, including continuous and discrete methods, such as TNA, and the development of novel open-source computational tools based on these methods has the potential to impact the field by increasing the number of assessment tools available to analyse heritage masonry structures.

Nevertheless, multiple challenges and open questions remain, such as how to overcome the specificity of the form diagram and provide engineers with adapted or automatically generated diagrams for the analysis or how to extend the analysis to the building scale, obtaining global structural safety factors. Future work should also ease working with scanned data obtained from existing buildings to leverage TNA's simple geometry-based input and enhance its applications in practical assessment scenarios.

References

Aita, D., Pedemonte, O., & Williams, K. (eds.) (2015). *Masonry structures: Between mechanics and architecture* (1st ed.). Springer International Publishing. ISBN 978-3-319-13003-3. https://doi.org/10.1007/978-3-319-13003-3.

Aita, D., Barsotti, R., & Bennati, S. (2019). A parametric study of masonry domes equilibrium via a revisitation of the Durand-claye method. In *Proceedings of the 7th International Conference on Computational Methods in Structural Dynamics and Earthquake Engineering (COMPDYN 2015), Crete, Greece, 2019* (pp. 663–672). Institute of Structural Analysis and Antiseismic Research School of Civil Engineering National Technical University of Athens (NTUA) Greece. ISBN 978-618-82844-5-6. https://doi.org/10.7712/120119.6947.19313.

Angelillo, M. (2014). Practical applications of unilateral models to Masonry Equilibrium. In F. Pfeiffer, F. G. Rammerstorfer, E. Guazzelli, B. Schrefler, P. Serafini, & M. Angelillo (Eds.), *Mechanics of masonry structures* (Vol. 551, pp. 109–210). Springer. ISBN 978-3-7091-1773-6 978-3-7091-1774-3. https://doi.org/10.1007/978-3-7091-1774-3_4.

Angelillo, M., Babilio, E., & Fortunato, A. (2013). Singular stress fields for masonry-like vaults. *Continuum Mechanics and Thermodynamics, 25*(2–4), 423–441. ISSN 09351175. https://doi.org/10.1007/s00161-012-0270-9.

Angelillo, M., Fortunato, A., Gesualdo, A., Iannuzzo, A., & Zuccaro, G. (2018). Rigid block models for masonry structures. International Journal of Masonry Research and Innovation, 3(4), 349, 2018. *ISSN, 2056–9459*, 2056–9467. https://doi.org/10.1504/IJMRI.2018.095701

Angelillo, M., Olivieri, C., & DeJong, M. J. (2021). A new equilibrium solution for masonry spiral stairs. *Engineering Structures, 238*, 112176. ISSN 01410296. https://doi.org/10.1016/j.engstruct.2021.112176.

Baker, S. J., Horne, M. R., & Heyman, J. (January 1956). *The steel skeleton* (Vol. 2). Plastic Behaviour and Design: Cambridge University Press. 978-0-521-04088-4

Bhooshan, S., Bhooshan, V., Dell'Endice, A., Chu, J., Singer, P., Megens, J., Van Mele, T., & Block, P. (2022). The Striatus bridge. *Architecture, structures and construction*. ISSN 2730-9894. https://doi.org/10.1007/s44150-022-00051-y.

Block, P. (2009). *Thrust network analysis: Exploring three-dimensional equilibrium*. Ph.D. thesis, Massachusetts Institute of Technology. https://dspace.mit.edu/handle/1721.1/49539.

Block, P., & Lachauer, L. (2014). Three-dimensional funicular analysis of masonry vaults. *Mechanics Research Communications, 56*, 53–60. ISSN 00936413. https://doi.org/10.1016/j.mechrescom.2013.11.010.

Block, P., & Ochesendorf, J. (2007). Thrust network analysis: A new methodology for 3D equilibrium. *Journal of the International Association for Shell and Spatial Structures, 48*(155), 1–7. ISSN 1028-365X.

Block, P., & Ochsendorf, J. (2002). Lower-bound analysis of masonry vaults (1892).

Block, P., Van Mele, T., Liew, A., DeJong, M., & Escobedo, D. (2018). Structural design, fabrication and construction of the Armadillo vault - the institution of structural engineers. *The Structural Engineer, 96*(5), 10–20. https://www.istructe.org/journal/volumes/volume-96-(2018)/issue-5/structural-design-fabrication-and-construction-of/.

Block, P., Calvo Barentin, C., Ranaudo, F., & Paulson, N. (2019). Imposing challenges, disruptive changes: Rethinking the floor slab

Block Research Group. RhinoVault 2 (2021). https://github.com/BlockResearchGroup/compas-RV2

Bruggi, M. (2020). A constrained force density method for the funicular analysis and design of arches, domes and vaults. *International Journal of Solids and Structures, 193–194*, 251–269. ISSN 00207683. https://doi.org/10.1016/j.ijsolstr.2020.02.030.

Chiozzi, A., Milani, G., & Tralli, A. (2017). A genetic algorithm NURBS-based new approach for fast kinematic limit analysis of masonry vaults. *Computers and Structures, 182*, 187–204. ISSN 00457949. https://doi.org/10.1016/j.compstruc.2016.11.003.

Culmann, K. (1875). *Die graphische statik*. Meyer and Zeller (A. Reimann), Zurich.

Cundall, P. (1971). A computer model for simulating progressive, large-scale movements in blocky rock systems. *International Journal of Rock Mechanics, Mining Sciences, and Geomechanics Abstracts, 25*(3), 107–116. https://www.semanticscholar.org/paper/A-computer-model-for-simulating-progressive%2C-in-Cundall/b185b3a85160c3a77ec30e6b6e8d2e2e6de82033.

D'Altri, A. M., Sarhosis, V., Milani, G., Rots, J., Cattari, S., Lagomarsino, S., Sacco, E., Tralli, A., Castellazzi, G., & de Miranda, S. (2019). Modeling strategies for the computational analysis of unreinforced masonry structures: Review and classification. *Archives of Computational Methods in Engineering*. ISSN 18861784. https://doi.org/10.1007/s11831-019-09351-x.

De Santis, S., de Felice, G., & Roscini, F. (2019). Retrofitting of masonry vaults by basalt textile-reinforced mortar overlays. *International Journal of Architectural Heritage, 13*(7), 1061–1077. ISSN 1558-3058. https://doi.org/10.1080/15583058.2019.1597947.

Dejong, M. J. (2009). *Seismic assessment strategies for masonry structures*. Ph.D. thesis, MIT, Cambridge.

Fantin, M., & Ciblac, T. (2016). Extension of thrust network analysis with joints consideration and new equilibrium states. *International Journal of Space Structures, 31*(2–4), 190–202. ISSN 20598033. https://doi.org/10.1177/0266351116661814.

Fraddosio, A., Lepore, N., & Piccioni, M. D. (2019). *Lower bound limit analysis of masonry vaults under general load conditions*. Springer International Publishing. ISBN 978-3-319-99440-6. https://doi.org/10.0.

Fraddosio, A., Lepore, N., & Piccioni, M. D. (2020). Thrust surface method: An innovative approach for the three-dimensional lower bound Limit Analysis of masonry vaults. *Engineering Structures, 202*(October 2019), 109846. ISSN 18737323. https://doi.org/10.1016/j.engstruct.2019.109846.

Fraternali, F. (2010). A thrust network approach to the equilibrium problem of unreinforced masonry vaults via polyhedral stress functions. *Mechanics Research Communications, 37*(2), 198–204. ISSN 00936413. https://doi.org/10.1016/j.mechrescom.2009.12.010.

Fugger, R., Maia Avelino, R., Iannuzzo, A., Block, P., & de Felice, G. (2022). A new numerical limit anaysis-based strategy to retrofit masonry curved structures with FRCM systems. In *The 8th European Congress on Computational Methods in Applied Sciences and Engineering*, 5–9 June 2022, Oslo, Norway.

Funari, M. F., Silva, L. C., Mousavian, E., & Lourenço, P. B. (2021). Real-time structural stability of domes through limit analysis: Application to St. Peter's dome. *International Journal of Architectural Heritage*, 1–23. ISSN 1558-3058. https://doi.org/10.1080/15583058.2021.1992539.

Grillanda, N., Chiozzi, A., Milani, G., & Tralli, A. (2019). Collapse behavior of masonry domes under seismic loads: An adaptive NURBS kinematic limit analysis approach. *Engineering Structures, 200*, 109517. ISSN 18737323. https://doi.org/10.1016/j.engstruct.2019.109517.

Heyman, J. (1966). The stone skeleton. *International Journal of Solids and Structures, 2*(2), 249–279. ISSN 00207683. https://doi.org/10.1016/0020-7683(66)90018-7.

Heyman, J. (1969). The safety of masonry arches. *International Journal of Mechanical Sciences, 11*(4), 363–385. ISSN 00207403. https://doi.org/10.1016/0020-7403(69)90070-8.

Heyman, J. (1995). *The stone skeleton*. Cambridge University Press. ISBN 978-0-521-47270-8. https://doi.org/10.1017/CBO9781107050310.

Huerta, S. (2008). The analysis of masonry architecture: A historical approach. *Architectural Science Review, 51*(4), 297–328. https://doi.org/10.3763/asre.2008.5136

Iannuzzo, A., Angelillo, M., De Chiara, E., De Guglielmo, F., De Serio, F., Ribera, F., & Gesualdo, A. (2018). Modelling the cracks produced by settlements in masonry structures. *Meccanica, 53*(7), 1857–1873. ISSN 0025-6455, 1572-9648. https://doi.org/10.1007/s11012-017-0721-2.

Iannuzzo, A., Van Mele, T., Block, P. (2020). Piecewise rigid displacement (PRD) method: A limit analysis-based approach to detect mechanisms and internal forces through two dual energy criteria. *Mechanics Research Communications, 107*, 103557. ISSN 00936413. https://doi.org/10.1016/j.mechrescom.2020.103557.

Lemos, J. V. (2007). Discrete element modeling of masonry structures. *International Journal of Architectural Heritage, 1*(2), 190–192. ISSN 1558-3058. https://doi.org/10.1080/15583050601176868.

Lemos, J. V. (2019). Discrete element modeling of the seismic behavior of masonry construction. *Buildings, 9*(2), 43. ISSN 2075-5309. https://doi.org/10.3390/buildings9020043.

Liew, A., Pagonakis, D., Van Mele, T., & Block, P. (2018). Load-path optimisation of funicular networks. *Meccanica, 53*(1–2), 279–294. ISSN 0025-6455. https://doi.org/10.1007/s11012-017-0714-1.

Liew, A., Avelino, R., Moosavi, V., Van Mele, T., Block, P. (2019). Optimising the load path of compression-only thrust networks through independent sets. *Structural and Multidisciplinary Optimization, 60*(1), 231–244. ISSN 1615-147X. https://doi.org/10.1007/s00158-019-02214-w.

LimitState Ltd. Limit State:RING - Industry Leading Masonry Arch Analysis Software | Limit State (2020). https://www.limitstate.com/ring.

Lourenço, P. B., & Rots, J. G. (1997). Multisurface interface model for analysis of masonry structures. *Journal of Engineering Mechanics, 123*(7), 660–668. ISSN 0733-9399. https://doi.org/10.1061/(ASCE)0733-9399(1997)123:7(660).

Maia Avelino, R. (2023). Compas TNO: Finding admissible thrust networks in vaulted masonry structures. https://doi.org/10.5281/zenodo.7895617.

Maia Avelino, R. (2023). Thrust network optimisation for the assessment of vaulted masonry structures. Ph.D. thesis, ETH Zurich, Zurich, 2023. https://doi.org/10.3929/ethz-b-000611351.

Maia Avelino, R., Iannuzzo, A., Van Mele, T., & Block, P. (2021a). Assessing the safety of vaulted masonry structures using thrust network analysis. *Computers & Structures, 257*, 106647. ISSN 00457949. https://doi.org/10.1016/j.compstruc.2021.106647.

Maia Avelino, R., Iannuzzo, A., Van Mele, T., & Block, P. (2021b). Parametric stability analysis of groin vaults. *Applied Sciences, 11*(8), 3560. ISSN 2076-3417. https://doi.org/10.3390/app11083560.

Maia Avelino, R., Iannuzzo, A., Van Mele, T., & Block, P. (2022a). An energy-based strategy to find admissible thrust networks compatible with foundation settlements in masonry structures. *Mechanics Research Communications, 125*, 103978. ISSN 0093-6413. https://doi.org/10.1016/j.mechrescom.2022.103978.

Maia Avelino, R., Olivieri, C., Donval, E., Fugger, R., Lai, M., Saretta, Y., Weichbrodt, A., & Sangirardi, M. (2022b). An integrated approach to the structural assessment of historic buildings: The case of St. Angelo Church in Anagni. *Structural, 241*. https://doi.org/10.12917/STRU241.13.

McInerney, J., & DeJong, M. J. (2015). Discrete element modeling of groin vault displacement capacity. *International Journal of Architectural Heritage, 9*(8), 1037–1049. ISSN 1558-3058. https://doi.org/10.1080/15583058.2014.923953.

Mehrotra, A., Liew, A., Block, P., & DeJong, M. J. (2023). An integrated modelling approach for the seismic collapse assessment of masonry towers. *International Journal of Architectural Heritage, 17*(1), 90–113. ISSN 1558-3058. https://doi.org/10.1080/15583058.2022.2139207.

Milani, E., Milani, G., & Tralli, A. (2008). Limit analysis of masonry vaults by means of curved shell finite elements and homogenization. *International Journal of Solids and Structures, 45*(20), 5258–5288. ISSN 00207683. https://doi.org/10.1016/j.ijsolstr.2008.05.019.

Moseley, H. (1843). On the theory of the arch. Practice and architecture of bridges. In J. Weale (Ed.), *Theory* (pp. 1–72). Architectural Library.

Nodargi, N. A.& Bisegna, P. (2022). Generalized thrust network analysis for the safety assessment of vaulted masonry structures. *Engineering Structures, 270*, 114878. ISSN 0141-0296. https://doi.org/10.1016/j.engstruct.2022.114878.

Obvis. (2016). ArchieM. Version 2.5.1. Obvis Ltd. http://www.obvis.com/archie-m/.

O'Dwyer, D. (1999). Funicular analysis of masonry vaults. *Computers & Structures, 73*(1–5), 187–197. ISSN 00457949. https://doi.org/10.1016/S0045-7949(98)00279-X.

Parisi, F., Balestrieri, C., & Varum, H. (2019). Nonlinear finite element model for traditional adobe masonry. *Construction and Building Materials, 223*, 450–462. ISSN 0950-0618. https://doi.org/10.1016/j.conbuildmat.2019.07.001.

Poleni, G. Memorie istoriche della gran cupola del tempio vaticano, e de'danni di essa, e de'ristoramenti loro, divise in libri cinque ... Stamperia del Seminario, Padova, 1748. https://doi.org/10.3931/e-rara-13403.

Prager, W. (1959). *An introduction to plasticity*. Addison-Wesley Publishing Company.

Rippmann, M. (2016). Funicular shell design: Geometric approaches to form finding and fabrication of discrete funicular structures. Ph.D. thesis, ETH Zurich, Zurich.

Rippmann, M., Lachauer, L., & Block, P. (2012). Interactive vault design. *International Journal of Space Structures, 27*(4), 219–230. ISSN 0266-3511. https://doi.org/10.1260/0266-3511.27.4.219.

Schek, H. J. (1974). The force density method for form finding and computation of general networks. *Computer Methods in Applied Mechanics and Engineering, 3*(1), 115–134. ISSN 00457825. https://doi.org/10.1016/0045-7825(74)90045-0.

Shin, H. V., Porst, C. F., Vouga, E., Ochsendorf, J., & Durand, F. (2016). Reconciling elastic and equilibrium methods for static analysis. *ACM Transactions on Graphics, 35*(2), 1–16. ISSN 0730-0301. https://doi.org/10.1145/2835173.

Smars, P. (2000). Sur La Stabilite Des Arcs et Voutes. Ph.D. thesis, KULeuven.

Smoljanović, H., Živaljić, N., & Nikolić, Ž. (2013). A combined finite-discrete element analysis of dry stone masonry structures. *Engineering Structures, 52*, 89–100. ISSN 0141-0296. https://doi.org/10.1016/j.engstruct.2013.02.010.

Tralli, A., Alessandri, C., & Milani, G. (2014). Computational methods for masonry vaults: A review of recent results. *The Open Civil Engineering Journal, 8*(1), 272–287. ISSN 1874-1495. https://doi.org/10.2174/1874149501408010272.

Ungewitter, G. (1890). *Lehrbuch Der Gotischen Konstruktionen*. Leipzig: T.O. Weigel Nachfolger.

Van Mele, T., & Block, P. (2014). Algebraic graph statics. *CAD Computer Aided Design, 53*, 104–116. ISSN 00104485. https://doi.org/10.1016/j.cad.2014.04.004.

Wächter, A., & Biegler, L. T. (2006). On the implementation of an interior-point filter line-search algorithm for large-scale nonlinear programming. *Mathematical Programming, 106*(1), 25–57. ISSN 0025-5610. https://doi.org/10.1007/s10107-004-0559-y.

Wolfe, W. S. (1921). *Graphical analysis: A text book in graphic statics*. New York: McGraw-Hill Book Company.

Zessin, J., Lau, W., & Ochsendorf, J. (2010). Equilibrium of cracked masonry domes. *Proceedings of the Institution of Civil Engineers: Engineering and Computational Mechanics, 163*(3), 135–145. ISSN 17550777. https://doi.org/10.1680/eacm.2010.163.3.135.

Zessin, J. F. (2012). Collapse analysis of unreinforced masonry domes and curving walls. Ph.D. thesis, MIT, Cambridge. http://dspace.mit.edu/handle/1721.1/70879.

Chapter 5
Discrete Element Modeling

José V. Lemos and Katalin Bagi

Abstract This chapter consists of two parts. First a general overview is given: the exact definition of the Discrete Element Method is provided, and the main types of existing models are introduced; then the theoretical fundaments of the most important approaches available today for masonry analysis are given. The second part of the chapter focuses on 3DEC, the most widespread commercial DEM code for masonry analysis. This second part begins with a theoretical overview. Application issues for masonry structures are then considered (block material behaviour, contact representation, structural elements like cables or bars etc.). Among the practical examples arches, barrel vaults, walls, coloumns and complete structures (like houses) are presented, including the issues of quasi-static versus dynamic analysis and of block fracture as well. The readers can find helpful advice on how to simulate their own problems most effectively.

5.1 Overview of DEM Models

5.1.1 Introductory Remarks

As already mentioned in Chap. 1.2, a numerical technique belongs to the Discrete Element Method if the following three criteria are satisfied:

(i) the model consists of finite-sized separate elements (rigid or deformable bodies) each having their own shape, size and position;
(ii) the elements are able for large displacements, independently of each other;

J. V. Lemos (✉)
LNEC—National Laboratory for Civil Engineering, Lisbon, Portugal
e-mail: vlemos@lnec.pt

K. Bagi
Budapest University of Technology and Economics, Budapest, Hungary
e-mail: bagi.katalin@emk.bme.hu

K. Bagi and M. Angelillo (eds.), *Discrete Computational Mechanics of Masonry Structures*, CISM International Centre for Mechanical Sciences 609,
https://doi.org/10.1007/978-3-031-32476-5_5

(iii) the modifications of the contact topology are automatically kept track by the code: in particular, the new contacts are automatically detected and taken into consideration after they are formed.

There exist several software packages and calculation tools today that satisfy these criteria. Most of them are timestepping methods: they simulate the behavior of the modelled system along time, in a series of small but finite timesteps. In general, they work in the following way:

→ The initial geometry and position of every element is defined, together with the material properties of the elements as well as of the contacts. (The elements can either be perfectly rigid, or have a deformable continuum model, e.g. linearly elastic isotropic, see below.)
→ Contacts are formed between suitably defined material points on the intersecting or contacting element surfaces. (The material properties of the contacts describe the relationship between the forces transmitted between the contacting elements and the relative displacements of those material points that form the contact.)
→ The forces that act on the elements (e.g. gravity) are also given; they may change throughout the process, i.e. they can vary in time and depend on position and velocity of their point of action.
→ Then, the equations of motion are numerically integrated, in a stepwise manner: the displacements of the elements are followed in small increments along time.

The commercial DEM software packages and most existing research codes belong to the timestepping methods. However, there exists an alternative to the timestepping methods: this is the traditional stiffness matrix approach (Bagi, 1993; Kishino, 1988). Here the stiffness matrix and the unbalanced force vector of the system is upgraded repeatedly throughout the state changing process, and from them, displacement increments are determined over and over again that lead to gradually changing approximations of the equilibrated position. Special solutions are applied for those situations when the stiffness matrix is singular (e.g. when unsupported blocks are falling out, or groups of blocks are sliding together along other blocks). Such software packages are not available today, neither commercially nor from researchers, hence this approach will not be discussed here in detail, but since they have a potential for future applications (Baraldi & Cecchi, 2017), they deserve being mentioned.

Returning to the time stepping methods now, according to the specific meaning of the equations of motion, there are basically three kinds of DEM models:

(1) **Perfectly rigid elements**

The elements are rigid bodies (their shape can be spherical, polyhedral, complex convex or concave), and the forces acting on them are reduced to their reference point (typically the mass centre): a reduced force and moment vector is prepared to every element this way, at the beginning of each time step. The equations of motion express the relations between translational and rotational accelerations and the reduced force and moment vectors. There are six scalar equations of motion for every element. Typical examples are the rigid-body version of UDEC/3DEC, the PFC code, EDEM, and the rigid-body version of Rocky-DEM and of the Non-Smooth Contact Dynamics method (Jean & Moreau, 1992).

(2) **Elements made deformable by an internal FEM mesh**

The polyhedral-shaped elements are subdivided into tetrahedral, practically a finite element mesh. These tetrahedral are mostly uniform-strain elements. In this case the translations of the nodal points are the basic kinematic unknowns (rotations are not considered independently); the mass of the Voronoi-cell around each node gives the "mass of the node", and the forces acting on each Voronoi-cell are reduced to the corresponding node (neglecting the moments). The equations of motion express the translational accelerations of the nodes. The deformable element version of UDEC/3DEC, Munjiza's FEM/DEM (Munjiza, 2004) and of the Non-Smooth Contact Dynamics method (Jean, 1999) are examples for this approach.

(3) **Deformable elements having a uniform strain field**

In this case the translation field is linear along the complete element, and the basic kinematic unknowns, in addition to the translation of the reference point and the rotation about it, are the strain components as well. There are 12 degrees of freedom for every element, and consequently, 12 scalar equations of motion. The theoretical basis for the equations of motion is the principle of stationarity of the potential plus kinetic energy. The DDA model of Shi (1992) works this way.

Classification according to the time integration technique

Another usual way to classify the existing timestepping discrete element models is according to the time integration method they use, i.e. from the positions and velocities known at the beginning of a time step, how those positions and velocities belonging to the end of the time step are determined.

Explicit methods are much easier to code and the calculation of a time step is significantly faster; however, because of the numerical stability problems typical for explicit methods, very small time steps have to be chosen and different unrealistic artificial damping has to be applied in order to solve quasi-static problems. The most widespread explicit scheme in DEM is the Central Difference method; it is applied in e.g. UDEC/3DEC, PFC, EDEM and Munjiza's FEM/DEM.

Implicit methods are more stable, and allow the user to apply longer time steps. The DDA method of Shi (1992) uses Newmark's beta-method, with a choice of the control parameters which corresponds to assuming the acceleration at the end of the time step being valid throughout the complete time step. The Non-Smooth Contact Dynamics method practically uses the implicit Euler method, with a Gauss–Seidel relaxation technique to find the contact forces, positions and velocities at the end of the considered time step. In both cases, however, convergence and uniqueness problems may still arise, due to the history dependent nature of the behavior of a system of solid blocks that can slide along, separate from and bounce into each other.

5.1.2 The Most Important Available Codes for Masonry Analysis

UDEC and 3DEC

UDEC (Universal Distinct Element Code) was the very first DEM model, developed by Cundall at the end of the 1960s (Cundall, 1971) as an alternative to FEM to simulate the behavior of fractured rock masses. Its three-dimensional version, 3DEC, is the most widely applied commercial software for masonry analysis today. Due to its importance, Sects. 5.2 and 5.3 are completely devoted to it: a detailed introduction is given, together with several applications. The main features can be summarized as follows:

→ The elements are convex polyhedra, or their any compositions. They can be rigid, or made deformable using an internal uniform-strain tetrahedral subdivision. Structural elements (cables, beams) can also be added to the blocks.
→ Contacts are formed by overlapping elements. Since the surfaces are discretized with a triangular mesh (in case of rigid as well as of deformable elements), the contacts are formed by several "sub-contacts" corresponding to the nodal points on the interpenetrated surfaces.
→ The material behavior of the elements and the contacts can be chosen by the user from a large variety, and the user can even code his or her version.
→ The elements are breakable only in such a sense that the user can set the material of the elements be "brittle", meaning that if the tension limit is reached in a tetrahedron, the tension limit drops to zero there.
→ The time integration method is explicit (the central difference scheme is applied).

DDA

DDA ("Discontinuous Deformation Analysis") was introduced by Gen-Hua Shi in his PhD dissertation (Shi, 1988) and in a journal paper (Shi, 1992). Shi suggested his model for the analysis of fractured rocks. The original version was a two-dimensional model, with uniform-strain polygonal elements. The contacts were undeformable (using a penalty function approach to achieve it) that transmitted normal and tangential forces between the elements.

In the next decades several versions of DDA were published. Three-dimensional models were developed (Shi, 2001), spherical elements were introduced (Zhao et al., 2000), higher-order translation field inside the elements were used (MacLaughlin, 1997) etc. The method has been widely applied for the analysis of masonry structures like arches, columns, stone bridges or brick buildings (Cocchetti & Rizzi, 2020; Fang et al, 2016; Jiang et al, 2014; Kamai & Hatzor 2005, 2008; Liu et al, 2018; Rizzi et al., 2014; Scheldt et al, 2002; Thavalingam et al., 2001).

DDA has the following main features:

→ The elements are deformable due to having a uniform-strain field throughout their interior. The elements are unbreakable.
→ In the original version, contacts are formed by those material points of the elements that form the first entrance when the two elements touch each other. A wide variety of contact constitutive behavior exists today.
→ The time integration method is implicit: Newmark's beta-method is applied, with a choice of control parameters that corresponds to having the accelerations valid at the end of the actual time step being constant already from the beginning of time step.
→ While in implicit methods much longer time steps can be applied than in explicit methods in general, in case of masonry structures "open-close iterations" can occur. Open-close iterations mean that the topology switches back-and-forth within a time step without finding an acceptable solution, and hence the calculations have to be repeated but with a shorter time step length. Consequently, for problems with significant rearrangements of the contact topology, the DDA method is usually computationally less efficient than the methods using explicit time integration.

Further details can be found in Bagi (2016a).

The Non-Smooth Contact Dynamics Method

The NSCD method was developed in the 1980s by Jean and Moreau (1992), Moreau (1988). Its original field of application was granular dynamics: in comparison to previous discrete element models, NSCD turned out to be very efficient when simulating granular flows, segregation, rapid avalanches, or vibration problems of granular assemblies.

While in the first NSCD versions the elements were spherical, polyhedral elements were soon included, opening the way to masonry mechanics applications. The deformability of the elements is also important to be reflected by a good numerical model (the Poisson effect, for example, can give a significant contribution to the strength of shells). Jean (1999) modelled masonry walls with deformable rectangular blocks, and developed the "LMGC" software that was proven to be able to provide realistic predictions for the quasi-static selfweight problem of a planar wall. Dubois, extending the method further, released and offered to the scientific community the open code "LMGC90" (Dubois & Jean, 2006), not only for using it by anyone interested, but also for further developments. The elements in LMGC90 are rigid or deformable, two- or three-dimensional, spheres or polyhedra.

There exists another available NSCD software ("SOLFEC", by Koziara & Bićanić, 2008). SOLFEC implements different (rigid, uniform-strain and finite element) element models, contact detection algorithms, and time integration techniques, aiming to be user-friendly as well. In order to achieve reasonable computation times for real problems, parallelization is also applied in SOLFEC.

The logic of the Non-Smooth Contact Dynamics method is very different from other discrete element techniques like 3DEC or DDA. In NSCD the basic unit of the analysis is the pair of two randomly chosen elements (contacting or non-contacting). The essence of the method is to find the contact forces transmitted between the two elements in such a way that during the analyzed time step the two elements should not overlap each other. The contact force is set to zero first, as if the two elements would not touch each other even at the end of the timestep, and if the two elements have to be slowed down in order to avoid overlap, a non-zero contact force is then calculated from the Newtonian equations of motion (also satisfying conditions corresponding to the mechanical behavior of the contact). So the motion of the system is numerically simulated in time through finite time steps, but in such a way that within the considered time step, an iterative process sweeps along randomly chosen pairs of the system over and over again, and gradually adjusts the estimated contact forces to get an improving approximation of a state that satisfies the dynamic equations of the system as well as the contact criteria.

The calculations correspond to the implicit version of Euler method: the accelerations at the end of the time step are assumed to be valid along the complete length of the time step.

The main disadvantage of NSCD is the non-uniqueness of the solution. The solver sweeps randomly along the pairs, and if the same problem is analyzed by starting the random number generator from two different initiations, the two contact force systems will be different. More details can be found in Bagi (2016b).

Munjiza's FEM/DEM

The FEM/DEM method (Munjiza, 2004; Munjiza et al, 1995, 1999) was introduced with the aim of modelling fragmentation phenomena of stones and other solid materials in situations when the individual solid bodies may form new contacts with each other while the stress field inside the bodies is also important (for example, stone blocks bounce into each other because of an explosion in a rocky neighborhood of an excavation, and break into pieces as a result of the impacts).

The elements in a FEM/DEM model are subdivided into uniform-strain tetrahedra, and the common faces between neighbouring tetrahedra can open up. These common faces in the interior of the discrete elements are breakable: in the intact state, multiple nodes are located at the same position, each of which belong to a different tetrahedron.

There is a choice for the type of stress and strain to be applied in the tetrahedra.

The contact treatment in FEM/DEM is unique and very sophisticated. Munjiza's contact recognition algorithm is linear, i.e. the computational time is proportional to the number of elements in the system. The mechanics of the contacts is described by Munjiza's "contact interaction algorithm": when two discrete elements get in contact, the contact forces (determined according to the gradient of a suitably defined potential function) are distributed over the complete overlapped domain, avoiding problems this way like unnatural stress concentrations or violation of the energy conservation principle.

The time integration is done using the central difference method, similarly to 3DEC.

In masonry mechanics, FEM/DEM is particularly advantageous when simulating seismic effects on structures whose voussoirs or bricks are endangered for fragmentation (Li et al., 2021; Nikolic et al., 2013; Smoljanovic et al., 2013).

Further methods

While a list of available DEM codes for masonry analysis can never be complete, there are a few more codes that must not be left out from this summary. One of them is the commercial code PFC version 7.0 from Itasca C.G., the developer of 3DEC (Itasca, 2019; www.itascacg.com) where rigid discrete elements can be defined in polyhedral shape as well as made from spherical elements being glued together. Another important code is Rocky-DEM (https://rocky.esss.co/software/; Fonte et al., 2015) whose main advantages are the extreme variety of available particle shapes, and the two available breakage models according to which the discrete elements can fall into pieces. The third one is the open source code YADE (Smilauer et al., 2021; https://yade-dem.org/doc/). YADE is useful basically for those researchers who have some programming experiences. YADE users share their subroutines for contact models, time integration, element shapes etc. Commercial and open-source DEM codes can be found in a wide variety today and an interested researcher can select the most appropriate one for his or her aims and possibilities.

5.2 Fundaments of the 3DEC Approach

5.2.1 Cundall's Distinct Element Method

The analysis of slope failures in jointed rock masses was the motivation for the initial development of the "distinct element method" by Cundall (1971). The original DEM codes allowed the analysis of systems of rigid bodies in mechanical interaction, either polygonal blocks or circular particles. The intended application was the analysis of the stability of slopes in jointed rock masses. The key requirements were the proper simulation of the nonlinear behavior of the joints, assuming no tensile strength and frictional sliding. A novel feature of the method was the ability to take into account the changes in geometry, including the complete detachment of the rock blocks, during the failure process. The main assumptions were:

- Polygonal blocks behave as rigid bodies
- Mechanical interaction represented by point contacts
- Contact behavior characterized by 3 parameters: normal stiffness, shear stiffness, and friction angle
- Solution based on the integration of the equations of motion by means of an explicit algorithm
- Static solutions obtained by dynamic relaxation using artificial damping
- Large displacement analysis, including automatic contact detection and update.

The conceptual model underlying this numerical representation corresponded to a medium composed of blocks of competent rock in which the deformation and failure modes were mostly governed by the discontinuities. In this 2D model, a rock mass was represented as a discontinuum, formed by an assembly of rigid polygonal blocks. The solution algorithm adopted was based on the integration of the equations of motion of the blocks by means of an explicit time stepping technique, the central-difference algorithm. Static solutions were obtained by dynamic relaxation, as in the classical finite-difference methods, in which artificial viscous damping was introduced to achieve convergence to static equilibrium, or to enable the smooth development of a failure mechanism (Otter et al., 1966). Dynamic analysis in the time domain can be performed with the same algorithm, using realistic values of damping.

The new model expanded the existing analysis capabilities. Limit equilibrium techniques employed in failure studies examined each potential unstable block independently. The DEM model could represent the interaction between all the blocks, taking into account joint deformability. A full mechanical analysis was carried out, considering kinematic compatibility and equilibrium, allowing the solution of boundary value problems.

In comparison to the FEM methods in use at the time, which approached discontinuum analysis by means of joint elements, the DEM model had definite advantages: a more versatile way of generating block systems, without mesh compatibility and numbering restrictions; the ability to carry the analysis into the large displacement regime; representation of failure modes involving complete block separation, which posed significant convergence difficulties for the iterative non-linear algorithms then available; finally, the solution algorithm adopted was simple and versatile, permitting static and dynamic problems to be solved with the same code.

Many of the features listed above would be retained in later codes, as they proved to be highly effective in the study of many other types of discrete systems. Among the main subsequent developments, we may include the consideration of deformation and failure of the blocks, general constitutive models for the discontinuities, and more efficient solution procedures.

5.2.2 3DEC Modelling of Masonry Structures. Basic Assumptions and Features

5.2.2.1 Equivalent Continuum and Discontinuum Idealisations

Masonry is a heterogeneous material composed of units, such as bricks, stones, or blocks of various materials and shapes, bonded together with or without mortar. The diversity of materials, construction techniques and the age of the structures makes for large differences in behaviour and performance. It is natural that different conceptual models can be used to analyze the mechanical response of masonry. The key distinction lies in the way the joints are represented (Lourenço, 2002).

The equivalent continuum or macro-models simulate the joint effects by adopting appropriate constitutive relations for the continuum elements. The discontinuum or micro-models represent explicitly the joints, either as joint or interface elements in FEM models, or as contacts between blocks in DEM models (Lemos, 2007). The purpose of the analysis, the size and type of structure, the available information, the computational requirements, as well as the engineer's experience, are all important factors in the selection of the most appropriate idealization for each case. The scale of the analysis remains a key aspect in the choice of model. The global response of large constructions is better suited for equivalent continuum models. DEM has been mostly employed in the study of structural components, such as panels, walls, arches, or smaller buildings, often viewed as an extension of the classical limit equilibrium methods. The progressive increase in computational power, however, has broadened the use of more detailed representations in masonry analysis.

5.2.2.2 Early Applications of Discrete Element Models to Masonry

The potential of DEM models developed for hard rock systems was recognized in other engineering fields dealing with complex, discontinuous structures displaying strongly nonlinear behaviour. Stone block structures were an evident example. Papastamatiou and Psycharis (1993) and Psycharis et al. (1993) applied the method to the nonlinear seismic analysis of the drum columns of the Apollo temple at Bassae, Greece. Afterwards, shaking table tests of a 1:3 marble model a Parthenon column, performed at the National Technical University of Athens, provided experimental data for comparison with the numerical simulations (Papantonopoulos et al., 2002), which supported the application to more complex geometries (Psycharis et al., 2003). Masonry arches were another structural type investigated with early DEM models, namely by Pagnoni and Vanzi (1995), Amadei et al. (1995), and Lemos (1995, 1998) (Fig. 5.1).

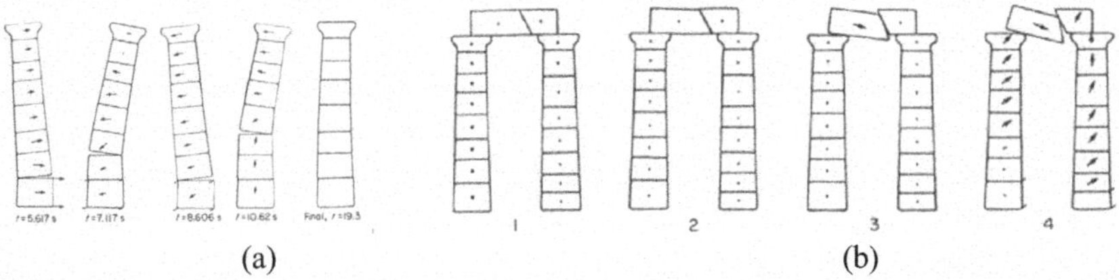

Fig. 5.1 **a** Single drum column undergoing rocking motion; **b** Collapse of broken architrave (Papastamatiou & Psycharis, 1993)

5.2.2.3 Representation of the Block Material Behaviour

Rigid Blocks

Rigid block models are appropriate for problems involving units of strong material, namely stone masonry. These models may be viewed as the numerical counterpart of the classical analytical approaches widely used in stability analysis of block structures. In DEM rigid block models, all system deformation is lumped at the discontinuities. If the actual block deformation is not negligible, the joint stiffness parameters may be modified to account for the global elastic deformation of the units. In the code 3DEC (Itasca, 2020) a rigid block is a convex polyhedron, non-convex shapes being created by rigidly assembling convex sub-blocks. The motion of a rigid block, or a rigid assembly of sub-blocks, is governed by the 6 degrees-of-freedom of rigid body mechanics.

Rigid blocks provide the simplest DEM models, with the minimum requirements in terms of material properties. For most static problems, a rigid block model has only a small computational advantage in comparison with a coarse-mesh deformable block model. However, for explicit dynamic algorithms, rigid blocks models tend to be more efficient, as will be discussed below.

Deformable Blocks

Cundall et al. (1978) extended the early rigid block models by including the deformation of the block material. This is achieved by discretizing each block into an internal finite element mesh, an approach sometimes designated by other authors as "finite-discrete elements" (Munjiza, 2004). In Cundall's codes, such as 3DEC, the internal mesh elements are generally referred to as zones, following the classical finite-difference terminology.

In 3DEC, the internal discretisation is based on uniform strain tetrahedral elements, as shown in Fig. 5.2. The choice of internal elements with a linear displacement field guarantees the block boundary to remain a polygonal line, in 2D, or a polyhedral surface formed by triangles, in 3D, which simplifies contact calculations. Triangular and tetrahedral meshes also have the advantage of fitting arbitrary block geometries, and there are many well tested procedures for automatic generation.

Deformable blocks allow the calculation of internal stresses, also providing a better simulation of structures with heterogeneous or softer materials. In stress analysis, the limitations of simple uniform-strain elements are well-known, for example for blocks undergoing bending deformation, thus they may require fine internal meshes if accurate local stresses are sought.

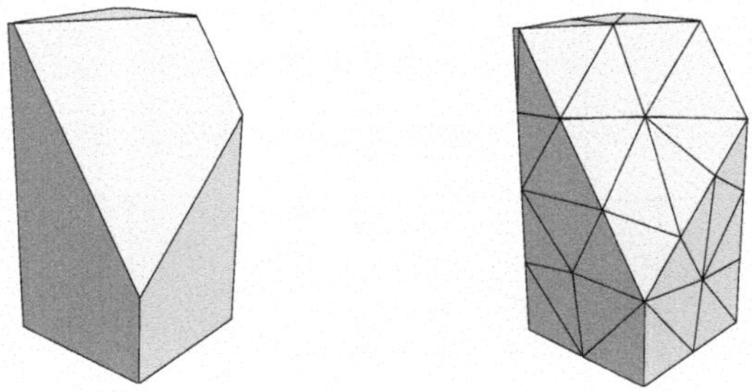

Fig. 5.2 **a** Rigid polyhedral block; **b** Deformable block with tetrahedral element mesh

Block Non-elastic Behaviour and Fracture

In most masonry analysis, deformable blocks are assumed elastic, as the joints are the main source of non-linear behaviour. However, deformable block models can also simulate the failure of the intact material by adopting elasto-plastic, damage or other nonlinear constitutive relations to the internal elements. User-defined models are also accepted by many codes.

Block fracture and splitting is also available in some DEM codes (e.g., Munjiza, 2004). A simpler alternative consists in the use of "bonded-block models", in which a random network of potential fracture paths is initially created, along which failure may take place during the analysis (Sarhosis & Lemos, 2018). The multi-block assemblies can be composed of rigid or deformable blocks, as will be discussed in Sect. 5.3.9.

5.2.2.4 Representation of Contact Between Blocks

Point Contacts and Joint Elements

The explicit representation of joints and interfaces is addresses in different ways by FEM and DEM models. FEM introduces zero-thickness joint or interface elements, derived from the original concept of Goodman et al. (1968). These elements are designed to be compatible with the adjacent continuum elements, using the same interpolation functions along the element tangent direction (Fig. 5.3). The joint element stresses are taken proportional to joint displacements, defined as the jump in normal and shear displacement across the two sides of the element. Therefore, a smooth distribution of stresses within the element is achieved. Note, however, that stresses in the continuum elements are based on strains, while in the joint elements they are obtained directly from displacement differences. Consequently, as the displacement interpolation functions are the same, the order of the stress polynomial will be different in the joint and the adjacent solid element.

In contrast, the representation of contact in most DEM models is based on the point contact approach: the interaction between blocks is represented by a set of

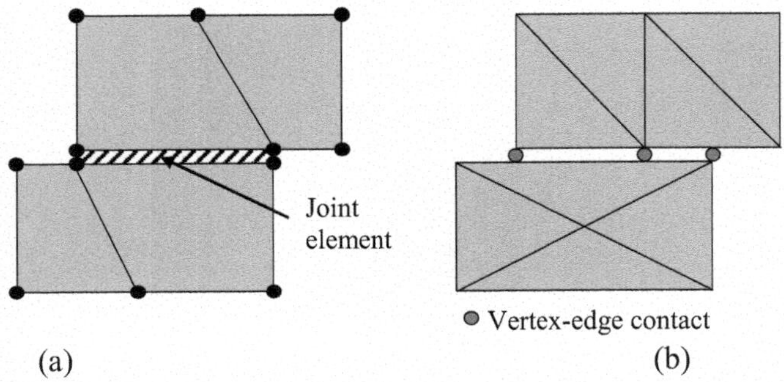

Fig. 5.3 Representation of contact between blocks by: **a** joint elements and **b** point contacts

contact points (Lemos, 2007). For example, an edge-edge contact is simulated by 2 point contacts (Fig. 5.3). These point contacts are assigned a length in 2D, or an area in 3D, so that stresses can be calculated resorting to the standard joint constitutive models, typically formulated in terms of stresses and relative displacements. The main advantage of this approach lies in its generality and simplicity, being able to handle the various types of geometric interaction and permitting large block movements. In particular, it allows for a smooth transition between the case of contact between faces and a true point contact, such as the interaction between a vertex and a face.

Contrary to the joint element approach, in point contacts the stress at a point depends solely on the joint displacement at that point, leading to a discontinuous stress distribution on the joint plane. As is to be expected, the accuracy of the stress distribution obtained on the joint plane is higher when joint elements are used. Therefore, DEM codes require a larger number of contact points for rigorous stress analysis. This drawback is compensated by their aptness for large displacement analysis and also for complex block configurations.

For deformable blocks, a vertex-edge contact may exist at any point along the edge. The appropriate interpolation functions are employed to find the relative block displacements at the contact point, and then to apply the contact forces to the 2 nodes that define the edge (in 2D). These procedures also allow adjacent blocks to have different discretisations, without matching of the nodes, as shown in Fig. 5.3. This provides a much greater flexibility in mesh design, namely the use of automatic generators for the internal mesh in the various deformable blocks. It should be noted, however, that the accuracy of the joint stress distribution improves if the nodes are matched.

Joint and Contact Stiffness

The mechanical deformability of masonry joints is described in terms of the normal and shear stiffness parameters, which express a relation between each joint stress component and the corresponding relative displacement between the blocks in contact. For mortared joints, the stiffness reflects the fill material deformability. For dry joints, they represent the increased deformation that occurs at the contact between rough or uneven block surfaces. Therefore, even in this case, joints stiffnesses have a physical significance which may be measured in the lab. As zero thickness is assumed in the numerical representation of the discontinuities, a small overlap is created when the joint is in compression, as in FEM joint elements. For the typical ranges of joint normal stiffness, this overlap is small and does not cause errors in the physics of the contact or numerical difficulties.

There are DEM formulations, however, which enforce the condition of non-interpenetration between the blocks (i.e., an infinite normal stiffness), which is also adopted in the classical analytical methods. The results of collapse analysis are generally independent of the joint stiffness parameters, thus in many cases, they are simply considered as numerical "penalty coefficients". However, when a rigid block model is intended to estimate the deformation of a stone masonry structure, the physical meaning of these parameters becomes evident. Measurements of the natural frequencies of the structure (e.g., under ambient vibration) often provide information to estimate the joint stiffnesses.

Contact Representation in 3DEC

The interaction between sharp block corners, for example of the diagonally opposite blocks created by an intersection of two joints, raises the question of defining the location of the contact and its normal vector. In the 2D code UDEC (Itasca, 2014), Cundall adopted an original logic that assumes corners to be rounded, i.e., approximated by circular arcs tangent to the adjacent edges. Therefore, corner-corner contacts are rigorously handled, with the contact normal uniquely defined by the line connecting the arc centres, evolving continuously as the two blocks rotate. This solution is very satisfactory in physical terms, but it becomes excessively complex in 3D.

For the 3DEC code, a different approach was devised (Cundall, 1988). Contact between polyhedral blocks may involve various types of interaction of faces, edges and vertices. In principle, all six combinations are possible: FF, EF, VF, EE, VE, and VV. A strategy is required to avoid checking all the possibilities. Cundall (1988) proposed the "common-plane" concept, applicable to the case of convex blocks, which provides an efficient way of detecting if two blocks in proximity are actually touching. The common-plane is defined as the plane that minimises the maximum overlap with either block. Figure 5.4 illustrates the concept in 2D, for simplicity. The decomposition of arbitrarily shaped polyhedra into convex sub-blocks still permits this method to be used, as shown in Fig. 5.4c.

Once two elements are found to be in contact, the actual point contacts, designated as "sub-contacts" in 3DEC, must be determined. For polyhedral blocks, there are several situations to be examined, depending on the interacting entities: faces, edges and vertices. In principle, two types of point contact are sufficient to describe the different geometrical possibilities: vertex-to-face (VF) and edge-to-edge (EE). The

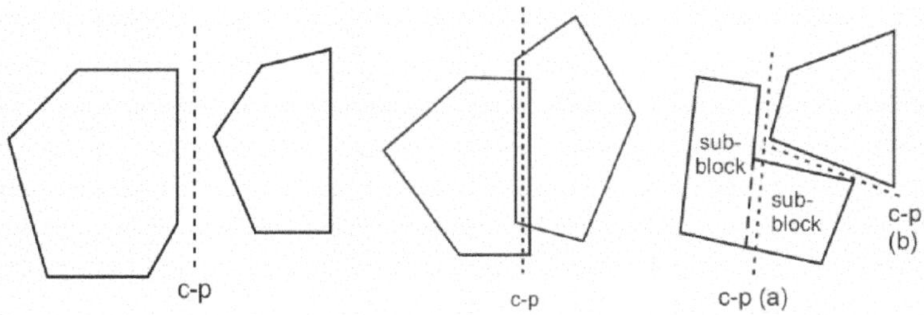

Fig. 5.4 Examples of common-plane between blocks

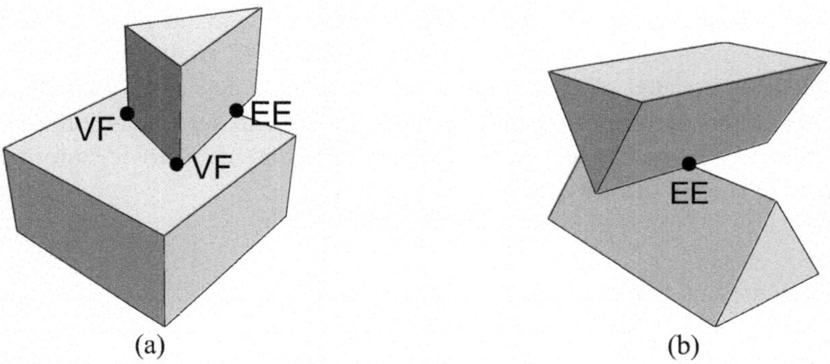

Fig. 5.5 Representation of block interaction by VF and EE point contacts: **a** face-to-face case; **b** edge-to-edge case

other combinations may be taken as particular cases. Figure 5.5 illustrates two cases: a face-to-face interaction described by 2 VF and 2 EE point contacts; and a true EE interaction that corresponds to a single point contact.

The representation of a face-to-face contact by a set of contact points is a simplified and efficient approach, well suited for systems of many blocks, either rigid or with coarse internal meshes. To achieve a more rigorous distribution of contact stresses, more sub-contacts can be created, either by dividing the faces of rigid blocks, or using a finer internal mesh for deformable blocks.

Contact Detection and Update

In principle, in the course of a large displacement analysis, each block in the assembly may move arbitrarily and come into contact with any other. The process of the detection of contacts must therefore be handled by the code without any user intervention. Even if this generality is not required in most masonry problems, the routines developed for this purpose are a great help to the user, as they can be used to create the contacts for the initial configurations, eliminating the need to define manually the system connectivity.

From a theoretical point of view, contact detection is a straightforward task, as only simple geometric calculations are required. The main problem is one of computational efficiency: for a system of many blocks, it is not possible to check all possible interactions between every pair, as the run times would increase rapidly with the number of blocks. Then, for each pair of blocks, every type of contact, for example, involving the various faces, edges and vertices that define the boundary of a polyhedron, would have to be considered in theory.

Special techniques for contact detection have been developed, and they are well reported in the literature (e.g., Cundall, 1988; Munjiza, 2004). In essence, the contact detection procedure may be divided into two parts: (i) the identification of the pairs that are in potential contact; (ii) the detailed resolution of the actual contact. In 3DEC, for the first task, a domain decomposition into a grid of 3D brick-shaped cells employed, so that only blocks that fall inside the same cell need to be checked against each other. For the second task, the verification of whether actual interaction exists, and the determination of its type, is based on the 'common-plane' algorithm mentioned above.

5.2.2.5 Reinforcement Elements

While the structural elements available in 3DEC were originally intended for rock mechanics applications, they are also useful for masonry analysis. The most elementary type is the local reinforcement element which is simply a spring attached to 2 points across a joint. This element may be composed of an axial spring and a shear spring (Fig. 5.6). Both springs have non-elastic behaviour, defined in terms of yield strength and rupture strain, allowing the element to break.

The "cable" elements are reinforcement elements inserted through the blocks and grouted, mainly for geotechnical applications. They can, however, be applied as

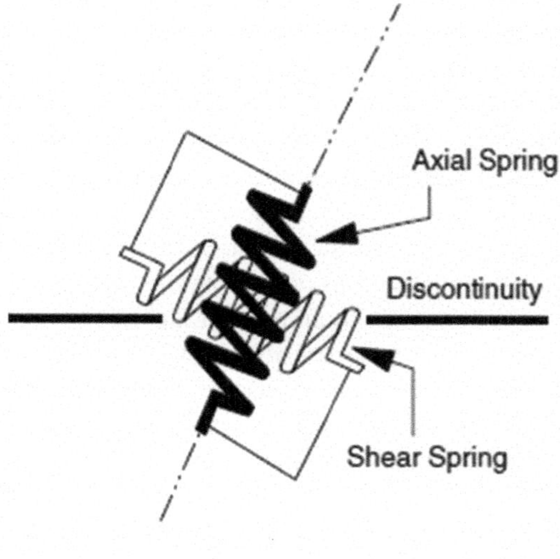

Fig. 5.6 Local reinforcement element, composed by a normal and a shear spring, both with nonlinear behaviour

ties, e.g. across an arch span, in masonry structures. The "beam" elements may also be useful in masonry models. They can be assembled into simple frame structures, connected to the block faces by means of breakable contacts. Local reinforcement elements only introduce stiffness components in the model. Cable and beam elements, however, involve adding structural nodes at the element end-points, for which the equations of motion have to be solved during the time stepping algorithm. The nodes of these elements interact with block faces through structural point contacts, with a Mohr–Coulomb type of nonlinear behaviour, so that they can break off the block or slide.

Standard local reinforcement elements were used by Psycharis et al. (2003), to simulate clamps and dowels connecting columns and architraves, Stefanou et al. (2011), and Catki et al. (2020). Beam elements were used by Gobbin et al. (2020) to represent externally bonded reinforcement of an arch.

It is also possible to create other types of reinforcement resorting to the internal programming language FISH in 3DEC. For example, ties connecting 2 arbitrary points in blocks at some distance, without links to intermediate blocks, can be implemented quite easily (Cismasiu et al., 2021). At every step, the function needs to retrieve the displacements on the 2 points, calculate the force based on the relative movement, possibly checking for element failure, and apply it at 2 anchor points.

5.2.2.6 Solution Procedures

Explicit time-stepping algorithms are the most common in DEM codes, as they are well suited for the strongly non-linear behaviour of systems of rapidly varying connectivity. This dynamic algorithm may also be used for static analysis, by means of "dynamic relaxation" (e.g. Underwood, 1983), which resorts high damping to eliminate vibration, thus allowing the convergence to equilibrium or the development of a failure mechanism. As the transient response is not sought, block and element masses may be scaled to improve convergence. The main options for damping will be reviewed in the next section.

The general calculation cycle of explicit time stepping algorithms is schematically represented in Fig. 5.7. At each time step, two sets of calculations are performed: (i) the equations of motion, either of rigid blocks or deformable block nodes, provide the new velocities and displacements, from which block and contact deformations can be obtained; (ii) application of the constitutive laws, to blocks elements and contacts, produces the new stresses, which can be assembled into the force terms for the motion equations of the next time step.

The time step of an explicit algorithm cannot be larger than a critical value in order to ensure numerical stability. This may be a significant drawback in dynamic analysis, as very stiff joints or thin elements imply small time steps. For arbitrarily shaped blocks with automatically generated tetrahedral meshes, the user has little control over the shape of elements, and the thinnest one will determine the time step. In this case, rigid block models may be much more efficient. For static problems, the procedure of mass scaling mitigates the problem, but large contrasts in stiffness or

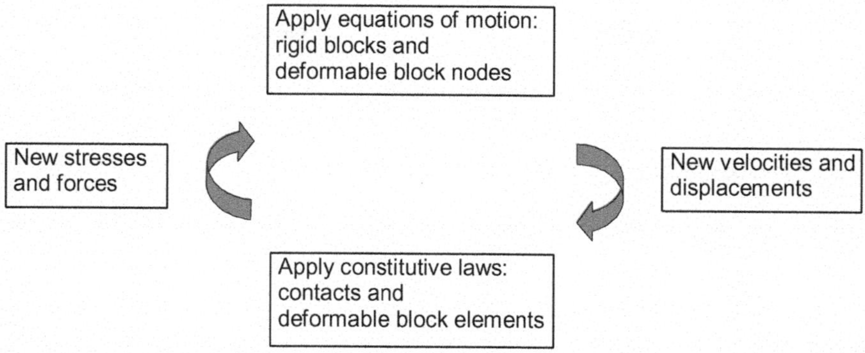

Fig. 5.7 Calculation cycle in explicit time-stepping algorithms

element size may still cause longer solution times. The elastic properties should not be unnecessarily high, for example, very stiff blocks are better represented as rigid, and very stiff elastic joints by attaching the adjacent blocks.

5.2.2.7 Large Displacement Analysis

Large displacement analysis is important in some masonry problems, for example, columns or other slender structures under horizontal seismic motion. In these problems involving block rocking, the response is strongly influenced by the motion amplitude, and the block overturning depends on the location of the mass centre. There are also cases of thin arches and vaults in which changes in geometry may have a strong impact on stability.

However, for many collapse analyses, usually performed under quasi-static conditions, it is possible to identify the failure mechanism while still in the small displacement range. Continuing the analysis into the large displacement regime may be instructive, as it helps to visualize and understand the structural behaviour, but it is often not essential for the determination of the collapse load. It should also be noted, however, that as the analysis progresses into the large displacement regime, the sources of inaccuracy increase. Firstly, there are uncertainties in the physical problem, as it becomes more dependent on geometric parameters, and initial or boundary conditions, which may not be well known. Secondly, there may be inaccuracies in the numerical models that are irrelevant in small displacement problems but affect the results of large deformation runs. Besides, the codes have often built-in assumptions to handle these situations, with implications regarding physical behaviour, not always well documented or known to the user.

In large displacement analysis, the geometry of the system, typically defined by the block vertices and mesh node, is progressively updated. The type and location of existing contacts between blocks are also updated, and new interactions may be detected. In most codes, these events are only triggered when some value of incremental displacement is attained, in order to avoid an excessive number of checks.

5.2.3 Formulation of a 3D DEM Block Model

5.2.3.1 Main Features

In this section, the fundamental expressions of the formulation of a 3D discrete element model for the mechanical analysis of block systems are presented, following the approach described by Cundall (1988) and Hart et al. (1988), and implemented in the code 3DEC (Itasca, 2020). The main features of the present formulation are:

- blocks may be assumed rigid, or deformable, discretised into an internal element mesh;
- the mechanical interaction between blocks is represented by sets of point contacts, and may be governed by general constitutive relations;
- an explicit solution algorithm is employed for static and dynamic analyses;
- a large displacement analysis is performed.

The equations of motion, of blocks and nodal points, the contact mechanical calculations, and the solution algorithm will be addressed in the following sections.

5.2.3.2 Rigid Block Motion

A rigid block is assumed to behave as a theoretical rigid body, with 6 degrees-of-freedom, 3 translations and 3 rotations. The 3 translational equations of motion of a rigid block centre of mass may be expressed, in the global axes, as:

$$m\ddot{u}_i + \alpha\, m\, \dot{u}_i = f_i \tag{5.1}$$

where

u_i denotes the displacement vector of the block centre,
m, the block mass, and:
α, the mass-proportional viscous damping parameter.

The force vector, where the elastic forces are included and is a function of the block displacements, is given by:

$$f_i = f_i^c + f_i^a \tag{5.2}$$

where

f_i^c is the sum of contact forces, and:

$f_i{}^a$, the sum of applied forces (including self-weight).

The 3 rotational degrees-of-freedom are governed by Euler's equations, expressed in the principal axes of inertia of the block as:

$$\begin{aligned} I_1\,\dot{\omega}_1 + \alpha\, I_1\,\omega_1 + (I_3 - I_2)\omega_3\omega_2 &= m_1 \\ I_2\,\dot{\omega}_2 + \alpha\, I_2\,\omega_2 + (I_1 - I_3)\omega_1\,\omega_3 &= m_2 \\ I_3\,\dot{\omega}_3 + \alpha\, I_3\,\omega_3 + (I_2 - I_1)\omega_2\,\omega_1 &= m_3 \end{aligned} \tag{5.3}$$

where

ω_i denotes the rotational velocity vector,
I_i, the principal mass moments of inertia, and:
α, the mass-proportional viscous damping parameter.

The moment term, on the r.h.s., is given by:

$$m_i = m_i^c + m_i^a \tag{5.4}$$

where

$m_i{}^c$ is the sum of moments produced by contact forces, and:
$m_i{}^a$, the sum of moments of applied forces.

For quasi-static analysis, mass and moments of inertia are scaled to improve the rate of convergence to the solution. The form of the translational Eq. (5.1) remain unchanged, with the real mass being replaced by a scaled mass m_s. In the rotational Eq. (5.3), a single scaled moment of inertia is adopted, and they can be expressed in the global axes in the simplified form:

$$I_s\dot{\omega}_i + \alpha\, I_s\,\omega_i = m_i \tag{5.5}$$

where I_s is the scaled moment of inertia, assumed equal in all directions.

5.2.3.3 Deformable Block Motion

The deformation of each element of a deformable block is described by the movement of its nodal points. Therefore, the motion of these blocks is calculated by integrating the equations of motion of each nodal point. The 3 equations of motion for a node I are:

$$m\,\ddot{u}_i^I + \alpha\, m\,\dot{u}_i^I = f_i^I \tag{5.6}$$

where

$u_i{}^I$ is the nodal displacement vector of node I,

m the nodal mass, and:
α the mass-proportional viscous damping parameter.

The nodal force vector is given by:

$$f_i^I = f_i^{cI} + f_i^{aI} + f_i^{eI} \tag{5.7}$$

where

f_i^{cI} are the contact forces transmitted to the node,
f_i^{aI}, the applied external forces, and:
f_i^{eI}, the internal nodal forces corresponding to the element stresses, obtained by integration according to standard FEM practice, employing the element deformation matrix.

5.2.3.4 Contact Mechanics

The present contact formulation is intended for both rigid and deformable blocks. As discussed above, the representation of mechanical interaction by point contacts in 3D can be achieved by two types of elementary contacts: vertex-to-face (VF) and edge-to-edge (EE) (see Fig. 5.5). In the following, the general designation contact will refer to any elementary point contact (termed sub-contacts in 3DEC).

Geometrical Parameters

Given 2 blocks A and B, the orientation of each contact is defined by the unit normal vector, n_i, defined as pointing from A to B. In general, if there are several contacts between the 2 blocks, they may be having different orientations. In the particular case of stiff convex blocks, it may be reasonable to employ a single normal for all contacts between the same 2 blocks. The normal of VF contacts will, in principle, be defined by the normal to the face, and for EE contacts by the normal to the plane defined by the 2 edges. Special cases, e.g., VE or VV, also need to be taken into account, for example by adopting Cundall's "common-plane" concept. A smooth, continuous variation of the contact normal, as the contact type and location are updated, is an important requirement to avoid numerical noise in the results.

Mechanical interaction between blocks is initiated as soon as physical contact is established, so for blocks in proximity the evolution of the contact gaps needs to be monitored. Nominal (or virtual) contacts are formed when the gap is within some fixed tolerance, and the distances between vertex and face or between contacting edges are evaluated to detect real interaction and trigger the development of contact forces.

Contact Velocities and Displacements

Once the contact position, x_i^c, is determined, the contact displacement is calculated as the relative displacement between blocks at that location. At each time step, the relative velocity at the contact point is evaluated as:

$$\dot{u}_i^c = \dot{u}_i^{cB} - \dot{u}_i^{cA} \tag{5.8}$$

where $\dot{u}_i^{cA}$ and $\dot{u}_i^{cB}$ stand for the velocities of blocks A and B at the contact position.

For a rigid block, the velocity at the point x_i^c is:

$$\dot{u}_i^{cA} = \dot{u}_i^A + e_{ijk}\, \omega_j^A (x_i^c - x_i^A) \tag{5.9}$$

where

x_i^A denotes the centre of mass of block A,
$\dot{u}_i^A$, its translational velocity,
ω_i^A, the rotational velocity,
e_{ijk} is the alternating symbol.

For deformable blocks, a different treatment is needed for VF and EE contacts. On the vertex side of a VF contact, the velocity is given by the vertex velocity. On the face side, the velocity is interpolated depending on the displacement field assumed within the block face. In the simplest case of a triangular face, a linear interpolation of the nodal velocities yields:

$$\dot{u}_i^c = W_J\, \dot{u}_i^{cJ} \quad (J = 1, 3) \tag{5.10}$$

The interpolation factors, W_J, are given by the area coordinates,

$$W_J = \frac{A_J}{A} \tag{5.11}$$

where A is the face area, and A_J the areas of the sub-triangles formed by the contact point and the face vertices. Note that a triangular face approximation is sometimes adopted for contact purposes even if the element shape functions are not linear.

For an EE contact, the velocities of the edge end-points are interpolated as:

$$\dot{u}_i^c = W_J^e\, \dot{u}_i^{cJ} \quad (J = 1, 2) \tag{5.12}$$

The interpolation factors are obtained from the length ratios:

$$W_J^e = \frac{\ell_J}{\ell} \tag{5.13}$$

where ℓ is the edge length, and ℓ_J the lengths of the 2 edge segments defined by the contact point.

For all types of blocks and contacts, once the relative contact velocity is determined, the increment in contact displacement is:

$$\Delta u_i^c = \dot{u}_i^c \, \Delta t \tag{5.14}$$

which can be resolved into a normal component, a scalar,

$$\Delta u_n = \Delta u_i^c \, n_i \tag{5.15}$$

and a shear component, a vector,

$$\Delta u_{si} = \Delta u_i^c - \Delta u_n n_i \tag{5.16}$$

which is expressed here in global coordinates (i = 1,3), but lies necessarily on the contact plane.

Contact Stresses and Forces

It is advantageous to assign an area to all point contacts, so that the same type of constitutive equations may be applied, in terms of tractions and relative displacements. For contacts that are part of a face-to-face interaction, the contact area of each point contact will be a fraction of the total area of the region of block interaction. For true point contacts, a nominal area is assigned, which in practice corresponds to the minimum contact stiffness allowed in the problem, namely intended to limit the maximum overlap between blocks.

The relation between contact stresses and forces will thus be:

$$F_n = A_c \, \sigma_n \quad \text{and} \quad F_{si} = A_c \, \sigma_{si} \tag{5.17}$$

where A_c is the contact area, and contact forces are resolved into normal and shear components in the same way as contact displacements.

At each step, trial elastic increments of the contact stresses are calculated as:

$$\Delta\sigma_n = k_n \, \Delta u_n \quad \text{and} \quad \Delta\sigma_{si} = k_s \, \Delta u_{si} \tag{5.18}$$

where the standard definition of joint stiffness as [stress/displacement] applies. An isotropic shear stiffness, k_s, was assumed in the above expression.

In large displacement analysis, the rotation of the contact normal has to be taken into consideration. Assuming this common rotation of the 2 blocks should not affect the existing contact forces in the local axes, it is necessary to correct the old shear force vector, to guarantee that it remains tangent to the new contact plane. The normal force, stored as a scalar, does not change. Denoting by n_i^{old} the unit normal in the previous step, and by n_i the new value, the existing shear force is corrected as:

$$F_{si}^{old} := F_{si}^{old} - e_{ijk}e_{kmn} F_{sj}^{old} n_m^{old} n_n \quad (5.19)$$

From this, the corrected existing shear stress is calculated, using the contact area. Then it is possible to add the trial stress increments to the existing stresses, and, after application of the selected joint constitutive model, formally expressed as:

$$\sigma_n, \sigma_{si} = f(\sigma_n^{old}, \sigma_{si}^{old}, u_n, u_{si}, \Delta u_n, \Delta u_{si}, \ldots) \quad (5.20)$$

the new contact stresses are obtained.

The normal and shear contact forces are obtained from the stresses, by multiplying by the contact area, and assembled into the vector:

$$F_i^c = F_n n_i + F_{si} \quad (5.21)$$

to be applied to the interacting blocks.

Application of Forces to the Blocks

The new contact forces must then be applied to rigid block or deformable block nodes. The total contact force vector, representing the action of block B on A, is:

$$F_i^c = F_n n_i + F_{si} \quad (5.22)$$

For rigid blocks, the contribution of this contact to the force and moment sums of block A is:

$$f_i^A = F_i^c \quad \text{and} \quad m_i^A = e_{ijk}(x_j^c - x_j^A)F_k^c \quad (5.23)$$

and the opposite forces are applied to block B

$$f_i^B = -F_i^c \quad \text{and} \quad m_i^B = -e_{ijk}(x_j^c - x_j^B)F_k^c \quad (5.24)$$

A loop over all the rigid block contacts yields the force and moment sums to be introduced in the equations of motion in the next time step.

For deformable blocks, the contact forces are distributed to the nodes by applying the reciprocal of the velocity interpolation procedure outlined above. For a VF contact, on the vertex block, the force is added directly to the nodal force. On the face block, it is distributed by the face nodes, according to the displacement interpolation assumed. For example, in a triangular face, each node J receives, from each contact, the force:

$$f_i^{cJ} = \pm W_J F_i^c \quad (J = 1, 3) \quad (5.25)$$

where the plus sign applies if the face block corresponds to block A, and the minus sign if it is block B. The force interpolation factors W_J are those defined for velocity interpolation, given by (5.16).

Similarly, for an edge to edge contact, the contribution to node J is:

$$f_i^{cJ} = \pm W_J^e F_i^c \quad (J = 1, 2) \tag{5.26}$$

using the same sign convention, and the previous definition of edge interpolation factors $W_J{}^e$ for the velocities in (5.18).

The forces from all contacts are summed into the nodal force vector, together with the internal element forces and applied loads, for use in the motion equations in the next step.

5.2.3.5 Solution Algorithm

The Central-Difference Algorithm

The equations of motion, either of the rigid blocks, (1) and (3), or of the deformable block nodes (6), are numerically solved by an explicit finite-difference algorithm, the central difference method (e.g., Hughes, 1987). In this method, accelerations and displacements velocities are evaluated at the time step end-points, while velocities correspond to mid-interval values. Therefore, the accelerations at time t are expressed in terms of the velocities at the mid-points of the time intervals, t+Δt/2 and t−Δt/2, by means of the finite-difference approximations. The velocities at time t are given by the averages of the mid-interval values. Inserting these expressions into the translational and the simplified rotational equations of motion, it yields:

$$\dot{u}_i(t + \Delta t/2) = [D_1\, \dot{u}_i(t - \Delta t/2) + \frac{f_i}{m}\Delta t]D_2 \tag{5.27}$$

$$\omega_i(t + \Delta t/2) = [D_1\, \omega_i(t - \Delta t/2) + \frac{m_i}{I_s}\Delta t]D_2 \tag{5.28}$$

where:

$$D_1 = 1 - \alpha\Delta t/2 \quad \text{and} \quad D_2 = 1/(1 + \alpha\Delta t/2) \tag{5.29}$$

Given the new velocities at $t + \Delta t/2$, then the displacement and rotation increments are calculated as:

$$\Delta u_i = \dot{u}_i(t + \Delta t/2)\Delta t \quad \text{and} \quad \Delta\theta_i = \omega_i(t + \Delta t/2)\Delta t \tag{5.30}$$

The new position and displacement of the block centre at time $t + \Delta t$ are:

$$x_i(t + \Delta t) = x_i(t) + \Delta u_i \quad \text{and} \quad u_i(t + \Delta t) = u_i(t) + \Delta u_i \tag{5.31}$$

The new block vertex velocities (see Eq. 5.9) and positions can thus be obtained. After updating the orientations of edges and faces, the contact calculations are executed, producing the new interaction forces for the next step.

For dynamic analysis, the general rotational equations apply, and the principal inertia axes have to be determined, and continuously updated during the calculation. Euler's Eq. (5.3) may be solved by an iterative procedure.

For deformable blocks, application of finite-difference approximations similar to Eq. (5.27) leads to the new values of nodal velocities, $\dot{u}_i^I$, displacements, u_i^I, and positions, x_i^I.

Numerical Stability

To ensure numerical stability of the central difference explicit algorithm, the maximum time step is given by (e.g., Belytschko, 1983):

$$\Delta t_{stab} = \frac{2}{\omega_{max}} \tag{5.32}$$

where ω_{max} denotes the highest system eigenfrequency.

When stiffness-proportional viscous damping is used, a lower limit applies:

$$\Delta t_{stab} = \frac{2}{\omega_{max}}\left(\sqrt{1 + \xi^2} - \xi\right) \tag{5.33}$$

where ξ is the fraction of critical damping at the highest frequency.

In practice, only an upper bound of the maximum eigenfrequency can be evaluated without excessive computational effort. A common procedure is to apply Gerschgorin's theorem (e.g., Underwood, 1983), to obtain an upper bound of the maximum eigenfrequency as the maximum ratio of stiffness over mass, for all the system degrees of freedom. The stiffness measure for each d.o.f. is given by the sum of the absolute values of the terms along the corresponding line of the system stiffness matrix. In explicit codes, the stiffness matrix is not assembled, and a loop over the contacts is performed to obtain the stiffness sums.

Therefore, the time step for stability is given by the minimum value, for all the system d.o.f.'s, of expressions of the form:

$$\Delta t = 2\sqrt{\frac{m^*}{k^*}} \tag{5.34}$$

where m^* stands for an inertial term, and k^* the corresponding stiffness term.

For the translational d.o.f.'s of a rigid block,

$$\Delta t = 2\sqrt{\frac{m}{k_t}} \tag{5.35}$$

where

m is the block mass, and
k_t is an upper-bound of the translational stiffness.

It is possible to evaluate k_t accurately for each d.o.f., considering the various contacts and their orientations. A simplified conservative assumption would be to use, for each block, the sum for all its contacts of the values:

$$k_{tc} = 2(k_n + k_s)A_c \tag{5.36}$$

where k_n and k_s are the normal and shear joint stiffness, and A_c the contact area.
For the simplified rotational equations of motion, it is:

$$\Delta t = 2\sqrt{\frac{I_s}{k_r}} \tag{5.37}$$

where

I_s is the scaled mass moment of inertia of the block, and
k_r is an upper-bound of the rotational stiffness.

The rotational stiffness term can be estimated, conservatively, by summing for all the block's contacts the terms:

$$k_{rc} = 2\,r^2(k_n + k_s)A_c \tag{5.38}$$

where r is the distance from the contact to the block centre.
In dynamic analysis, it is advisable to find a more accurate estimate of the rotational time step, considering separately the 3 principal moments of inertia, and the stiffness that correspond to rotations about each axis.
For deformable blocks, the time step for each node is calculated:

$$\Delta t = 2\sqrt{\frac{m}{k_n}} \tag{5.39}$$

where m is the nodal mass, and the nodal stiffness term is a sum of two components, involving the contact and element stiffnesses:

$$k_n = k_{nc} + k_e \tag{5.40}$$

The contact component, k_{nc}, is evaluated in the same way as the translational stiffness of a rigid block, but taking into account only the contacts that involve the vertex, or the edges or faces converging on it.

The element stiffness component, k_e, may be obtained from the element stiffness matrix, e.g. by adding the absolute values of the stiffness coefficients of the line corresponding to the d.o.f. under consideration. More expedite estimates are available for specific types of elements. For a tetrahedron, 3DEC uses:

$$k_e = (K + \frac{4}{3}G)\frac{3A_{max}}{h_{min}} \tag{5.41}$$

where

K	is the material bulk modulus,
G,	the shear modulus,
A_{max},	the maximum element face area,
h_{min},	the minimum height of the tetrahedron.

In what concerns contact stiffness, for any type of block, there are 2 alternative ways to apply the formulas presented. The first is to loop through the existing contacts, in order to get good estimates of the current stiffness. This implies a time step verification, and possible update, whenever new contacts are created or deleted. The second option is to get a conservative estimate of the maximum contact stiffness that may develop, and to perform time step updates only at regular intervals, or when changes are introduced in the model.

Damping in Quasi-Static Analysis

For quasi-static analysis, damping is intended to improve convergence to the solution. For this purpose, mass-proportional viscous damping is often used, included in the equations of motion (1) or (11), which effectively corresponds to a damping force of the type:

$$f_i^d = -\alpha\, m_s\, \dot{u}_i \tag{5.42}$$

This form of damping is more effective when the parameter α is chosen to provide critical damping at the lowest system frequency (e.g., Hughes, 1987). As calculation of eigenvalues is too costly, adaptive procedures to control α have been developed (e.g., Cundall, 1987; Underwood, 1983).

An alternative to viscous damping, designated as local damping, was proposed by Cundall (1987), based on a different type of energy dissipation. The damping force is defined as:

$$f_i^d = -\lambda |f_i| \mathrm{sign}(\dot{u}_i) \tag{5.43}$$

where λ is a non-dimensional damping constant, and f_i is the unbalanced force.

This type of damping is intended to approach the behaviour of hysteretic damping, in which the energy loss per cycle is independent of frequency. The damping force always opposes motion, but it is scaled to the force term, unlike viscous damping, which is scaled to the magnitude of velocity. A value of the parameter λ in the range 0.7–0.8 is suitable for most quasi-static problems, so it does not have to be changed in the course of the analysis. The effective damping force varies from point to point, depending on the magnitude of the unbalanced forces. As these tend to zero, the damping force vanishes, so there is no resistance to oppose steady-state failure modes.

Damping in Dynamic Analysis

For dynamic analysis, Rayleigh damping is commonly employed. This corresponds to a damping matrix with mass-proportional and stiffness-proportional components:

$$C = \alpha M + \beta K \tag{5.44}$$

The mass-proportional component, which affects mostly the lower frequencies, is included in the equations of motion. In an explicit algorithm, the stiffness-proportional component, responsible for damping the higher frequencies, is added to the contact forces and the element stresses. For the contacts, at each step, the viscous force is given by:

$$F_i^v = \beta \frac{\Delta F_i^{ce}}{\Delta t} \tag{5.45}$$

where ΔF_i^{ce} is the elastic contact force increment.

For deformable blocks, the viscous stress term is:

$$\sigma_m^v = \beta \frac{\Delta \sigma_m^e}{\Delta t} \tag{5.46}$$

where $\Delta \sigma_m^e$ is the elastic stress increment.

The reduction of the time step due to the inclusion of the stiffness-proportional damping, given by Eq. (5.33), may be substantial, leading to prohibitive the run times, so this term is often left out, which increases the high frequency noise in the response.

Mass-proportional damping has an effect inversely proportional to frequency. When setting its value in dynamic failure analyses, some caution has to be exercised. The reduction of effective system stiffness as joints separate and slide implies that frequencies estimated for the initial "elastic" situation may be too high, and therefore the amount of damping may be excessive and actually condition or even stop the failure mechanism. Alternative damping models for explicit algorithms have been proposed, e.g. Maxwell damping (Lemos et al., 2022), which overcome some of the difficulties with Rayleigh damping.

Mass Scaling

In quasi-static analysis, scaling of the inertial masses is an essential technique to improve the convergence to the solution, given the contrasts in block or element size and stiffness. It is a common procedure to assign the time step an arbitrary value, and then to invert the time step expressions given above, in order to calculate the scaled mass and moment of inertia for each rigid block, and the scaled nodal masses of deformable blocks.

5.3 Application Topics and Examples

5.3.1 Masonry Arches and Shells

Masonry arches and arch bridges are commonly analyzed by the classical methods using limit equilibrium concepts, which provide an assessment of the observed collapse modes based on mechanisms. Numerical techniques, such as DEM, extend the analysis capabilities by addressing special shapes and support conditions, dynamic loads, and the representation of the fill material.

Straight arches were analyzed by Meng (2018) using 3DEC. The most important outcome of the investigations was about the concept of load bearing capacity. Usually, load bearing capacity of an arch is understood as the largest load that can still be resisted by the structure when the load is placed directly to the critical position of the arch. (This critical position depends on the arch thickness and on the angle of embrace.) However, in reality, live load arrives to the structure from outside: the position of the load is gradually changing as the vehicle expressing it proceeds from the springing inwards along the structure. It was pointed out by Meng (2018) that for certain combinations of arch geometries and contact friction angles, the critical position cannot be reached by a travelling load that is gradually coming inwards from the springing: the distortions caused by the load when it is still far from the critical position may already be so significant that the structure fails because of the deformed shape deviating too much from the intact, unloaded shape of the arch. Hence, a modified definition for the load bearing capacity can be suggested: The quasi-static load bearing capacity should be, by definition, the largest load that can safely travel quasi-statically along the whole arch without causing collapse.

The bond pattern and the brick shape have a definite mechanical effect on the behavior of masonry shells. Discrete element simulations using 3DEC (Chen & Bagi, 2020; Simon & Bagi, 2016) and NSCD (Beatini et al., 2018) proved that a significant hoop-directional tensile resistance can develop in shells under meridional

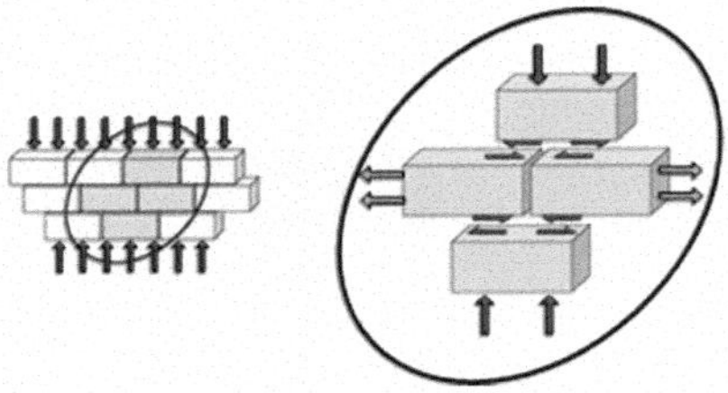

Fig. 5.8 Explanation of the crosswise tensile resistance (Simon & Bagi, 2016)

Fig. 5.9 3DEC models of barrel vaults with (left) transversal; (right) longitudinal arrangement of bricks (Chen et al., 2021)

compression. The phenomenon is explained in Fig. 5.8; Chen and Bagi (2020) quantified it for different herringbone and running bond patterns. The effect can be so large that the orange-slice-like cracking of domes can easily be prevented with it.

Straight barrel vaults were studied by Chen et al. (2021): several overall geometries were formed by longitudinal and transversal arrangements of bricks (Fig. 5.9 presents two examples), and the load bearing capacities for different vertical loads were determined. Two competing effects were pointed out: barrels with longitudinal arrangements distribute the loads along the complete structure while in transversal vaults the arches work nearly independently of each other (causing a lower resistance to loads), while larger size of the voussoirs along the arch direction increases the load bearing capacity.

Skewed bridges have been object of several DEM studies. Forgacs et al. (2018) have compared the effect of construction techniques on the ultimate capacity of the arch (Fig. 5.10). The effect of the backfill materials can be represented as a dead load, a conservative assumption, often adopted in safety assessment studies. Explicitly including the fill in the model may provide a more realistic estimate, but the fill material non-elastic behavior has to be considered. A deformable block with elasto-plastic behavior has been used by several authors (Gobbin et al., 2020; Pulatsu et al., 2019a; Toth et al., 2009). The arch itself may be created with rigid or deformable blocks. A more refined representation of the fill by means of random block systems was investigated by Sarhosis et al., (2019a, 2019b), which removes numerical constraints on the development of extrados hinges. The dynamic response of arch bridges was modelled by Saygili and Lemos (2021), using rigid block models. In situ measurements of natural frequencies allowed the calibration of the joint stiffnesses. Masi et al. (2020) addressed the effects of blast loading on arched structures.

Fig. 5.10 Models of skewed bridges with different construction methods: **a** False skew method; **b** Helicoidal method (Forgacs et al., 2018)

These 3DEC simulations above were conducted on perfect geometries. However, real masonry shells contain several inaccuracies due to fabrication and construction, which may significantly affect the occurring internal force distributions. Dell'Endice et al. (2020) analyzed this issue, by comparing 3DEC models of a cloister vault with small-scale physical models. They pointed out that indeed, the internal force distributions (and, consequently, the load bearing) were considerably altered by the imperfections.

5.3.2 *Walls and Panels*

Laboratory tests of masonry walls and panels, either under static loading or on the shaking table, have provided invaluable data to test and validate numerical models. For these relatively small specimens, it is usually possible to achieve a close representation of the block pattern, thus DEM idealizations are an appropriate choice. Bui et al. (2017) presented a series of comparisons with experiments, in which the large displacement capabilities of DEM are manifest. Figure 5.11 shows the results of the test of brick panel with dry joints under an applied partial settlement. These authors also present comparisons with tests of stone block assemblies subject to out-of-plane loads, performed by Restrepo-Velez et al. (2014).

Fig. 5.11 Experimental and DEM failure mode of brick panel subject to partial settlement (Bui et al., 2017)

Experiments of brick panels under in-plane shear and compression loading were successfully simulated with DEM models by Malomo et al. (2019) and Pulatsu et al. (2020). Dynamic tests of panels subject to out-of-plane loads were analyzed by Godio and Beyer (2018). Galvez et al. (2022) discuss the influence of damping in the rocking behaviour of blocks and walls.

DEM models are also being applied to more complex masonry structures, for example to investigate the seismic response and safety of archeological remains (Valluzzi et al., 2022).

5.3.3 Columns and Tall Structures

The dynamics of drum columns were among the first DEM applications to masonry (Fig. 5.1). Rocking phenomena are critical in the development of collapse mechanisms. Shaking table tests of columns (Papantonopoulos et al., 2002) and single blocks (Pena et al., 2007) have allowed the validation of DEM models to approach these problems. The experiments also demonstrated the significant variability to be expected in these nonlinear dynamic problems involving large amplitude movements. Numerical simulations confirm that small variations in properties or dynamic input can led to different outcomes (Psycharis et al., 2003). Therefore, multiple analysis and parameter studies are required in safety assessment studies.

A 3DEC model of the roman temple in Évora, Portugal, was developed by Nayeri (2012), who represented accurately the actual shapes of drum columns and architraves measured by laser scanning. Rigid blocks were used, with the joint stiffness parameters calibrated by means of in situ dynamic characterization tests, in order to approximate the measured frequencies and the lower mode shapes. Figure 5.12 shows 3 stages of collapse under seismic loading.

DEM models have been applied in many studies involving the dynamics of columns (e.g., Giamundo et al., 2014; Sarhosis et al., 2019a, 2019b; Stefanou et al., 2011) and other tall structures, such as spires (DeJong & Vibert, 2012), obelisks (Lemos et al., 2015), minarets (Catki et al., 2020), and towers (Sarhosis et al., 2017).

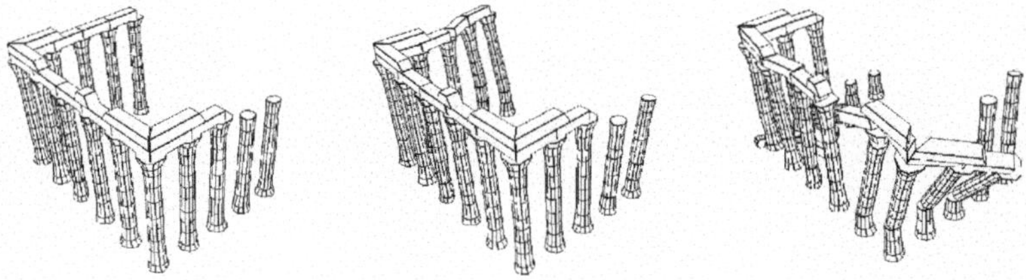

Fig. 5.12 Rigid block model of the Roman temple of Évora. Three stages of collapse under seismic action (Nayeri, 2012)

5.3.4 *Analysis of Natural Frequencies Using Rigid Block Models*

Natural frequencies of vibration govern the linear response of a structure under dynamic loads. Ambient vibration measurements are routinely used to identify the natural frequencies of masonry structures, providing important information on its deformability, the interaction of its parts and the general state of conservation. Therefore, they can bring useful data for the validation and calibration of numerical models intended to evaluate seismic response. The collapse of tall structures under strong impulsive loads is often not much affected by the eigenfrequencies. However, the evaluation of the response, as well as the estimate of irreversible displacements, caused by less severe ground motion requires models that represent well the dynamic characteristics of the structure.

Eigenvalue analysis is a standard task in FEM codes. As DEM codes typically do not use matrix solutions, it is not always available. 3DEC, however, has a simple built-in procedure for eigenvalue calculations, which can be applied to rigid or deformable block models. In both cases, a stiffness matrix of the structure has to be assembled, as well as a mass matrix, assumed diagonal. For rigid block models, the variables are the 6 degrees-of-freedom of the block, while the stiffness matrix represents the contact springs, assumed elastic. Therefore, the normal and shear stiffness parameters and all sub-contacts, considering their areas and orientation, are used to build the global stiffness matrix. Under very low amplitude vibration, as is the case of in situ ambient noise characterization tests, the assumption of elastic response is acceptable. Therefore, these tests give helpful information for the choice of joint stiffnesses for dynamic analysis with a rigid block model, so that the global deformability of the structure is adequately represented.

The accuracy of natural frequencies given by rigid block has been shown by comparisons with elastic analytical solutions. Increasing the number of rigid blocks leads to the convergence of frequencies and eigenmodes to the solutions, for example for the case of bending of beams and plates (Lemos, 2017). For an accurate solution, the number of point contacts must be sufficient. For example, for a rectangular section under bending, 3×3 contact points are advisable.

Figure 5.13 displays a DEM model of a stone masonry minaret (Cakti et al., 2020). Rigid blocks were used to represent the outer shell and the inner staircase. The joint stiffnesses were calibrated with the in situ measurements of the eigenfrequencies of the 3 modes shown in the figure, the first two bending modes and the torsional mode.

5.3.5 *Simulation of Displacements Caused by Earthquakes*

Reproducing the observed effects of earthquakes, whether total or partial collapse patterns, or irreversible displacements, is an important test for numerical models. In parallel, researchers are also using numerical models as a simulation tool to support

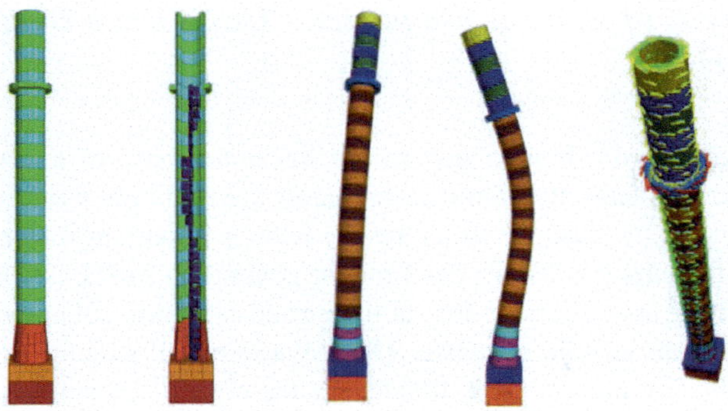

Fig. 5.13 Model of minaret and eigenmodes. **a** Rigid block model; **b** Internal structure; **c** First bending mode; **d** Second bending mode; **e** Torsional mode (Cakti et al., 2020)

the discussion on the historical seismic events that caused those effects and to test explanatory hypotheses. Schweppe et al. (2017) used a DEM model to investigate the ruins of the Roman Temple of Kedesh, Israel, as a way to gain knowledge on the historical seismicity of the region. Young et al. (2015) developed a model of the ancient temple of Nemea to investigate if an earthquake was a plausible cause of collapse.

Testing and validating numerical models is a critical task to improve the predictive capabilities of numerical models. While laboratory tests provide more reliable data, the observation of the effects of earthquakes on structures gives invaluable data on the actual performance of the historical constructions. The main difficulty is typically the lack of records of the seismic excitation at the location. Parameter studies are often necessary to evaluate the influence of less well-known data.

Saifullah and Wittich (2022) employed a DEM model to reproduce the overturning of 2 statues during the 2014 South Napa earthquake. Lemos et al. (2015) sought to understand the rotational movements observed in a stone block obelisk after the 2015 Lorca earthquake, Spain (Fig. 5.14). Seismic records of the event were available. Natural frequencies were also measured, which allowed the calibration of the joint stiffnesses. A notable feature was that the frequencies of the first bending modes were different in the 2 orthogonal directions. Given the apparent symmetry of the structure, this fact was ascribed to the foundation conditions. The foundation joint was thus partitioned in areas of different stiffness in order to reproduce the measurements. This calibration allowed a good agreement with the rotational displacements of the blocks observed after the earthquake, as shown in the figure.

Fig. 5.14 Stone block obelisk. **a** DEM model; **b**, **c** Final displacements after earthquake, model and observed (Lemos et al., 2015)

5.3.6 Shaking Table Test of a Stone Masonry Building Model

Shaking table tests offer a major contribution to the data available for validation of numerical codes. Mendes et al. (2017) described the results of blind prediction and post-test analysis of experiments on a traditional stone masonry construction. One of the DEM models employed in the simulations is shown in Fig. 5.15 (Lemos & Costa, 2017). A rigid block representation was used, but the geometry was much simplified, adopting brick shaped blocks. The block sizes were in the same order of the real ones, and the general pattern was simulated, but the actual shapes were not reproduced, thus missing the irregularity of the joints. Before the test, a pushover analysis gave a reasonable prediction of the collapse mode, but obviously a much lower failure acceleration. The dynamic analyses before the tests were not successful in predicting the outcome. After the test, the measured natural frequencies were supplied, allowing a new calibration of the model deformability. This model provided a fair approximation of the near-collapse mode and permanent displacements, mostly involving the façade and a lateral wall, as shown in the figure.

5.3.7 Modelling Structural Reinforcement

Clamps and dowels between stone blocks are easily represented in DEM models by local reinforcement elements, essentially breakable springs acting in tension or shear (Catki et al., 2020; Psycharis et al., 2003; Stefanou et al., 2011).

Structural beam elements were used by Gobbin et al. (2020) to represent externally bonded reinforcement applied to a brick arch tested in the laboratory. These are standard 2-node truss elements which are connected to the blocks by contact elements with limited tensile and shear strength, thus allowing the simulation of debonding.

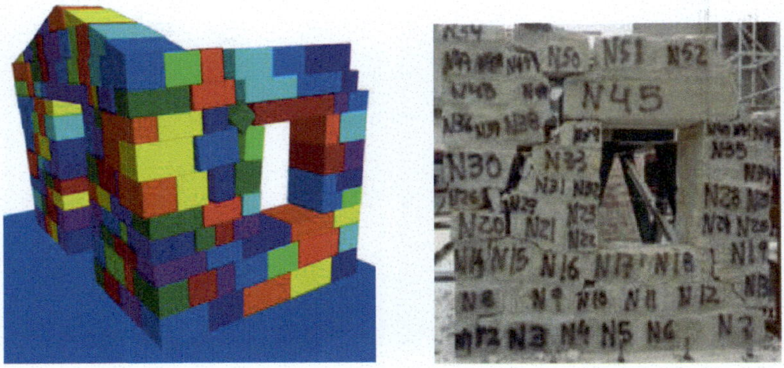

Fig. 5.15 **a** Rigid block model of stone masonry house, final stateafter shaking table test; **b** Experimental results (Lemos & Costa, 2017)

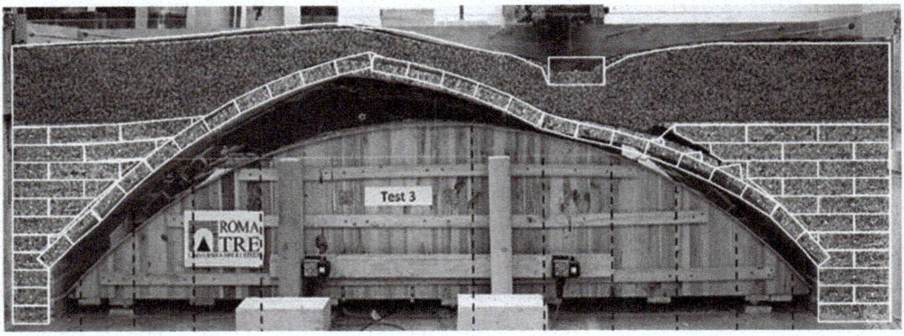

Fig. 5.16 Comparison of the numerical (white outline) and experimental failure mode of reinforced arch (Gobbin et al., 2020)

The DEM model provides a good agreement of the collapse loads and failure modes observed in the experiments (Fig. 5.16).

5.3.8 Irregular Block Geometries

Historical and traditional masonry constructions often include blocks with irregular shapes or rubble assemblies of various types, which influence the mechanical behaviour. In practical analysis, there is seldom information to represent them in the models, which in any case would lead to excessive detail. However, in research projects, several authors have taken advantage of DEM to quantify the effects of irregular, uneven or damaged components.

De Felice (2011) simulated a 3-leaf cross section of a traditional wall with DEM, simulating closely the observed irregular blocks geometries, namely in the very

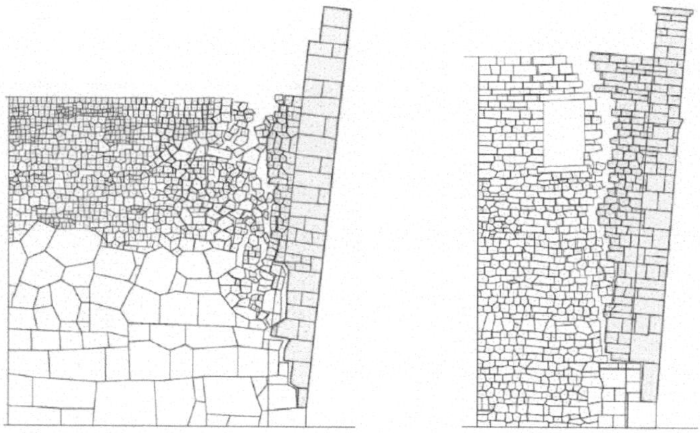

Fig. 5.17 Detailed DEM models of façade collapse of 2 churches considering the interaction with the lateral walls (Gobbin et al., 2021)

detailed representation of the inner leaf of rubble masonry. Static and dynamic analyses were performed, and the results highlighted the importance of the connection between the outer wall leaves in maintaining the integrity of the section. Meyer et al. (2007) also addressed the 3-leaf wall problem, focusing on the internal collapse under dynamic input with different frequencies.

Gobbin et al. (2021) analyzed the collapse of church façades under seismic loading taking into account the contribution of the lateral walls (Fig. 5.17). These were represented by the actual irregular block geometries acquired in detailed surveys. The numerical failure modes were compared with crack patterns observed after the 2009 L'Aquila earthquake.

5.3.9 Block Fracture

Nonlinear behaviour of the block material can be addressed in deformable block models by assigning appropriate constitutive relations to the internal elements, as done in FEM models. The consideration of block cracking and splitting poses a greater challenge. Munjiza (2004) and co-workers have developed strategies to allow the blocks to fracture during the simulation and thus divide into smaller blocks. Rigorous fracture propagation analysis is computationally intensive, requiring a fine local discretization which is not possible in large block models. A more practical solution in the DEM environment is the approach designated as "bonded-block model" by Garza-Cruz and Pierce (2014). In this case, a random network of joints is generated, defining the potential fracture paths. These joints are initially bonded, with tensile and cohesive strengths corresponding to the block material. During the analysis they may break, based on a given fracture criterion, thus creating real joints that separate the

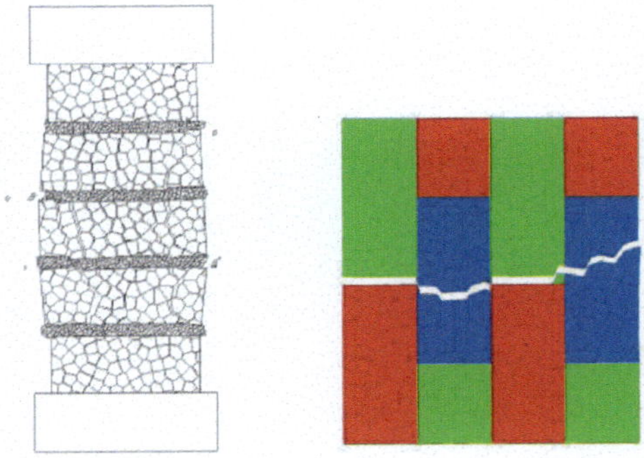
(a) (b)

Fig. 5.18 Models considering block fracture. **a** Compression test of brick stack (Sarhosis & Lemos, 2018); **b** Direct tension test parallel to bed joints (Pulatsu et al., 2020)

sub-blocks. The sub-blocks may be rigid or deformable, as the nonlinear behaviour is circumscribed to the potential fractures. The random sub-block system may be generated by several procedures, namely Voronoi polyhedral patterns or Delaunay tetrahedral assemblies.

The bonded-block approach, also designated as detailed micro-modelling, has been applied in masonry, mostly to simulate tests on laboratory specimens. Figure 5.18a shows an example of a 2D model of a brick stack under compression, in which both brick and mortar are divided into bonded assemblies of Voronoi polygons (Sarhosis & Lemos, 2018). Chen et al. (2018) applied the same concept to triplet tests on brick specimens. Pulatsu et al., (2019a, 2019b, 2020) developed 3D models using Voronoi polyhedra sub-blocks inside each unit. Figure 5.18b shows brick cracking in a test in which a tensile stress is applied parallel to the direction of the bed joints. Bonded-particle models, using circular or spherical particles, often employed in fracture studies of geomaterials, are an alternative approach. Azevedo et al. (2016) applied it to the analysis of traditional stone masonry panels composed of irregular blocks.

5.4 Closing Remarks

Masonry structures display a multiplicity of designs, materials, or construction techniques. A wide range of tools are available to analyze their behaviour and predict its evolution, from the classical analytical methods to a variety of complex numerical models. Discrete elements, a numerical approach based on the mechanics of discontinuous media, have particular strengths and practical limitations which have been discussed in this chapter. Each established methodology has an optimal range of application and its own set of requirements. It is clear that more elaborate analysis methods are more demanding in terms of input data, computational power and expertise of the analyst. Therefore, a critical judgement of these variables should assist the selection of the most appropriate approach for a given problem.

Laboratory testing of materials and structures has played a major role in assessing and validating the analysis methods. The progressive growth of monitoring data from real structures, involving either long term behaviour or the response to specific events, such as earthquakes, will further improve the means to test new concepts and models, enhancing our confidence on their predictive capabilities.

References

Amadei, B., Mirabella Roberti, G., & Binda, L. (1995). Modeling the stability of masonry structures with the discontinuous deformation analysis. In J. Middleton & G. N. Pande (Eds.), *Computer methods in structural masonry—3* (pp. 1–11). Books & Journals International.

Azevedo, N. M., Lemos, J. V., & Almeida, J. R. (2016). Discrete element particle modelling of stone masonry. In Sarhosis, et al. (Eds.), *Computational modeling of masonry structures using the discrete element method* (pp. 146–169). IGI Global. https://doi.org/10.4018/978-1-5225-0231-9.ch007

Bagi, K. (1993). A quasi-static numerical model for micro-level analysis of granular assemblies. *Mechanics of Materials, 16*(1–2), 101–110.

Bagi, K. (2016a). The DDA method. In V. Sarhosis, K. Bagi, J. V. Lemos, & G. Milani (Eds.), *Computational modeling of masonry structures using the discrete element method* (pp. 90–102). IGI Global.

Bagi, K. (2016b). The contact dynamics method. In V. Sarhosis, K. Bagi, J. V. Lemos, & G. Milani (Eds.), *Computational modeling of masonry structures using the discrete element method* (pp. 103–122). IGI Global

Baraldi, D., & Cecchi, A. (2017). A full 3D rigid block model for the collapse behaviour of masonry walls. *European Journal of Mechanics – A/Solids, 64*, 11–28. https://doi.org/10.1016/j.euromechsol.2017.01.012

Beatini, V., Royer-Carfagni, G., & Tasora, A. (2018). The role of frictional contact of constituent blocks on the stability of masonry domes. *Proceedings of Royal Society A, 474*, 20170740.

Belytschko, T. (1983). An overview of semidiscretization and time integration procedures. In T. Belytschko & T. J. R. Hughes (Eds.), *Computational methods for transient analysis* (pp. 1–65). North Holland.

Bui, T. T., Limam, A., Sarhosis, V., & Hjiaj, M. (2017). Discrete element modelling of the in-plane and out-of-plane behaviour of dry-joint masonry wall constructions. *Engineering Structures, 136*, 277–294.

Cakti, E., Saygili, O., Lemos, J. V., & Oliveira, C. S. (2020). Nonlinear dynamic response of stone masonry minarets under harmonic excitation. *Bulletin of Earthquake Engineering, 18*(10), 4813–4838. https://doi.org/10.1007/s10518-020-00888-y

Chen, W., Konietzky, H., Liu, C., Fu, H., & Zhang, J. (2018). Prediction of brickwork failure using discrete-element method. *Journal of Materials in Civil Engineering ASCE, 30*(9), 06018012. https://doi.org/10.1061/(ASCE)MT.1943-5533.0002431

Chen, S., & Bagi, K. (2020). Crosswise tensile resistance of masonry patterns due to contact friction. *Proceedings of Royal Society A, 476*(2240), 20200439.

Chen, Sh., Ferrante, A., Clementi, F., & Bagi, K. (2021). DEM analysis of the effect of bond pattern on the load bearing capacity of barrel vaults under vertical loads. *International Journal of Masonry Research and Innovation, 6*(3), 346–373.

Cismasiu, C., Silva, P. B. S., Lemos, J. V., & Cismasiu, I. (2021). Seismic vulnerability assessment of a stone arch using discrete elements. *International Journal of Architectural Heritage*. https://doi.org/10.1080/15583058.2021.1963506

Cocchetti, G., & Rizzi, E. (2020). Static upper/lower thrust and kinematic work balance stationarity for least-thickness circular Masonry arch optimization. *Journal of Optimization Theory and Applications, 187*, 707–757. https://doi.org/10.1007/s10957-020-01772-0

Cundall, P. A. (1971). A computer model for simulating progressive large scale movements in blocky rock systems. In *Proceedings of symposium rock fracture (ISRM)*, Nancy (Vol. 1). Paper II-8.

Cundall P. A., Marti J., Beresford P., Last, N., & Asgian, M. (1978). *Computer modeling of jointed rock masses*. Waterways Experiment Station Report WES-TR-N-78-4.

Cundall, P. A. (1987). Distinct element models of rock and soil structure. In E. T. Brown (Ed.), *Analytical and computational methods in engineering rock mechanics* (pp. 129–163). George Allen and Unwin.

Cundall, P. A. (1988). Formulation of a three-dimensional distinct element model—Part I: A scheme to detect and represent contacts in a system composed of many polyhedral blocks. *International Journal of Rock Mechanics and Mining Sciences, 25*(3), 107–116.

Dubois, F., & Jean, M. (2006). The non smooth contact dynamic method: Recent LMGC90 software developments and application. In P. Wriggers, & U. Nackenhorst (Eds.), *Analysis and simulation of contact problems. Lecture notes in applied and computational mechanics* (Vol. 27, pp. 375–378).

De Felice, G. (2011). Out-of-plane seismic capacity of masonry depending on wall section morphology. *International Journal of Architectural Heritage, 5*, 466–482.

DeJong, M. J., & Vibert, C. (2012). Seismic response of stone masonry spires: Computational and experimental modeling. *Engineering Structures, 40*, 566–574.

Dell'Endice, A., Iannuzzo, A., DeJong, M. J., Van Mele, T., & Block, Ph. (2020). Modelling imperfections in unreinforced masonry structures: Discrete element simulations and scale model experiments of a pavilion vault. *Engineering Structures, 228*, 111499.

Fang, J., Qin, X., Cai, Y., Wang, Q., & Lei, J. (2016). Earthquake response analysis on masonry structural by discontinuous deformation analysis and finite element method. *Journal of Geodesy and Geodynamics, 36*(6), 520–524.

Fonte, C., Oliveira Jr, J., & de Almeida, L. (2015). DEM-CFD coupling: Mathematical modelling and case studies using ROCKY-DEM® and ANSYS Fluent®. In *Eleventh international conference on CFD in the minerals and process industries CSIRO*, Melbourne, Australia.

Forgacs, T., Sarhosis, V., & Bagi, K. (2018). Influence of construction method on the load bearing capacity of skew masonry arches. *Engineering Structures, 168*, 612–627.

Galvez, F., Sorrentino, L., Ingham, J., & Dizhur, D. (2022). Damping considerations for rocking block dynamics using the discrete element method. *Earthquake Engineering Structural Dynamics, 51*, 935–957. https://doi.org/10.1002/eqe.3598

Garza-Cruz, T. V., & Pierce, M. (2014). A 3DEC model for heavily veined massive rock masses. In *Proceedings of 48th US rock mechanics/geomechanics symposium*, ARMA, Alexandria, Paper 14-7660.

Giamundo, V., Sarhosis, V., Lignola, G. P., & Cosenza, E. (2014). Discrete element modelling of the archaeological colonnade in Pompeii. In *9th international masonry conference, Guimarães*, Portugal.

Gobbin, F., De Felice, G., & Lemos, J. V. (2020). A discrete element model for masonry vaults strengthened with externally bonded reinforcement. *International Journal of Architectural Heritage*. https://doi.org/10.1080/15583058.2020.1743792

Gobbin, F., De Felice, G., & Lemos, J. V. (2021). Numerical procedures for the analysis of collapse mechanisms of masonry structures using discrete element modelling. *Engineering Structures, 246*, 113047. https://doi.org/10.1016/j.engstruct.2021.113047

Godio, M., & Beyer, K. (2018). Evaluation of force-based and displacement-based out-of-plane seismic assessment methods for unreinforced masonry walls through refined model simulations. *Earthquake Engineering Structures Dynamics*, 1–22. https://doi.org/10.1002/eqe.3144

Goodman, R. E., Taylor, R. L., & Brekke, T. L. (1968). A model for the mechanics of jointed rock. *Journal of the Soil Mechanics Foundations Division. ASCE, 94*(3), 637–659.

Hart, R. D., Cundall, P. A., & Lemos, J. V. (1988). Formulation of a three-dimensional distinct element model—Part II: Mechanical calculations. *International Journal of Rock Mechanics and Mining Sciences, 25*(3), 117–125.

Hughes, T. J. R. (1987). *The finite element method—Linear static and dynamic finite element analysis*. Prentice-Hall.

Itasca. (2014). *UDEC—Universal Distinct Element Code*. Itasca Consulting Group, Minneapolis, MN.

Itasca. (2019). *PFC 6.0 documentation*. Itasca Consulting Group, Minneapolis, MN.

Itasca. (2020). *3DEC—Three-dimensional distinct element code*. Itasca Consulting Group, Minneapolis, MN.

Jean, M., & Moreau, J. J. (1992). Unilaterality and dry friction in the dynamics of rigid body collections. In *Proceedings of contact mechanics international symposium* (pp. 31-48), Lausanne, Switzerland: Presses Polytechniques et Universitaires Romandes.

Jean, M. (1999). The non-smooth contact dynamics method. *Computer Methods in Applied Mechanics and Engineering, 177*, 235–257.

Jiang, H., Wang, L., Li, L., & Guo, Z. (2014). Safety evolution of an ancient maonry seawall structure with modified DDA method. *Computers and Geotechnics, 55*, 277–289.

Kamai, R., & Hatzor, Y. H. (2005). Dynamic back analysis of structural failures in archeological sites to obtain paleo-seismic parameters using DDA. In M. MacLaughlin & N. Sitar (Eds.), *Proceedings of ICADD-7*, December (pp. 121–136). Honolulu.

Kamai, R., & Hatzor, Y. H. (2008). Numerical analysis of block stone displacements in ancient masonry structures: A new method to estimate historic ground motions. *International Journal for Numerical and Analytical Methods in Geomechanics, 32*, 1321–1340.

Kishino, Y. (1988). Disc model analysis of granular media. In M. Satake, & J. T. Jenkins (Eds.), *Micromechanics of granular materials* (pp. 143-152). Elsevier.

Koziara, T., & Bićanić, N. (2008). Semismooth Newton method for frictional contact between pseudo-rigid bodies. *Computer Methods in Applied Mechanics and Engineering, 197*(33–40), 2763–2777.

Lemos, J. V. (1995). Assessment of the ultimate load of a masonry arch using discrete elements. In J. Middleton & G. N. Pande (Eds.), *Computer methods in structural masonry—3* (pp. 294–302). Books & Journals International.

Lemos, J. V. (1998). Discrete element modelling of the seismic behaviour of stone masonry arches. In J. Middleton, G. N. Pande, & B. Kralj (Eds.), *Computer methods in structural masonry—4* (pp. 220–227). E&FN Spon, London.

Lemos, J. V. (2007). Discrete element modeling of masonry structures. *International Journal of Architectural Heritage, 1*(2), 190–213.

Lemos, J. V., & Costa, A. C. (2017). Simulation of shake table tests on out-of-plane masonry buildings. Part (V): Discrete element approach. *International Journal of Architectural Heritage, 11*(1), 117–124. http://doi.org/10.1080/15583058.2016.1237587

Lemos, J. V., Oliveira, C. S., & Navarro, M. (2015). 3D nonlinear behavior of an obelisk subjected to the Lorca May 11, 2011 strong motion record. *Engineering Failure Analysis, 58*, 212–228.

Lemos, J. V. (2017). Contact representation in rigid block models of masonry. *International Journal of Masonry Research and Innovation,* 2(4), 321–334. https://doi.org/10.1504/IJMRI.2017.087445

Lemos, J. V., Dawson, E. M., & Cheng, Z. (2022). Application of Maxwell damping in the dynamic analysis of masonry structures with discrete elements. *International Journal of Masonry Research and Innovation, 7(6)*, 663–686. https://doi.org/10.1504/IJMRI.2021.10043266

Li, W., Chen, X., Wang, H., Chan, A. H. C., & Cheng, Y. (2021). Evaluating the seismic capacity of dry-joint masonry arch structures via the combined finite-discrete element method. *Applied Sciences, 11*(18), 8725. https://doi.org/10.3390/app11188725

Liu, S., Li, Z., Zhang, H., & Wu, W. (2018). A 3-D DDA damage analysis of brick masonry buildings under the impact of boulders in mountainous areas. *Journal of Mountain Science, 15*(10), 657–671. https://doi.org/10.1007/s11629-017-4453-5

Lourenço, P. B. (2002). Computations of historical masonry constructions. *Progress in Structural Engineering and Materials, 4*, 301–319. https://doi.org/10.1002/pse.120

MacLaughlin, M. M. (1997). *Discontinuous deformation analysis of the kinematics of landslides* (117 p.). PhD Dissertation, Department of Civil and Environment Engineering, University of California Berkeley.

Malomo, D., DeJong, M. J., & Penna, A. (2019). Distinct element modelling of the in-plane cyclic response of URM walls subjected to shear-compression. *Earthquake Engineering & Structural Dynamics, 48*, 1322–1344.

Masi, F., Stefanou, I., Maffi-Berthier, V., & Vannucci, P. (2020). A discrete element method based-approach for arched masonry structures under blast loads. *Engineering Structures, 216*, 110721. https://doi.org/10.1016/j.engstruct.2020.110721

Mendes, N., Costa, A. A., Lourenço, P. B., Bento, R., Beyer, K., Felice, G., Gams, M., Griffith, M., Ingham, J., Lagomarsino, S., et al. (2017). Methods and approaches for blind test predictions of out-of-plane behavior of masonry walls: A numerical comparative study. *International Journal of Architectural Heritage, 11*, 59–71.

Meng, Y. (2018). *Frictional sliding in limit state analysis codes of masonry arches.* MSc Thesis, supervised by K. Bagi, Department of Structural Mechanics, Budapest University of Technology and Economics. https://doi.org/10.13140/RG.2.2.19377.94565

Meyer, P., Ochsendorf, J., Germaine, J., & Kausel, E. (2007). The impact of high-frequency/low-energy seismic waves on unreinforced masonry. *Earthquake Spectra, 23*(1), 77–94.

Moreau, J. J. (1988). Unilateral contact and dry friction in finite freedom dynamics. In J. J. Moreau, & P. D. Panagiotopoulos (Eds.), *CISM courses and lectures* (Vol. 302). Springer, Vienna.

Munjiza, A., Owen, D. R. J., & Bicanic, N. (1995). A combined finite discrete element method in transient dynamics of fracturing solids. *Engineering Computations, 12*(2), 145–174.

Munjiza, A., Andrews, K. R. F., & White, J. K. (1999). Combined single and smeared crack model in combined finite-discrete element method. *International Journal for Numerical Methods in Engineering, 44*, 41–57.

Munjiza, A. (2004). *The combined finite-discrete element method.* John Wiley.

Nayeri, S. A. (2012). *Seismic assessment of the roman temple in Évora, Portugal.* MS thesis, University of Minho, Guimarães, Portugal.

Nikolic, Z., Smoljanovic, H., & Zivaljic, N. (2013). A combined finite-discrete element analysis of dry stone masonry structures. *Engineering Structures, 52*, 89–100.

Otter, J. R. H., Cassell, A. C., & Hobbs, R. E. (1966). Dynamic relaxation. *Proceedings of the Institution of Civil Engineers, 35*, 633–665.

Pagnoni, T., & Vanzi, I. (1995). Experimental and numerical study of the seismic response of block structures. In J. Middleton & G. N. Pande (Eds.), *Computer methods in structural masonry—3* (pp. 213–222). Books & Journals International, Swansea.

Papantonopoulos, C., Psycharis, I. N., Papastamatiou, D. Y., Lemos, J. V., & Mouzakis, H. P. (2002). Numerical prediction of the earthquake response of classical columns using the distinct element method. *Earthquake Engineering Structural Dynamics, 31*, 1699–1717.

Papastamatiou, D., & Psycharis, I. (1993). Seismic response of classical monuments—A numerical perspective developed at the temple of Apollo Bassae, Greece. *Terra Nova, 5*, 591–601.

Peña, F., Prieto, F., Lourenço, P. B., Costa, A. C., & Lemos, J. V. (2007). On the dynamics of rocking motion of single rigid-block structures. *Earthquake Engineering Structural Dynamics, 36*(15), 2383–2399.

Psycharis, I. N., Papastamatiou, D. Y., Cundall, P. A., & Lorig L. (1993). Numerical and analytical modeling of Greek classical columns. In *Proceedings of 3rd Pan American congress of applied mechanics—PACAM III* (pp. 487–490). São Paulo, Brasil.

Psycharis, I. N., Lemos, J. V., Papastamatiou, D. Y., Zambas, C., & Papantonopoulos, C. (2003). Numerical study of the seismic behaviour of a part of the Parthenon Pronaos. *Earthquake Engineering Structural Dynamics, 32*, 2063–2084.

Pulatsu, B., Erdogmus, E., & Lourenço, P. B. (2019a). Comparison of in-plane and out-of-plane failure modes of masonry arch bridges using discontinuum analysis. *Engineering Structures, 178*, 24–36. https://doi.org/10.1016/j.engstruct.2018.10.016

Pulatsu, B., Erdogmus, E., Lourenço, P. B., & Quey, R. (2019b). Simulation of uniaxial tensile behavior of quasi-brittle materials using softening contact models in DEM. *International Journal of Fracture, 217*, 105–125. https://doi.org/10.1007/s10704-019-00373-x

Pulatsu, B., Erdogmus, E., Lourenço, P. B., Lemos, J. V., & Hazzard, J. (2020). Discontinuum analysis of the fracture mechanism in masonry prisms and wallettes via discrete element method. *Meccanica, 55*, 505–523. https://doi.org/10.1007/s11012-020-01133-1

Restrepo-Vélez, L. F., Magenes, G., & Griffith, M. C. (2014). Dry stone masonry walls in bending—Part I: Static tests. *International Journal of Architectural Heritage., 8*(1), 1–28. https://doi.org/10.1080/15583058.2012.663059

Rizzi, E., Rusconi, F., & Cocchetti, G. (2014). Analytical and numerical DDA analysis on the collapse mode of circular masonry arches. *Engineering Structures, 60*, 241–257.

Saifullah, M. K., & Wittich, C. E. (2022). Seismic response of two, freestanding statue-pedestal systems during the 2014 South Napa earthquake. *Journal of Earthquake Engineering, 26*(10), 5086–5108. https://doi.org/10.1080/13632469.2020.1859004

Sarhosis, V, Milani, G., Formisano, A., & Fabbrocino, F. (2017). Evaluation of different approaches for the estimation of the seismic vulnerability of masonry towers. *Bulletin of Earthquake Engineering*. https://doi.org/10.1007/s10518-017-0258-8

Sarhosis, V., & Lemos, J. V. (2018). A detailed micro-modelling approach for the structural analysis of masonry assemblages. *Computers & Structures, 206*, 66–81. https://doi.org/10.1016/j.compstruc.2018.06.003

Sarhosis, V., Baraldi, D., Lemos, J. V., & Milani, G. (2019a). Dynamic behaviour of ancient freestanding multi-drum and monolithic columns subjected to horizontal and vertical excitations. *Soil Dynamics and Earthquake Engineering, 120*, 39–57. https://doi.org/10.1016/j.soildyn.2019.01.024

Sarhosis, V., Forgacs, T., & Lemos, J. V. (2019b). A discrete approach for modelling backfill material in masonry arch bridges. *Computers & Structures, 224*, 106108.

Saygili, O., & Lemos, J. V. (2021). Seismic vulnerability assessment of masonry arch bridges. *Structures, 33*, 3311–3323. https://doi.org/10.1016/j.istruc.2021.06.057

Scheldt, T., Lu, M., & Myrvang, A. (2002). Numerical analysis of Gjovick cavern. In Y. Hatzor (Ed.), Proceedings ICADD-5 (pp. 125–132). Abingdon, Balkema

Schweppe, G., Hinzen, K.-H., Reamer, S. K., Fischer, M., & Marco, S. (2017). The ruin of the Roman Temple of Kedesh, Israel; example of a precariously balanced archaeological structure used as a seismoscope. *Annals of Geophysics, 60*(4), S0444. https://doi.org/10.4401/ag-7152

Shi, G. H. (1992). Discontinuous deformation analysis: A new numerical model for the statics and dynamics of deformable block structures. *Engineering Computations, 9*(4), 157–168.

Shi, G.-H. (1988). *Discontinuous deformation analysis – A new model for the statics and dynamics of block systems*. PhD thesis, University of California Berkeley, USA.

Shi, G.-H. (2001). Three dimensional discontinuous deformation analysis. In N. Bicanic (Ed.), *Proceedings of ICADD-4* (pp. 1–21), 6-8 June 2001, Glasgow.

Simon, J., & Bagi, K. (2016). DEM analysis of the minimum thickness of oval masonry domes. *International Journal of Architectural Heritage, 10*(4), 457–475.

Smilauer, V., et al. (2021). *Yade documentation* (3rd ed.). The Yade Project. https://doi.org/10.5281/zenodo.5705394. http://yade-dem.org/doc/

Smoljanovic, H., Zivaljic, N., & Nikolic, Z. (2013). Nonlinear analysis of engineering structures by combined finite-discrete element method. *Gradevinar, 65*, 331–334.

Stefanou, I., Psycharis, I., & Georgopoulos, I. O. (2011). Dynamic response of reinforced masonry columns in classical monuments. *Journal of Construction and Building Materials, 25*(1), 4325–4337.

Thavalingam, A., Bicanic, N., Robinson, J. I., & Ponniah, D. A. (2001). Computational framework for discontinuous modelling of masonry arch bridges. *Computers and Structures, 79*(19), 1821–1830.

Toth, A. R., Orban, Z., & Bagi, K. (2009). Discrete element modelling of a stone masonry arch. *Mechanics Research Communications, 36*(4), 469–480.

Underwood, P. (1983). Dynamic relaxation. In T. Belytschko & T. J. R. Hughes (Eds.), *Computational methods for transient analysis* (pp. 245–265). North Holland.

Valluzzi, M. R., Salvalaggio, M., Lorenzoni, F., Politi, M., & Boaga J. (2022). The engineering approach to conservation of massive archaeological structures in seismic areas: The Apollo Nymphaeum in Hierapolis of Phrygia. *International Journal of Architectural Heritage*. https://doi.org/10.1080/15583058.2022.2056545

Young, M. P., Schultz, A. E., & Lemos, J. V. (2015). Seismic analysis of the Panhellenic Sanctuary of Nemea, Greece. In *Proceedings of. 12th North American masonry conference*, Denver, Colorado.

Zhao, Sh., Salami, M. R., & Rahman, M. Sh. (2000). Three dimensional spherical DDA model for granular media. In J. L. Tassoulas (Ed.), *Proceedings of the 14th Engineering Mechanics Conference*, 21–24 May 2000, Austin, Texas.

The manufacturer's authorised representative in the EU is Springer Nature Customer Service Centre GmbH, Europaplatz 3, 69115 Heidelberg, Germany. If you have any concerns regarding our products, please contact ProductSafety@springernature.com

Printed and bound by CPI Group (UK) Ltd, Croydon, CR0 4YY
05/10/2026
02228133-0001